Oxygen, the Breath of Life: Boon and Bane in Human Health, Disease, and Therapy

Authored by

Olen R. Brown

Dalton Cardiovascular Research Center
134 Research Park Drive
University of Missouri
Columbia, MO 65201
United States

advertisements or ideas contained in the Work.

Limitation of Liability:

In no event will Bentham Science Publishers, its staff, editors and/or authors, be liable for any damages, including, without limitation, special, incidental and/or consequential damages and/or damages for lost data and/or profits arising out of (whether directly or indirectly) the use or inability to use the Work. The entire liability of Bentham Science Publishers shall be limited to the amount actually paid by you for the Work.

General:

1. Any dispute or claim arising out of or in connection with this License Agreement or the Work (including non-contractual disputes or claims) will be governed by and construed in accordance with the laws of the U.A.E. as applied in the Emirate of Dubai. Each party agrees that the courts of the Emirate of Dubai shall have exclusive jurisdiction to settle any dispute or claim arising out of or in connection with this License Agreement or the Work (including non-contractual disputes or claims).
2. Your rights under this License Agreement will automatically terminate without notice and without the need for a court order if at any point you breach any terms of this License Agreement. In no event will any delay or failure by Bentham Science Publishers in enforcing your compliance with this License Agreement constitute a waiver of any of its rights.
3. You acknowledge that you have read this License Agreement, and agree to be bound by its terms and conditions. To the extent that any other terms and conditions presented on any website of Bentham Science Publishers conflict with, or are inconsistent with, the terms and conditions set out in this License Agreement, you acknowledge that the terms and conditions set out in this License Agreement shall prevail.

Bentham Science Publishers Ltd.
Executive Suite Y - 2
PO Box 7917, Saif Zone
Sharjah, U.A.E.
Email: subscriptions@benthamscience.org

BENTHAM
SCIENCE

CONTENTS

FOREWORD

We are constantly barraged with scientific publications and news media regarding issues pertaining to our health in regard to prevention and cures, product defects, food benefits and dangers, renewable energy, fossil fuels, global warming, climate change, and other concerns. The contradictions and reversals in opinions on many of these subjects make us question where the truth lies. This book is an unbiased overview and synopsis of the benefits and detriments of oxygen culminating from more than 56 years of theoretical and laboratory research by the author.

As delineated in the title of this book, the effects of oxygen in its multiple forms and compounds can be a benefit or detriment to our way of life especially to our health and to the world as we know it. These are important but very complex topics. In fact, even though the oxygen element was isolated and discovered in or around 1774 it will be obvious after reading this book that our understanding of the influence of this element has made great strides but in reality, we have only uncovered the tip of the iceberg for what lies before us in the future.

Mastering a subject or concept throughout decades of one's professional career is one thing but taking the time and then having the patience and ability to assimilate and record the obtained wisdom and experience in a clear and concise document that contains details that can be read and understood by an outsider to the particular field is a totally different matter. This book provides such readily understandable sections for readers who are looking for information to initiate research on a particular issue that may be related to or affected by oxygen. But, just as important, the book also contains historical facts, theoretical and experimental scientific data, and detailed explanations with illustrations for a reader more experienced with chemistry, physics and/or biology who is seeking an in depth understanding in order to formulate his or her own opinions and conclusions regarding the benefits, detriments, and/or dangers associated with oxygen as related to a specific topic of interest.

For myself as one with a limited background in chemistry and biology obtained from college level freshman courses, it was refreshing to read an unbiased informative comprehensive treatise on what the scientific and medical communities have learned about oxygen up to this time in history. I believe anyone seeking knowledge on any of the many topics associated with oxygen will agree that this is a very comprehensive and well written book. So, read, enjoy, and become informed.

David A. Hullender
The University of Texas at Arlington
Texas
United States

PREFACE

The facts of science are published daily in a bewildering flurry of information that arrives in scientific papers and is immediately summarized (often incorrectly) by the press and on the internet. This book was written with a broader view: to try to make sense of what we know, or think we know, and from my desire to understand and explain things. As a scientist through most of six decades, it has been a great pleasure to teach students at all college levels and my interests include microbiology, biochemistry, molecular biology, toxicology, and the history of science.

The specific focus of my research is oxygen and its role in health and disease. This includes oxygen free radicals, cellular defenses against the toxicity of oxygen, and the central role of oxygen in cellular metabolism. Oxygen's dual nature was referred to as oxygen's boon and bane by a colleague, Irwin Fridovich, and I have chosen these apt words in the title of this book. Contributions from my laboratory include: evidence that specific enzymes, but not enzymes in general, are sites of oxygen toxicity, and with Richard Seither, while a student, the discovery that hyperoxia induces genetic stringency in microbes. Stringency stops grown and metabolism and protects cells from damage.

I began laboratory research in the 1960s, an exciting time for scientists because of NASA and the stated objective by President John F. Kennedy to "Go to the moon and do it in this decade". The US Space Program chose pure oxygen at low pressure for manned space flight. The Russian Space Program used larger rockets and chose a two-gas system of nitrogen and oxygen at atmospheric pressure because of the toxicity and fire hazard of pure oxygen. NASA's choice ultimately led to a tragic fire on the ground in a test capsule filled with low-pressure, pure oxygen and caused the death of three Apollo 1 astronauts in 1967. My research on the mechanisms of oxygen toxicity was relevant to NASA's Space Program and also to the Office of Naval Research where the toxicity of oxygen was a concern for deep sea diving.

This eBook was written to bring together in a popular science version, while remaining true to science, the many and varied aspects of oxygen, without which we cannot live for more than a few minutes, but which is lethal at concentration only a few times that found in the ordinary air we breathe.

Olen R. Brown
The University of Missouri
Columbia- MO
United States

ACKNOWLEDGEMENTS

Scientific study, experimentation, and discovery brought together many individuals in my laboratory and interactions with students; and generous and helpful colleagues contributed to and helped me prepare to write this eBook. I thank them all with a gratitude borne out of being fellow travelers in science and discovery. One of the great joys in science for me has been the attempt to understand our world in the broadest sense that is reasonably possible. Science is about asking the "how" questions. Answers are found in individual science publications and other writings by scientists in fields from astronomy to xenobiotics (chemicals, foreign to nature). In my lectures and study to understand science and especially in writing this book, I have freely drawn on this body of work by many scientists from many fields and obviously I make no claim of originality except for the research I have published.

I thank many talented individuals who created the images I have chosen to use from those made available on the internet licensed as "free to modify, share, and use commercially", and those who generously provided other images I have acknowledged in figure legends.

I thank Cameron Brown for his assistance, especially for his computer skills with manuscript preparation, and John Allen and Emily Brown, especially for their critical reading for content and clarity, of this e-book.

CONFLICT OF INTEREST

The author confirms that author has no conflict of interest to declare for this publication.

DEDICATION

To my former students and research associates.

CHAPTER 1

Oxygen: Origin in the Universe and Brief Chemistry

Abstract: Oxygen is the 3rd most abundant element in the known universe. Only hydrogen and helium, in that order, are more abundant. The big bang theory of creation asserts that all the elemental oxygen on earth was created late in the life of a dying star. On earth, oxygen is the most abundant element in the lithosphere (land), hydrosphere (water), and atmosphere (air). More precisely, the lithosphere is all the crust or solid, upper portions of the Earth; the hydrosphere includes all the rivers, lakes, seas, and oceans; and the atmosphere is the gas-filled space above and near the earth. Atomic oxygen is chemically and biologically reactive and primarily exists as molecular oxygen (O_2, two like atoms combined), or in combination with certain elements (primarily metals). Oxygen forms reactive intermediates and free radicals including peroxide, superoxide and hydroxyl radical. The latter is said to be the most reactive species known in chemistry. Oxygen has a unique arrangement of electrons that is conducive to one-electron transfer reactions that can produce oxygen free radicals and cause biological oxidant stress. However, oxygen is especially suited to serve as the terminal acceptor of electrons in the biological process of electron transfer that is linked *via* coupled reactions to oxidative phosphorylation that creates adenosine triphosphate (ATP), the universal storage and transfer form of energy for all aerobic life on earth.

Keywords: Atmosphere, Aufbau order, Big bang, Burning, Christ's last breath, Combustion, Dmitri Mendeleev, Double bond, Elemental oxygen, Final electron-acceptor, Free radicals, Hydrogen peroxide, Hydrosphere, Hydroxyl radical, Inflation, Lithosphere, Nucleosynthesis, Ozone, Periodic table, Photosynthesis, Sir Fred Hoyle, Superoxide dismutase, William A. Fowler.

Olen R. Brown

INTRODUCTION

Oxygen is like Goldilocks' porridge, chair and bed in the children's story *The Three Bears*: oxygen is "just right"; just right for life.

By analogy with Goldilocks' porridge (neither too hot nor too cold), oxygen for living systems is neither too reactive nor to inert; its chemical and biological reactivity is just right. Oxygen has many roles in cells. It must oxidize (burn in a controlled way) the food in our bodies to release energy in a manner that does not consume us. Energy must be stored in a transferable, accessible form and adenosine triphosphate (ATP) performs this function uniquely for growth, repair, moving, and thinking. ATP's stored energy, made possible by biological reactions involving oxygen, is just right. It is somewhat like the dynamite discovered by Alfred Nobel that so tamed the energy of gunpowder that it literally could move mountains but was safe enough for untrained men to handle because it required a lighted fuse to detonate. Oxygen has just the "right stuff" to snatch electrons from chemicals like the sugar glucose. Glucose is derived metabolically in cells by enzymes and is made abundantly by plants with energy ultimately derived from the light of the Sun, a nuclear furnace that is just the right distance away to be safe (93 million miles). Thus, it is a property of electrons of oxygen (mystically having both particle and wave-like properties) that undergirds an exquisite system that efficiently stores energy in ATP that is transported for use everywhere in our bodies.

By further analogy with Goldilocks' chair and bed (just the right size for sitting and softness for sleeping), oxygen is correct in size and electronic structure to readily diffuse into cells and fit energetically into complex metabolic schemes as substrate, intermediate, and product that are easily inspired, reacted, and excreted. A gaseous element, it can freely diffuse in the atmosphere where it maintains a near constant concentration of approximately 21%. It can diffuse from air in the lungs into red blood cells, requiring only hemoglobin and a few regulatory molecules to serve as subway trains and ticket conductors. Oxygen thus has free entry into cells and it requires no special passport to cross their membrane borders, partly because of its small size. Once inside cells, oxygen freely diffuses around the cytoplasm and into organelles called mitochondria. Mitochondria work like factories to make ATP, which functions like charged batteries to do work everywhere in the cell. Oxygen participates in biosynthesis and in oxidation-reduction reactions and combines with carbon to form carbon dioxide that is readily excreted at low cost in energy. To paraphrase Descartes: "I have ATP, therefore I am".

Each molecule of atmospheric oxygen has two atoms of oxygen. Oxygen forms various structural components in cells and important metabolic intermediates with many different functions and many different oxygen contents. For example, each molecule of ATP has 13 atoms of oxygen. Whenever oxygen burns an organic substance, whether in a forest fire, a log in your fireplace, or in the mitochondria within cells of your body, the process is vastly different, but the primary end products are simple molecules of water and carbon dioxide. They are non-toxic (safe) wastes and are easily disposed of by diffusion out of cells and into the lungs where they are exhaled with minimal expenditure of energy. Water, by mass, is approximately 89% oxygen; carbon dioxide is approximately 73% oxygen; and the entire human body is approximately 65% oxygen. This near universal presence of oxygen throughout the biochemistry of cells is further, strong evidence for the essential nature of oxygen for life. In the form of water, oxygen is considered almost universally to be the signature of life by searchers who hope to discover life on other planets. To further paraphrase Descartes: "I have oxygen, therefore I am".

OXYGEN IN THE UNIVERSE

To continue our journey toward an understanding of the role of oxygen in Nature, it will be helpful to start at the beginning, or what many scientists now believe to be the beginning.[1] The dominant, current theory in astrophysics about the origin of the universe is the "big bang". The term big bang was coined by Sir Fred Hoyle [1] as a perhaps derisive term (he later denied it was meant to be derisive) for the event that is generally claimed in science to be the origin of the Universe.[2] An origin for the Universe implies an Originator and some preferred the idea that the Universe had always existed.

Hoyle, about 70 years ago, disagreed with the theory that the universe had a beginning. Indeed, even today, based on recent discoveries widely reported in the popular press, the big bang may not be a valid description to account for the origin of everything. Nonetheless, the big bang theory has led to an enormous body of research and theory that proposes that everything in the Universe resulted from the expansion of an almost indescribably small, hot, dense source (some say nothing). This resulted in the formation of all the fundamental quarks, protons, neutrons and electrons, and later also formed a portion of the lightest elements, primarily hydrogen and helium. Hoyle [2] also wrote the first papers describing stellar nucleosynthesis of the elements heavier than helium. Hoyle became popular but often stated controversial opinions on various topics in astronomy. He never won the Nobel Prize which many felt he deserved. A co-researcher, William Alfred Fowler who won a share in the Nobel Prize for Physics in 1983, bluntly stated that Hoyle was deserving also [3]. Indeed, the grand concept of nucleosynthesis in stars was first definitively established by Hoyle in 1946 and remains the most accepted, basic understanding of the mechanism for how oxygen came to be. Fowler's acknowledgement to Hoyle is best credited by citing from Fowler's published work [3]:

> "After Whaling's confirmation of Hoyle's ideas I became a believer and in 1954/1955 spent a sabbatical year in Cambridge, England, as a Fulbright Scholar in order to work with Hoyle. There Geoffrey and Margaret Burbidge joined us. In 1956 the Burbidges and Hoyle came to Kellogg and in 1957 our joint efforts culminated in the publication of 'Synthesis of the Elements in Stars' in which we showed that all of the elements from carbon to uranium could be produced by nuclear processes in stars starting with the hydrogen and helium produced in the big bang."

Hoyle also wrote popular science and science fiction books. During WWII he helped to develop countermeasures against radar-guided guns, like those of the German battleship the Graf Spree [4]. In astronomy, Hoyle favored the hypothesis that the origin of life on Earth was by panspermia, the thesis that life came to earth *via* comets or by other means after originating on other worlds. His strongly held controversial views about the origin of life led him to make statements critical of evolution [5]. It is my personal opinion that his critical views about elements of the evolutionary doctrine contributed strongly to his omission for the Nobel Prize award which I believe he clearly deserved. A few excerpts on this topic, quoted from Hoyle's paper published in 1981 [5] suffice:

> "The big problem in biology, as I see it, is to understand the origin of the information carried by the explicit structures of biomolecules... then let evolution in some chemical environment cause the simple enzymes to change gradually into complex ones we have today. The deceit comes from omitting to explain what is in the environment that causes such an evolution... Wasn't it even more ridiculous to suppose that the vastly more complicated system of biology had been obtained by throwing chemicals at random into a wildly chaotic astronomical stewpot? Would you not say to yourself... Some super calculating intellect must have designed the properties of the carbon atom, otherwise the chance of my finding such an atom through the blind forces of nature would be utterly miniscule."

The big bang and subsequent nucleosynthesis in stars has been described elegantly, and far more completely than my brief account here, by Lawrence Krauss in a fascinating book [6]. Krauss and many others have written that the big bang took several minutes to form any elements heavier than hydrogen. Oxygen, however, was not created until stars were formed and it is the second most abundant product formed by fusion reactions (helium is the most abundant). Indeed, oxygen, based on analysis and conclusions of many physicists, was not formed in the universe until stars, somewhat more massive than our Sun were born, burned their store of hydrogen, grew old, and in their later stages created oxygen and heavier elements. The eventual explosion of a dying star of this type scattered all the elements (as "stardust" into the surrounding space), including oxygen that were synthesized in the furnace of nuclear reactions in the star. Current physics dictates that these elements were the building blocks of our solar system including the earth and all that it contains. Many have expressed the thought-provoking idea that a star had to die for us to be born.

Thus, all of the atomic oxygen and other stable, more massive elements that are present on earth today, theoretically, were produced by a dying star that exploded before the Earth was formed. Molecular oxygen (diatomic oxygen, O_2, I shall use the terms interchangeably) is created and destroyed by chemical processes. It is created by photosynthesis of green plants and certain other life forms (primarily phytoplankton). The air we breathe today contains about 20.9% oxygen. Indeed, it is a general principle, agreed to in physics, astronomy and chemistry that chemical and biochemical processes do not create more elemental oxygen but can add and subtract oxygen by combining and releasing both elemental oxygen and molecular oxygen. Based on estimates of global rates of photosynthesis, and turn-over rates, an amount of oxygen equal to all of Earth's global atmospheric oxygen today could have been generated within the last 2,000 years [7]. Atomic oxygen is quite reactive chemically and it rapidly combines with itself to form O_2 or combines with (oxidizes) other substances. Thus, geochemists and atmospheric scientists conclude that oxygen exists in the lithosphere primarily combined as silicates; in the hydrosphere combined with hydrogen as water; and in the atmosphere combined with itself as molecular oxygen [8].

Nucleosynthesis of Light Elements

Astrophysics teaches that temperatures in the first microseconds after the big bang were too high for atoms to exist. Indeed, the predominant theory concludes that in its earliest moments the Universe was mostly radiation (a form of energy). Subsequently with cooling and expansion (very rapid expansion, known as "inflation" is essential for the theory) of the universe, the light elements hydrogen and helium formed and the elements deuterium, lithium and perhaps traces of beryllium were created by fusion reactions as described by Claus Grupen [9].

The subject of "inflation" of the Universe has recently made news with the awarding of the Kavli Prize in Astrophysics "for work in developing the theory of cosmic inflation, the theory that explains the origin and structure of the universe", as announced in a report from Stanford University on May 28, 2014 [10]. The prize was shared with Alan Guth at MIT, and Alexei Starlbinsky, Cosmologist at the Landau Institute for Theoretical Physics at the Russian Academy of Sciences. The report also says that researchers including Chao-Lin Kuo, also of Stanford, had announced support for the theory with the (at the time, unconfirmed) detection of gravity waves that would have been created in the first moments of the formation of the universe [11]. Paul Halpern [1] recently has written eloquently about this discovery including the insight that the big bang needs to have two parts (events): the singularity (traditional big bang) and the inflation with the release of a huge amount of energy that was converted into "a deluge" of particles that formed the matter of the universe. This would have included the basic particles used in the creation of oxygen.

I will omit the details, necessary for a more complete understanding of origins, in favor of a briefer treatment that is helpful in perceiving theoretically how oxygen came to be. The process of expansion and cooling of the universe predicts that at a particular temperature about 30,000 years after the big bang (properly called a singularity), the universe was sufficiently cooled and electrons, protons, and neutrons could form stable atoms [12]. Photons of light were set free and could travel through the now transparent universe: "let there be light". Thus, the theory explains the formation of the simplest elements beginning about 30,000 years after the singularity that was the big bang. None of this, of course, is original thought by me but is generally accepted and usually expressed in similar ways by scientists since the late 1940s.

Oxygen, Theoretically, Was Created in Stars

Accepted theory says that stellar nucleosynthesis created oxygen by physical not chemical processes. Stars, including our Sun, are described to have life cycles; they are said to be "born, grow old, and die". Stars similar to our Sun are relatively young and are composed primarily of hydrogen which they fuse in nuclear reactions to make helium. Toward the end of their life cycles, some stars above a certain mass range, begin to run out of hydrogen, but continue fusion reactions to form heavier and heavier elements. A good source for this type of information is Donald Clayton's book: *Principles of Stellar Evolution and Nucleosynthesis* [13]. Von Baeyer also has written elegantly about this process and he recounts that it has been calculated that at least one atom of elemental oxygen that was in the last breath exhaled by an historical figure 2,000 years or so ago is present in each breath you take today [14]. Von Baeyer attributes this calculation to Enrico Fermi (1901-1954), credited as being "the father of the atomic bomb", who was noted for his "Fermi solutions". These calculations were amazingly accurate but were based on intelligent, logical estimates from very simple starting information. Von Baer [14] recounts that in one such estimate, Fermi mentally calculated (based on his estimation of the effect on the movement of paper he released from his hand into the air gust resulting from the Trinity Test) that the first atomic bomb was equivalent to 10 megatons of TNT. The blast was eventually calculated to be about twice Fermi's estimate, a verification of his simple, quick estimate[3].

This fact that there are oxygen atoms (as parts of diatomic oxygen molecules in each breath we take) that likely were in the lungs of a historical person who lived approximately 2,000 or more years ago informs us about oxygen's: small size, enormous abundance, stability, permanence and rapid mixing throughout the atmosphere. Note that I refer to elemental oxygen not the molecule dioxygen which may acquire elemental oxygen from many sources in 2,000 years.

Let us make a calculation, using Von Baer's procedures [14] and some generally-accepted data. Let us estimate the likelihood that an atom of oxygen that was exhaled in the last breath of Christ will be present in the next breath you take. Several generally-accepted assumptions and approximations are required: complete mixing of molecular oxygen– which is reasonable because of the time of approximately 2,000 years and the nature of the gaseous atmosphere; the permanence of oxygen atoms– oxygen atoms are the same atoms since cosmological creation because no new atoms have been created or destroyed; a constant concentration of oxygen in the atmosphere of approximately 21%; the same volume for Christ's last breath and your breath (assume 1 liter); and 1.777×10^{20} moles of atmospheric air– a total of approximately 1×10^{44} (100,000,000,000,000,000,000,000,000,000,000,000,000,000,000) air molecules. Oxygen atoms are permanent; oxygen molecules are not because they exchange atoms of oxygen. The calculation is: 2.836 atoms of oxygen present in Christ's last breath will be in the next breath you take. For a different breath volume use the equation: atoms oxygen in 1 breath = 2.836 L^2 (breath volume).

In this brief overview of the origin of atoms, particularly focused on oxygen, the generally-accepted information from many sources has been summarized as a basis for better understanding of oxygen's roles in chemistry and biology. Knowledge and theories of origins currently are expanding and details will change or perhaps even require extensive revisions. However, the information as currently known and required to understand the basic chemistry and biology of oxygen likely will remain in place. Life requires an intricate and unique set of very special attributes, attributes designed into all matter and energy. An alternative, favored by many, is that the design is only apparent, not real. These are very similar ideas except for interpretation. The lesson to be learned shouts out the smallness, abundance, permanence and uniqueness of the element oxygen.

Oxygen's Place in the Periodic Table of the Elements

The history of science teaches that chemistry began in alchemy with mystical attempts to turn base metals into gold and as vain searches for the elixir of life, perhaps by aid of the philosopher's stone. Dictionaries differ but many attribute the word alchemy to be derived from Greek, Latin, French and Arabic sources and it may refer to "the art of transforming metals". Basic elements of the material world were categorized as earth, air, fire and water. The efforts of alchemists were mostly hidden in mystery and little real "chemistry" was done. Before chemistry could develop, it was necessary to institute careful weighing and measuring of the starting and ending products of chemical reactions. In this transition from alchemy it was repeatedly noted that elements had properties that "repeated". Thus, elements could be arranged in rows and columns based on these repeating properties. The validity of such arrangements was presaged by the observation that there were gaps in their charts that strongly suggested the presence of unknown, undiscovered elements. This notion that theories can be validated by their predictive powers is a continuing theme in science. A milestone in practical understanding of the repeating properties of known elements was the work of Dmitri Mendeleev, also spelled Mendeleyev (1834-1907). Mendeleev (Fig. **1**) arranged the then known 67 elements based on their atomic masses in a table which he published in *Principles of Chemistry,* volume 1, p. 27, in 1869 [15].

Oxygen was element number six (based on properties, not order of discovery) and was placed in the second of six columns in his table. He claimed that the basis for his ordering, which related to a periodicity in occurrence of chemical properties, came to him in a dream [15][4].

Fig. (1). Dmitri Mendeleev.

Mendeleev's mother had great faith and was a great inspiration to him. Daniel Q. Posin [16] has written a wonderfully detailed book about Mendeleev and says that he was reared with three chief influences from significant people: "*Everything in the world is science* (Bessargin, husband of his sister Olga); *everything in the world is art* (Timofer, the glassblower); *and everything in the world is love* (his mother, Maria). Maria's dying words to him were said to have been that he should "*Patiently search divine and scientific truth*".

It is difficult to imagine that the extensive knowledge of the chemistry of oxygen (and the other elements) could have progressed without the insight and direction provided by the Periodic Table. The modern form of the Periodic Table, extended and extensively revised, is in the inside cover of many chemistry books and on the walls of most chemistry labs.

OXYGEN AND OXYGEN RADICAL CHEMISTRY IN BRIEF

The properties of oxygen essential to understanding the chemistry and biology of oxygen are best simplified and summarized by referring briefly to the place of oxygen in the modern Periodic Table.[5] Oxygen is in group 16 (of 18), and period 2 (of 7). Groups are arranged in vertical columns and periods in horizontal rows. Oxygen is toward the top right in the Table. Atomic oxygen has 8 electrons in "orbitals". The word orbital customarily is used instead of orbit to indicate that it is orbit-like. The electrons surround the nucleus composed of eight protons and eight neutrons in the most common oxygen isotope which makes up more than 99% of earthly oxygen. It is useful to think of oxygen as having electrons that spend most of their time in restricted regions, sometimes described as "shells" or "clouds" of electrons. Oxygen's position in the Periodic Table toward the top and right reveals its electronegativity which is second only to fluorine. Neon does not count because its valence orbitals are completely filled. Recall that electronegativity is a calculated (not measured) characteristic and describes the propensity of an element to gain or lose electrons. More specifically, fluorine is above oxygen, but oxygen is to the right of fluorine in the Periodic Table. Linus Pauling (1901-1994) made calculation, considered to be as accurate as any, of electronegativity [17]. Specifically, however, it is a predictor of reactivity of atomic fluorine and oxygen, but not for diatomic fluorine and dioxygen. A higher number for electronegativity indicates a greater propensity to gain electrons. This is a function of the number of protons in the atomic nucleus and the diameter of the outer orbital (a larger atom attracts its electron less avidly). Suffice it to say, that oxygen wins out in having the right balance of reactivity to serve life's functions and fluorine does not.

Diatomic oxygen (O_2) is depicted in a space-filling model in Fig. (**2**). The shape of dioxygen is a consequence of the interactions of the valence electrons and the protons in the nuclei of the two atoms, and the shape is produced by the space occupied by the "cloud" of 16 electrons (8 from each atom of oxygen) that move about the positively charged protons in the nucleus which is very small and not seen in this diagram. The merged spheres result from the double bond that arises from four "shared" outer (valence) electrons. The two lighter spots indicate the site of the nucleus of each oxygen atom. The small size can be appreciated by the fact that the distance between the nuclei of the two bonded atoms is approximately 66 pm (one pm, picometer, is one trillionth of a meter). Another measure of its size is to note that 32 grams of oxygen have a total number of 6.02 x 10^{23} oxygen molecules. A modest inhalation (single breath) of air contains approximately 5 x 10^{21} oxygen molecules (five with 21 zeros, sometimes expressed as a thousand-billion-billion). As a further comparison, astronomers estimate there are approximately 10^{22} to 10^{24} stars in the known Universe; however, on a dark, clear night only a few thousands can be seen as separate points of light by the unaided eye.

Fig. (2). Model of the oxygen molecule (O_2).

The mathematical description of the atom and the nature of the position and movement of electrons around the nucleus of atoms are beyond our purpose and needs here. However, to understand the role of oxygen in the process of biological oxidation, a general discussion of the role of electrons in forming chemical bonds is useful. Using our crude terminology, the first (inner) shell of oxygen's electrons is filled to capacity with two electrons. The second shell, proceeding outward from the nucleus of oxygen can hold up to eight electrons, but oxygen has only six electrons. To achieve stability, oxygen will tend to capture two additional electrons or to share two additional electrons from outside the atom. Oxygen can acquire a share in two electrons from "reaction" with elements such as another oxygen atom, with hydrogen or with carbon. When four electrons are shared a "double" bond is formed. Oxygen can also react by capturing two electrons from another element, typically a metal, to form an oxide. In the chemistry of life, oxygen also forms various functional groups including alcohol (OH), aldehyde (HC=O) and acid (HOC=O) which are successively oxidized groups on organic molecules that contain principally hydrogen, oxygen and carbon.

Oxygen, by Analogy

As we have briefly stated, the recurring (periodic) properties of the chemical elements exist because of recurring structure in the make-up of atoms. We will be concerned only with the parts of the Periodic Table having to do with the lower molecular-weight elements which have fewer orbitals. This is sufficient to understand the character of oxygen.

Electrons are subatomic components with a dual nature that is both particle and wave-like. Electrons have one unit of negative charge and a very small mass. Electrons are present as a cloud surrounding the central part of the atom, called the nucleus which contains protons with a positive charge of 1. The electron's negative charge is precisely equal and opposite to the charge of the proton. In all atoms except hydrogen, there are also neutrons in the nucleus. Neutrons are neutral in charge and precisely equal in mass to the proton. Being neutral, neutrons have essentially no effect on electrons in atoms.

Outside the atomic nucleus, all elements have electrons. The words cloud, shell and sub-shell are used by scientists to help us understand the nature of electrons. None of these words accurately describe the permitted positions of the electrons in atoms but they are useful analogies. The atom exists because of a delicately balanced set of forces between the nucleus and the electrons in the cloud surrounding it. If these forces were not set almost precisely at the values they have, no atoms could form or persist.

Because of elaborately interconnected attributes, the atom is stable in certain configurations and can be formed from subatomic particles. The permitted states of atoms are set by innate properties of the atomic nucleus including the number of protons and neutrons and the strong and weak forces, and by the number and energies of the orbiting electrons. Perhaps there is a fundamental particle that is "pregnant with everything".

Thus, all types of matter are made up of a nucleus containing protons and neutrons (almost always) and a "cloud" of electrons outside the nucleus. Other particles, including quarks, have been discovered in matter, and are important in physics. However, to understand the chemistry of a molecule of oxygen, formed by the union of two oxygen atoms, we need be concerned only with the outer part (involving six electrons) of each oxygen atom's cloud of electrons. Chemistry occurs with, and because of, these clouds of electrons and their behaviors. Chemistry involves both free trade in and sharing of electrons. Electrons, by analogy, are the glue that holds parts of molecules together.

The chemistry of oxygen is dominated by the effects of the arrangement of valence electrons which permit oxidation of other chemicals. The gain in electrons by oxygen eventually reduces it to water. In this reduction process, gain of electrons one at a time results in some negatively charged radicals with unpaired electrons that have consequences for life that we shall explore further in Chapter 7. However, further brief comment is required about the manner in which oxygen atoms bond with other oxygen atoms to form molecular oxygen, and the way molecular oxygen reacts biologically, including the generation of reactive "free" radicals. The nature of the chemical bond is complex and since the work of Linus Pauling [17], it has been understood in quantitative terms involving mathematical descriptions of the behavior of electrons (the "stuff" of bonds). For our purposes here, a simplification is possible that is useful in explaining: the free radicals formed from oxygen, the nature of its double bond, and oxygen's propensity to undergo transfer of electrons one-at-at-a-time.

Oxygen Double Bond: Formation and Dissolution

Some additional explanation of the nature of the double bond in oxygen is useful. Dioxygen is composed of two oxygen atoms joined by a double bond. A double bond is created by the sharing of 4 electrons between two atoms (a single bond involves 2 electrons). Atomic oxygen has a total of 8 electrons and molecular oxygen has 16 electrons. A simple depiction of these electrons is *via* a "box" diagram (Fig. **3**) that includes unpaired electrons and accounts for the fact that oxygen is paramagnetic. Electrons have a charge of negative 1 and are attracted by protons in the atomic nucleus, each of which has a charge of plus 1, for a total of plus 16. The electrons spend most of their time in what is sometimes simply described (not using complex mathematics) as shells or orbitals. Electrons in orbitals closest to the nucleus have less energy and the inner shell is filled first when considering the placement/location of electrons. This closest shell to the nucleus is not involved in bond formation by oxygen and is not shown in Fig (**3**). When considering the electrons, they are placed in an Aufbau order (*i.e.*, lower energy orbitals filled before higher energy orbitals). The lowest orbital (not shown) can contain a maximum of 2 electrons which is the case for atomic oxygen. In the diagram (Fig. **3**), the arrows indicate electrons and the boxes indicate orbitals. The 12 electrons in the s and p orbitals from two atoms of oxygen combined in molecular oxygen are shown. Paired electrons are shown pointing up and down to indicate that they have the required opposite spin states.

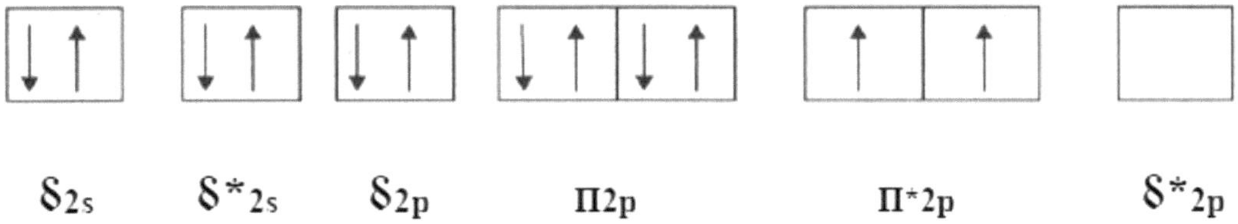

Fig. (3). Standard representation of the configuration of the electrons in the 2s and 2p orbitals of oxygen.

There are two unpaired electrons in the pi*2p orbital. An unpaired electron is necessary to account for the fact that oxygen is paramagnetic. There are also eight total valence electrons in the bonding orbitals (sigma 2s, sigma 2p, pi 2p) and two in the antibonding orbitals (sigma*2s) which accounts for the double bond character of dioxygen (the outermost electrons of 2 atomic oxygen atoms are shared to form the double bond of molecular oxygen). Note that the molecule contains bonding and antibonding (*) orbitals. Electrons in bonding orbitals promote bond formation and electrons in antibonding orbitals tend to weaken bonds. Addition of electrons can lead to dissolution of bonds. In a crude analogy, the electrons being shared by the two atomic nuclei, and spending time in an orbital that involves both, act somewhat like glue to hold the molecule together.

The outermost orbital can accept two electrons; dioxygen has 2 such orbitals and can accept 4 electrons. These electrons can be gained one at a time with an effect on the double bond as follows. Loss of one electron leaves 1-½ bond; loss of a second electron leaves 1 bond; and loss of a third electron leaves ½ bond, which is no bond at all, and the atoms of oxygen separate. The process of gain of electrons by oxygen can be depicted as follows:

$O_2 + 1e = O_2^{\cdot -}$ (superoxide ion or superoxide free radical)

$O_2^{\cdot -} + 1e = H_2O_2$ (peroxide)

$H_2O_2 + 1e = OH^{\cdot -}$ (hydroxyl radical)

$OH^{\cdot -} + 1e = H_2O$ (water)

For simplicity, the reactions shown are not balanced. Balancing requires H^+ which is freely available in cells. Electrons are denoted by e. The purpose is to show the contributions of electrons which are required to form the radicals superoxide and hydroxyl, and the non-radical peroxide *via* 1-electron transfer. Significant to our interests, formation of the reactive radicals superoxide, peroxide and hydroxyl radical occur biologically and accommodations must be made by cells to ameliorate their toxic effects. We shall have more to say about these radicals in Chapter 7.

Oxygen is a colorless, odorless and tasteless gas under ordinary pressures and temperatures on earth. It has some properties shared with sulfur and selenium. Oxygen is second only to fluorine in its electronegativity (tendency to attract electrons); however, it is slower to react than predicted because the bond energy is very high and reactions that require that it be broken occur only at very high temperatures. Most reactions of oxygen are those that do not require the oxygen-to-oxygen bonds to be completely broken, typical of which is the formation of peroxide as previously noted. All metals except silver, platinum and gold react directly with oxygen to form oxides.

Generally without thinking about it, we all utilize the oxidizing capacity of molecular oxygen. For example, burning is a type of oxidation and requires oxygen. The oxygen concentration is a critical factor that determines whether a substance will burn. For example, in a closed container filled with air (21% oxygen) ordinary ignited paper will not continue to burn until the oxygen is exhausted. Rather, the paper fire will go out when the concentration of oxygen falls to about 13% to 15% [18]. This is not a fair test, however, because in the closed space rapid oxidation (burning) is impaired by the rising concentration of

CO_2, produced from carbon in the paper. Oxidation will continue, but it is slow and does not sustain a flame. The differences among oxidation, combustion, and explosion are matters of rate. Oxidation is a chemical term and can occur at rates from imperceptible to explosive; an explosion is violent, rapid combustion. However, within the cells of our bodies a "cool combustion" called biological oxidation occurs. It is defined as a loss of electrons with the electrons ultimately transferred to oxygen. Oxidation and reduction occur by coupled reactions and no net overall change occurs in oxidation state of the system. This process can be catalyzed by enzymes where it is controlled and results in energy storage in usable forms, with only a part of the energy "lost" as heat. Enzymatically-controlled oxidations allow myriad and abundant unicellular and multicellular life to flourish; this would be remote or impossible if oxygen's fundamental chemical properties were only slightly different.

An interesting example of oxidation is the rapid rusting of an iron nail exposed to air and water. Instead of saying the surface of the nail has burned, we more properly say that the outer surface of the iron has oxidized (Fe^{+2} converted to Fe^{+3}) with transfer of an electron to molecular oxygen. Oxidation is speeded up when the iron is covered by a film of water, and when the iron has more exposed surface. Recall the last time you scrubbed a grill with a pad made of iron and soap and returned only a short time later to find the rusty mess that was the scrubbing pad. We can take advantage of the oxidization-reduction relationship of iron and oxygen by shorting the poles of a charged battery with a small wire or with a tuft of steel wool. This will produce heat sufficient to ignite wood shavings and it can be used to start a camp fire.

The chemical reactivity of oxygen causes it to find stable unions with other elements, primarily with metals in earth's crust. Were molecular oxygen not being produced by green plants on Earth, it is logical that the Earth's original oxygen now would be combined with other elements and no free oxygen (or only traces) would be present in our atmosphere today [19]. Probably all, or nearly all, of the oxygen on Mars is now combined and perhaps only a trace is free in the atmosphere [20].

Under special conditions, three (not the usual two) molecules of oxygen combine to form ozone (O_3), a more reactive, stronger oxidant than O_2. Ozone that forms near the Earth's surface is considered to be an air pollutant and it can harm our respiratory systems. It also is a component of the smog that forms in cities, especially during hot weather. Ozone is also formed high up near the top of the atmosphere by the energy of sunlight reacting with certain gases. This ozone forms what has been called the ozone layer, which is just a region of higher ozone concentration. Ozone has a high ability to absorb harmful radiation present in sunlight and thus it protects life. This ozone shield is another miraculous property of oxygen working in coordination with many factors to result in a protective system above the Earth. The ozone layer is not a perfect shield. Obviously a complete shield would not allow Earth to benefit from the Sun's useful rays which are essential for the balance required for the life we know on Earth. More information about ozone is provided in Chapter 3.

Biologically, peroxides and superoxide (previously described chemically) are important oxidizing agents. Superoxide is a reactive form of oxygen formed when molecular oxygen (dioxygen) gains a single electron. Peroxide contains an oxygen-to-oxygen single bond. The simplest peroxide is hydrogen peroxide, a common chemical found in the typical medical cabinet. The oxygen-to-oxygen bond of peroxide is relatively unstable and only small concentrations are usually found in nature. Hydrogen peroxide can form in cells as a short-lived toxic product or intermediate. It is capable of oxidizing proteins, lipids in membranes of cells and sub-cellular organelles, and DNA.

Of great significance for our understanding of the effects of oxygen on the chemistry of life molecules is the ability of oxygen to form the previously referred to toxic intermediates called oxygen "free radicals". Because of the reactivity of these forms of oxygen, there is a need for, and the presence of, enzymes that rapidly "destroy" oxygen free radicals. These enzymes are present within nearly all cells in the body and they greatly increase destruction rates of the oxygen radicals. Each enzyme reacts specifically with only one type of oxygen radical. One such enzyme (Fig. **4**) stimulates the dismutation (simultaneous oxidation and reduction) of superoxide anions. It is called superoxide dismutase (SODase, later called SOD 1), an enzyme that was discovered in the cytosol of cells by Joe McCord while a graduate student in the laboratory of Irwin Fridovich [21, 22].

Fridovich later discovered the mitochondrial form of SODase (SODase 2) which contains Manganese (Mn) instead of copper (Cu) and Zinc (Zn) found in SOD 1 [23]. Both McCord and Fridovich are pioneers but still contributors in the field of the biology and chemistry of oxygen radicals. SODase converts the superoxide radical ($O_2^{\cdot -}$) into O_2 and H_2O_2. Catalase is abundantly present in cells and functions to convert toxic hydrogen peroxide (H_2O_2) into water and dioxygen.

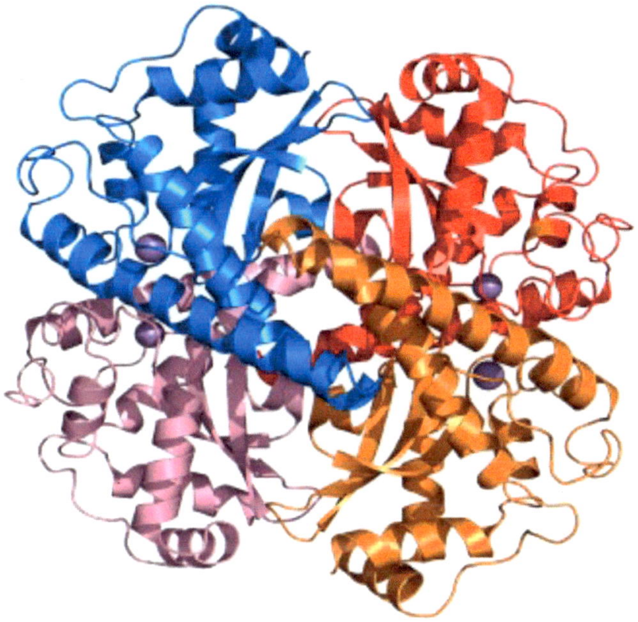

Fig. (4). Structure of human Mn superoxide dismutase 2 tetramer.

Thus, SODase is an important factor in the balance of processes required to make use of oxygen without intolerable damage to the fabric of life. The hydroxyl radical is more difficult for cells to deal with in some ways. Because it is extremely oxidizing, it has great potential to damage essential cell molecules. However, this extreme reactivity actually limits its potential for cell damage because it is destroyed by reaction with the first molecule it happens to meet and it reactivity is said to be "diffusion limited". As an aside, hydroxyl radicals are also formed in the upper atmosphere and as evidence of their reactivity, they are said to have an average life-time of less than one second even in this gas phase with large distances between molecules [24]. In cells, an hydroxyl radical will react with the first molecule it contacts after formation; and this is most likely to be water and not a vital substance such as: DNA, or even a protein enzyme, or a component in the cell plasma membrane, or membrane of an internal organelle whose destruction could be most damaging or even lethal. For example, the peroxidation of lipids and its prevention either by a therapeutic or by body processes is important for understanding atherosclerosis, heart disease and stroke. There is evidence that oxygen free radicals are involved in ageing and degenerative brain diseases, and in cancer. Oxidant stress needs to be balanced by anti-oxidant mechanisms; however, little success has been achieved with supplementation of the body's natural defenses with anti-oxidants including water-soluble vitamin C, and lipid-soluble vitamin E, or synthesized variants of superoxide dismutase. We shall have more to say about this in several following chapters.

Ultimately, our knowledge of cosmology and specifically of the creation of oxygen itself rests on theories and we cannot perform experiments that recreate galaxies, stars and planets. With current "colliders" particle physics has been able to probe the make-up of atoms but a unification of the forces of nature still escapes us. The chemistry of oxygen rests on firmer and more fully-discovered grounds and it appears that there no big surprises awaiting discovery at the fundamental level. Biology, however, has not outgrown its descriptive roots and yields only grudgingly to grand theories and of course, hardly at all to mathematics.

NOTES

[1] We dance round in a ring and suppose, But the secret sits in the middle and knows. Robert Frost (1874-1963). *The Secret Sits;* first appeared in *The Witness Tree* (1942).

[2] His comments occurred on the BBC radio network during the *Third Programme* broadcast on March 28, 1949.

[3] "There is something fascination about science. One gets such wholesale returns of conjecture out of such a trifling investment of facts." Mark Twain, *Life on the Mississippi.*

[4] The first principle is that you must not fool yourself and you're the easiest person to fool. Richard Feynman, *Surely You're Joking, Mr. Feynman!"*

[5] Everything should be made as simple as possible, but not simpler. Albert Einstein (1879-1955) attributed- See p. 314 *The Expanded Quotable Einstein*, Alice Calaprice, Princeton University Press (2000).

REFERENCES

[1] Halpern P. The Big Bang's identity crisis - The nature of reality. PBS. Available from: http://www.pbs.org/wgbh/nova/blogs/physics/2014/05/the-big-bangs-identity-crisis/ 2014.

[2] Burbidge E, Burbage G, Fowler W, Hoyle F. Synthesis of the elements in stars. Rev Mod Phys 1957; 24(4): 546.

[3] Anon. Fowler - Biographical. Available from: http://www.nobelprize.org/nobel_prizes/physics/laureates/1983/fowler-bio.html 1983.

[4] Mitton S. Fred Hoyle a life in science. Cambridge University Press 2011. [http://dx.doi.org/10.1017/CBO9780511852107]

[5] Hoyle F. The Universe: Past and present Reflections. Eng Sci 1981; (Nov): 8-12.

[6] Krauss L. An odyssey from the big bang to life on earth... and beyond. 1st ed. Boston, New York, London: Little, Brown and Company 2001; p. 9.

[7] Wallace B. Features: Atmospheric oxygen--an endangered resource? Available from: http://www.columbia.edu/cu/21stC/issue-2.1/broecker.htm 2014.

[8] Evolution of the atmosphere: Composition, structure and energy. Available from: http://www.globalchange.umich.edu/globalchange1/current/lectures/Perry_Samson_lectures/evolution_atm/ 2014.

[9] Grupen C. Big bang nucleosynthesis. Astroparticle physics. Berlin: Springer 2005; pp. 213-28.

[10] Mcgregor B. Alan Guth receives prestigious $1 million Kavli prize in astrophysics. Available from: http://frenchtribune.com/teneur/1422686-alan-guth-receives-prestigious-1-million-kavli-prize-astrophysics 2014.

[11] Moskowitz C. Gravity waves from big bang detected. Available from: http://www.scientificamerican.com/article/gravity-waves-cmb-b-mode-polarization/ 2014.

[12] Anon . Mysteries of deep space. Available from: http://www.pbs.org/deepspace/ 2014.

[13] Clayton DD. Principles of stellar Evolution and Nucleosynthesis. New York: Mc-Graw Hill 1968; p. 9.

[14] von Baeyer HC. The fermi solution: Essays on science. 1st ed. New York: Random House 1994; p. 9.

[15] White K. Mendeleev and the periodic table. New York: Rosen Publishing Group, Inc. 2005.

[16] Posin DQ. Mendeleyev: The Story of a great scientist. New York: Mc-Graw Hill 1948.

[17] Pauling L. The Nature of the Chemical Bond. IV. The Energy of Single Bonds and the Relative Electronegativity of Atoms. J Am Chem Soc 1932; 54(9)

[http://dx.doi.org/10.1021/ja01348a011]

[18] Lundgren P, Backström S, Chiti S. Fire prevention through oxygen reduction at Forsmark - Nuclear engineering international. Available from: http://www.neimagazine.com/features/featurefire-prevent-ion-through-oxygen-reduction-at-forsmark-4271282/ 2014.

[19] Ward PD. The life and death of planet Earth : how the new science of astrobiology charts the ultimate fate of our world. USA: Times Books 2003; p. 240.

[20] Mahaffy PR, Webster CR, Atreya SK, *et al.* Abundance and isotopic composition of gases in the martian atmosphere from the curiosity rover. Science (80-) 2013; 341(6143): 263-6. Available from: http://www.sciencemag.org/content/341/6143/263.abstract retrieved 06-04-2014. [http://dx.doi.org/10.1126/science.1237966]

[21] McCord JM, Fridovich I. Superoxide dismutase: The first twenty years (1968–1988). Free Radic Biol Med 1988; 5(5-6): 363-9. Available from: http://www.sciencedirect.com/science/article/pii/08915849-88901098.

[22] Fridovich I. The trail to superoxide dismutase. Protein Sci 1998; 7(12): 2688-90. Available from: https://www.ncbi.nlm.nih.gov/pmc/articles/PMC2143889/pdf/9865966.pdf. [http://dx.doi.org/10.1002/pro.5560071225] [PMID: 9865966]

[23] Weisiger RA, Fridovich I. Mitochondrial superoxide simutase. Site of synthesis and intramitochondrial localization. J Biol Chem 1973; 248(13): 4793-6. [PMID: 4578091]

[24] Isaksen IS, Dalsøren S. Gettinga better estimate of an atmospheric radical. Science (80-) 2011; 331(6013): 38-9.

CHAPTER 2

Oxygen: Essential Role in Life

Abstract: Oxygen is especially suited to serve as the terminal acceptor of electrons in the biological process within mitochondria of electron transfer. This process is linked *via* coupled reactions to oxidative phosphorylation that creates adenosine triphosphate (ATP), the universal storage and transfer form of energy for all aerobic life on earth. However, the unique arrangement of oxygen's electrons is conducive to one-electron transfer reactions that can produce oxygen free radicals and cause biological oxidant stress. Hans Krebs was the genius who uncovered the complex, biological cycle that we know as the Krebs cycle which revealed the long-hidden secrets of how the energy of consumed food is made available for cellular functions. Oxygen is the only element with the special combination of unique attributes required for cellular bioenergetics, the process of "cold" burning of food molecules to power the cell's energy needs. A special example of oxygen's role in "cold burning" is the firefly's luminescence.

Keywords: Apollo 13, ATP synthesis, Cellular respiration, Citric acid cycle, Coupled reactions, Cytochromes, Essential oxygen, Final electron-acceptor, Free radicals, Futile cycles, Hydrogen peroxide, Hydrosphere, Hydroxyl radical, Krebs cycle, Luciferase, Luciferin, Mitochondrial membranes, Oxidative metabolism, Ozone, Photosynthesis, Two-electron transfer, Valence electron.

INTRODUCTION

Oxygen has unique features, some of which are described in Chapter 1. As a molecule (diatomic oxygen) and as an element (monatomic oxygen) it is capable of undergoing reactions unique to living cells. Metabolism of cells is conveniently thought of in two categories: catabolism and anabolism. Catabolism refers to processes that break down chemicals to serve as nutrients for cells. Anabolism refers to the processes that create more complex molecules from simpler molecules. Growth, multiplication and repair of cells would not be possible

Olen R. Brown

without anabolism, and anabolism requires coupled reaction (subsequently to be described) to provide the necessary energy to create new bonds between atoms and parts of molecules.

Catabolism is a complex process involving many stages and many different organelles inside cells. Of interest to us here, is the fact that oxygen serves as the terminal receptors for hydrogen ions and electrons that are derived from chemicals like glucose. To better explain this process it is useful to consider how glucose brings energy into a process that occurs primarily in special organelles called mitochondria (singular mitochondrion). Organelles are intracellular structures that are bounded by membranes and designed for specific functions. With the exception of certain anaerobes, life as we know it on earth cannot exist without oxygen. This derives primarily from the fact that oxygen serves as the terminal acceptor of electrons in a long sequence of reactions whereby energy is transferred and eventually "stored" in adenosine triphosphate (ATP). It is this ATP that serves as the source of energy for life processes. Lawrence Krauss has written about this [1][1]. Respiration adds oxygen to the process as the terminal receptor of hydrogen ions and electrons removed from intermediates derived from sugar and taken through the cytochromes system where energy gets stored in ATP.

It is possible to roughly estimate the amount of oxygen consumed annually by Earth's human population. Of course, this requires assumptions and the estimate's accuracy depends on the validity of the assumptions[2]. There are data from various sources that differ. I shall make a few estimates and leave detailed calculation and evaluation to Chapter 4. A reasonable approximation of the quantity of oxygen consumed annually by the world's human population is 1.4 petagrams (a petagram is 10^{15} grams). Oxygen is produced by photosynthesis which generates one molecule of oxygen for each molecule of carbon dioxide. Photosynthesis is reported to produce approximately 550 to 642 petagrams (average = 596) of carbon dioxide annually [2]. Thus, the mass of molecular oxygen produced is 32/44 times 596 petagrams = 433 petagrams (calculated based on the molecular masses of oxygen and carbon dioxide, 32 and 44 grams). I estimate that humans use is about 1.4/433 which is 0.0032 (0.32%; approximately 1 out of every 300). Consider that most life forms on earth consume oxygen. This includes all the other land mammals, the fish and crustaceans, the arthropods, the insects, the aerobic bacteria; algae, and the plants themselves. The oxygen concentration appears to be remaining approximately steady in today's atmosphere. Humans are only a fraction of the total world's biota that consume oxygen. Is this calculation wrong that 1 of every 300 molecules of oxygen is used by human respiration? Could the data about atmospheric oxygen and carbon dioxide concentrations be incorrect? Are we missing something here that is important to our conclusions about the global balance of carbon dioxide and oxygen? I will address this important issue again in Chapter 4.

CELLULAR RESPIRATION

Cellular respiration is the process of oxidative phosphorylation within cells which results in formation of ATP. There is the intake of oxygen and the release of carbon dioxide. This is such a broad subject that I must limit content, and I shall focus on outlining biological processes at the cell level that are helpful in providing an understanding of the specific role of oxygen.

Consider a molecule of oxygen that has arrived at a typical cell. (In Chapter 8 oxygen transport is described, including the intricate mechanisms for "loading" of oxygen on red blood cells in the lungs and "unloading" at the cellular site). Oxygen is less soluble in water as temperature increases and the amount that can dissolve in blood plasma that bathes cells is relatively low at the partial pressures of oxygen normally experienced when breathing air. Calculations show that there are about 3 ml of oxygen per liter of blood at the partial pressure of 100 mmHg which approximates the concentration near to but outside of cells that are well-oxygenated. There are about 5 liters of blood in the average person and a total of approximately only 15 ml of oxygen dissolved in blood plasma; far too little to supply the needs of the body and explanations are in order (see Chapter 8).

There are no special means (such as binding to membrane proteins) for oxygen and it gets inside cells by simple diffusion which is temperature-and concentration-dependent (oxygen moves from higher to lower concentration). Oxygen can move through interstices in the lipid bi-layer of the membrane or through plasma filled channels in membranes. Thus, a never-ending supply of oxygen at the outer membrane surface of cells is required.

Once inside cells, oxygen's primary role is to receive electrons and become reduced. These electrons are transferred in a specific sequence along molecules that are collectively part of a system called the cytochrome system (more thoroughly described later is this chapter). The cytochromes are linked with, and derive electrons from, a cyclic series of enzymatic conversions known as the citric acid cycle or Krebs cycle (Fig. **1**). This cycle repeatedly accepts a two-carbon compound (as an acetyl group attached to coenzyme A) derived in most cases from glucose. In this cyclic metabolic process, the two-carbon compound is oxidized to carbon dioxide and water. The transfer of the electrons results in the coupled creation of ATP, a process called oxidative phosphorylation for reasons that may be obvious. This occurs within the membranes of mitochondria in animal cells and in the cell membrane of most bacteria by similar processes involving oxygen.

For many years, how food in the body was "converted into energy" was a mystery. It required an enormous amount of careful laboratory work, ingenuity and insight to eventually understand the cyclic nature of specific enzyme pathways with biological catalysts that not only speeded up reactions but made reactions specific and conducted in specific sequences. This provided the means for break-down products of glucose to be fed into cyclic reactions that once started, could continually regenerate the cycles. I recall first learning about the details of these pathways under the influence of a demanding, but fair, biochemical professor, I recall remarking to my fellow students before a final exam: "I expect we shall get only one question: start with glucose and make a man; show all the substrates, reactants, enzymes, coenzymes, and metals". This was not considered sexist in my day.

A most important contributor to dispelling the mystery of biological respiration was the German scientist Hans Krebs (1900-1981) (Fig. **2**). The cycle we have referred to Fig. (**1**) is known as the citric acid cycle but more usually called the Krebs cycle, after Nobel Laureate Hans Krebs [3].

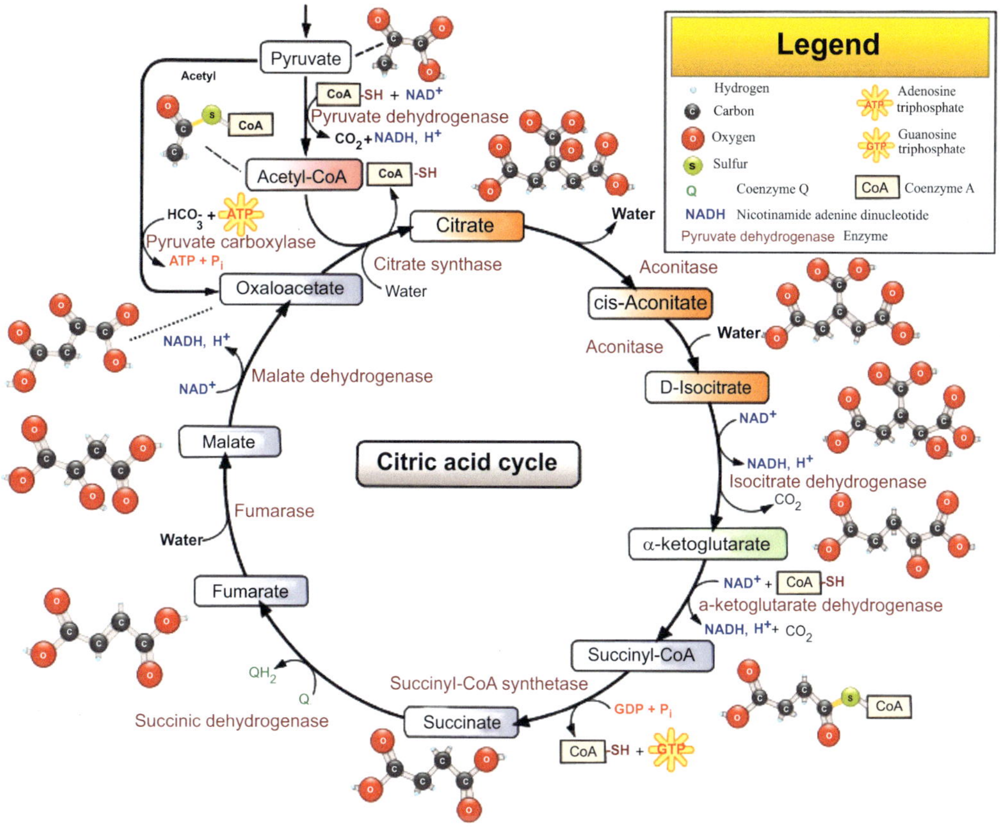

Fig. (1). Citric acid cycle (Krebs cycle).

Fig. (2). Sir Hans Krebs with his signature.

The total number of ATP molecules obtained by the complete oxidation of one molecule of glucose to carbon dioxide and water theoretically is 38. Two of these ATPs come from glycolysis (which does not require oxygen) and 36 from aerobic metabolism *via* the Krebs cycle and the cytochrome system. Thus, aerobic metabolism is 16 times as effective as the anaerobic process for energetics.

The Krebs cycle to which we have referred appropriately bears his name because by his hard effort he uncovered the mystery of how food is metabolized by cells to produce energy, ultimately by the mitochondrial system that synthesizes ATP from ADP. For this effort he was awarded a share of the Nobel Prize in Physiology or Medicine for 1953, and his award was announced as being for: "… his discovery of the citric acid cycle" [3]. He shared this prize with the great Fritz Lipman (1899-1986) who's Nobel Prize (also for Physiology or Medicine,1953) was for: "… his discovery of co-enzyme A and its importance for intermediary metabolism" [4]. Co-enzyme A is integral to the production of ATP. Krebs, for a time up to 1930 was an assistant to Otto Warburg (1883-1970) who also won the Nobel Prize in Physiology or Medicine in 1931 for "his discovery of the nature and mode of action of the respiratory enzymes" [5]. All of these discoveries were intimately related to oxygen as the biological oxidant essential to energetics.

In 1931 Krebs moved to Freiburg, Germany where his research was interrupted when the Nazis came to power in 1933 [6]. Krebs was Jewish and was fired without cause. He went to England and joined the biochemistry school at Cambridge. Krebs worked in applied science including nutrition and during WWII "he developed a bread that helped to keep the British people nourished at a time of food shortage". More detail is provided in a "remembrance" when Krebs died in 1982 [7]. He was able to emigrate before the situation for Jews became intolerable. In his memoriam [8], it is stated that the paper reporting the Krebs cycle was published in 1937 in *Enzymologia*. However, it had been previously rejected by *Nature* which recently acknowledged that Krebs cherished the rejection slip from *Nature* for his work which ultimately was recognized as worthy of the Nobel Prize [9].

Krebs married Margaret Cicely Fieldhouse and they had 2 sons. Krebs contributed to Britain's war effort during WWII by supervising research for the British

Medical Council [10]. The origin of the "bread" statement (see above) was that he was credited with helping to develop the national whole meal loaf that "kept the English people well-nourished through the war years despite food shortages" [6, 10, 11]. Krebs also studied ways to preserve liver for transplantation and contributed to the field of inborn errors of metabolism [12]. He was interested in many subjects and even published on juvenile delinquency [6, 12]. Krebs' papers have been cited tens of thousands of times [6, 13]. During one period, more than a dozen British chairs of biochemistry were occupied by his students [6, 8]. In 1958 he was knighted by England, his adopted country [7].

In researching the life of Krebs[14], I came across an interesting newspaper article from 1953 [15]. It was published in the newspaper *Association of Jewish Refugees in Great Britain* (AJR) in London[3]. The writer was Marie Sackinand and the article was entitled: "Hans A. Krebs, The Nobel Prize Winner from Hildeshelm". It was, obviously written for the occasion of his Nobel Prize, and Hildeshelm refers to his birthplace in England. A passage by Sackinand will be quoted at length since no transcription would do it justice:

> "When I first heard the news of his having won the award my mind turned back to April 1, 1933, Boycott Day... two S.A. men were standing outside the surgery of Professor Krebs's father, an eminently efficient surgeon in Hildesheim, our common home-town. The S.A. men were carrying a poster saying *"Don't consult this Jewish"*. It was on the same day that the weekly German Zionist paper published a leading article entitled *"Tragt ihn mit Stolz, den gelben Fleck"* (Wear the Yellow Star with pride). If I then felt proud to have my lot thrown in with Dr. Krebs rather than with the two S.A. men I feel prouder still to-day that the highest award should have been granted to one whom the Nazis disdained. In Professor Krebs the two factors which might make a man great are combined: genius and a passion for work... Professor Krebs' industry is boundless. It is not known at what time he begins work in the morning– nobody has ever arrived in the laboratory before him– but his wife tells me that he continues working at home every night till 11 p.m. His interest in his work

> makes him immune from noises and it does not disturb him in the least if his own three children as well as a number of their friends play round his desk. When they were smaller he is said to have so completely forgotten their presence that on crossing the room he occasionally stepped on some of them… His only relaxation is gardening… he grows Bavarian 'Bierrettich' vegetable which is hardly eatable but of which he is prouder than of the "Krebs-Cycle". It is only natural for a man like Professor Krebs not to have to smoke or drink… His assistants treat him as if he were the absent-minded kind of professor and tell him that his tie is not straight or that he forgot to comb his hair."

Thus, Krebs and others contributed to discovering important aspects of the production of ATP in cells for which they received Nobel Prizes. This is a testament to the essentiality of ATP for life, and to the importance of the oxygen required for its biochemical generation.

The production of ATP results in storage of some of the energy of glucose in a form universally available for energy-requiring processes which include synthesis of components needed for growth and repair, and for the energy requirements of all the cells of the body including the muscles and the brain. Because of the significant advantage of aerobic respiration in energy production it is hard to imagine that large, multicellular animals, including humans could exist anaerobically, *i.e.* without oxygen. We shall address limitations that oxygen overcomes further here and in Chapter 7 and Chapter 10. We will do this in relation to limitations imposed by multicellular status and the benefits of oxygen as a small gaseous molecule to serve as source of electrons and as CO_2 to rid the cells of carbon wastes.

OXIDATION-REDUCTION AND FUTILE CYCLES

Oxidation, which we have described with preliminary details, cannot occur in the absence of reduction: a molecule is oxidized for each molecule that is reduced. There must be a balance and electrons that are lost from one compound must be gained by another: oxidation is a loss in electrons and reduction is a gain in electrons [15, 16]. Therefore, oxygen *via* reduction creates oxygen radicals, and metals can be good sources of electrons for creating these radicals. Metals, including iron, copper and manganese have significant roles in biological oxidation and in the toxicity of some chemicals. For example, the herbicide paraquat can have oxidized and reduced forms. In the presence of oxygen, electrons can be transferred to oxygen with formation of toxic, oxygen radicals.

The term futile cycle has been used in several contexts including synthesis and degradation of glucose with a net consumption of ATP and generation of heat. Here, I use the term futile redox cycle in a restricted context where it relates to a reducing reaction and an oxidizing reaction that are coupled with no net change in the chemical component. For example, compounds such as quinones can cycle between radical and non-radical species with the consumption of oxygen and production of oxygen radicals and reactive oxygen species. In my own work I have investigated this as contributing to the mechanism of toxicity of the herbicide paraquat and the cancer drug doxorubicin (trade name Adriamycin©). In a reaction-couple involving oxidation and reduction (redox reaction), there can be a continuous reaction that seems to produce no net results (futile cycle). However, there may be an input to the cycle and an output. The input can be a compound that will be consumed, and the output can be an oxygen radical that is toxic or it can be linked to formation of various other oxygen radicals including the hydroxyl radical – said to be the most oxidizing substance in chemistry. The relevance here is that the hydroxyl radical can oxidize any biological chemical. The reactivity of the hydroxyl radical is ameliorated somewhat by being limited to its diffusion radius (the distance it likely travels before it contacts some molecule). When the encounter is with DNA the damage of free radical oxidation can well be a mutation of a single base, or even damage to the structure of the chromosome. It is possible that this could result in cancer. When a large number of oxygen free radicals are formed, there can be significant damage to membranes, enzymes and other structures. It is more likely, however, that free radical reactions will occur with abundant, nonessential molecules (including water) where the damage may be so slight and with consequences so minor as to be of no biological consequences and even be non-detectable.

ENERGY CONVERSION AND STORAGE

Coupled Reactions

It is a simple fact that life could not exist without ATP [17]. There must be a means for energy to be stored and moved about in cells. This is required because many biological reactions will not occur at appreciable rates without an input of energy. To permit energy "releasing" reactions to "pull" energy "requiring" reactions, coupled reactions are utilized.

Coupled reactions are two reactions that share an intermediate in common that links the reactions. It is important as a concept because it allows linking of a reaction that releases energy with one that requires energy. Life would not be possible without coupled reaction. For example, glucose that is derived from consumed food contains energy that exists in the bonds holding the hydrogen, oxygen and carbon of glucose together. When these bonds are broken, energy is released. This energy is usually released as heat, but when there is a coupled reaction, some of the energy can be conserved in the bonds of other chemicals. A reaction that will not occur at a measurable rate (an "uphill" reaction) can be made to occur by "coupling" it with an energy-releasing ("downhill") reaction. The coupling occurs *via* a chemical compound (an intermediate) that occurs in both reactions. Thus, the energy-requiring reaction (uphill) can be made to occur. This is the basis of the ability of living cells to synthesize all of their biochemicals and structures from simpler molecules. A useful analogy is to think of a stream flowing downhill with waterfalls at various points. The energy in the flowing water can be captured by waterwheels at the waterfalls and used to generate electrical energy. We have repeatedly referred to the molecule adenosine triphosphate (ATP). A space-filling model of its chemical structure is shown in Fig. (**3**). In this model, as is traditional, the spheres represent individual atoms and in this space-filling approach, spheres are merged.

Colors depict different atoms. Traditionally, black is carbon, hydrogen is white, nitrogen is blue, oxygen is red and phosphorous is orange. ATP has a molecular mass of 507 grams per mole (oxygen's atomic mass is 16) and the formula (where subscripts indicate the number of atoms) for ATP is: $C_{10}H_{16}N_5O_{13}P_3$. The 13 atoms of oxygen are scattered about the structure where they make up approximately 41% of the mass of ATP. There are three groupings of atoms (called phosphate groups) that contain phosphorous, hydrogen and oxygen, and they are at the left-hand end of the molecule in this model. These phosphate groups can be lost one at a time, leaving ADP (adenosine diphosphate), and then AMP (adenosine monophosphate). Of course, the reverse (gaining phosphate groups) occurs during the generation of ATP.

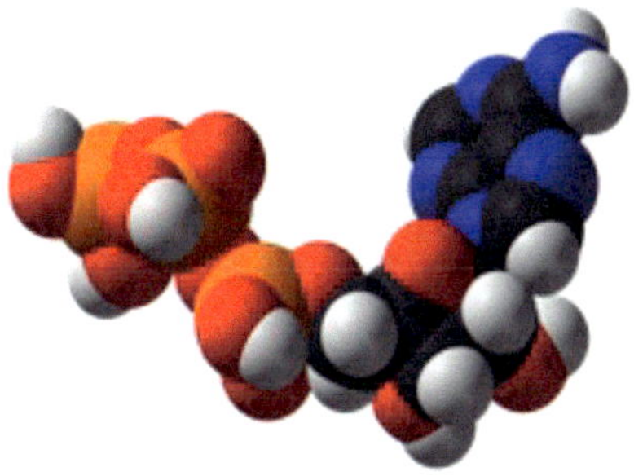

Fig. (3). Space-filling model of adenosine triphosphate (ATP).

The molecules of ADP and AMP, by analogy, are often described as partially uncharged "batteries" compared to the ATP which is a fully-charged battery. Although the bond between the last phosphorous in the chain and oxygen is often called a "high energy" bond (for understandable reasons) there is really nothing special about the energy in this bond. However, when it is broken there is much less energy left in the cleaved parts and some of this energy can be used to do the work of coupled reactions to create chemical bonds as needed for the assembly of larger molecules. ATP readily diffuses about inside the cell from its site of formation by the system of membrane-bound enzymes and the cytochrome chain in mitochondrial respiration to sites where it is utilized. This is significant: energy does not have to be "generated" at specific sites where it may be required, but energy can be transferred *via* coupled reactions to energy-requiring reactions in cells. By this means these reactions, which must be otherwise thermodynamically permitted reactions, are accelerated greatly. For life to function efficiently, this is an absolute requirement: that the ATP thus generated can diffuse from its site of formation to sites requiring energy throughout cells. If ATP did not exist, something essentially like ATP would have to take its place for life as we know it to exist on the Earth.

THE CYTOCHROME SYSTEM IN MORE DETAIL

The cytochrome system, as shown in Fig. (**4**), takes its hydrogen (protons, hydrogen ions) from the coenzyme NADH generated by the citric acid (Krebs) cycle. These hydrogen ions and associated electrons are removed enzymatically from metabolites derived ultimately from (primarily) glucose. The electrons can be viewed as flowing down an energy gradient of cytochromes embedded in the inner mitochondrial membrane where at each of three places there is sufficient energy to add phosphate groups to ADP and have energy left over to satisfy the 2nd law of thermodynamics. Oxygen is shown interacting with cytochrome IV with reduction to water. This is where and how oxidative phosphorylation occurs in our cells: in the inner membranes of mitochondria which are organelles present in most cells of our body. An exception is the mature red blood cell which has no mitochondria; its mitochondria having been extruded during later stages of maturation, apparently to save space for carrying oxygen, it's essential function. Mature red blood cells then must make-do with the less efficient anaerobic glycolysis to synthesize their ATP. The mitochondrial inner membrane provides the architecture for imbedding the cytochromes themselves in the required sequence.

In cellular biosynthesis, self assembly of intricate cell components is required. This is true for the cytochromes. Although all cytochromes have similar properties each is unique and functions in sequence to direct the flow of electrons from, cytochrome b, then to c, then to a, and finally to a_3. The cytochromes have just the right chemical properties to permit efficient transfer of electrons down an energy gradient.

A model of cytochromes c with monomeric protein surrounding the heme group, containing iron inside, is shown in Fig. (**5**). In the final steps, cytochrome c transfers electrons to a complex called cytochromes IV that has associated cytochrome c oxidase, the last enzyme in the respiratory electron transport chain.

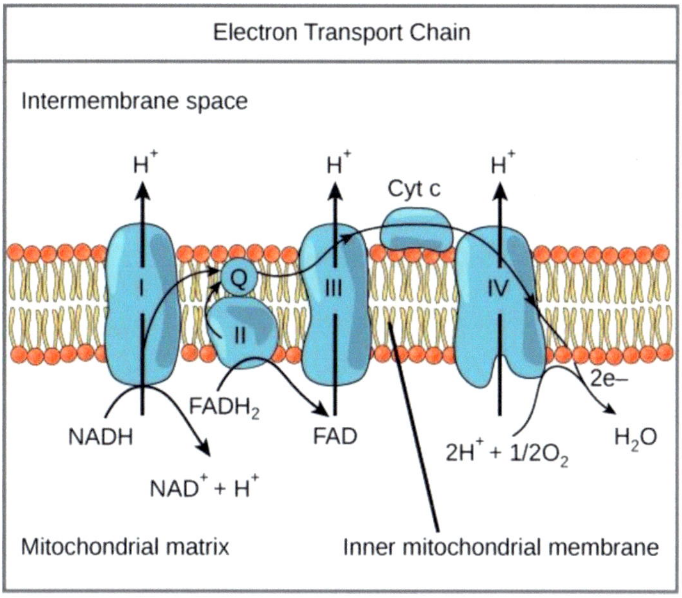

Fig. (4). The cytochromes system synthesizes ATP using O_2 as the final acceptor of hydrogen and electrons.

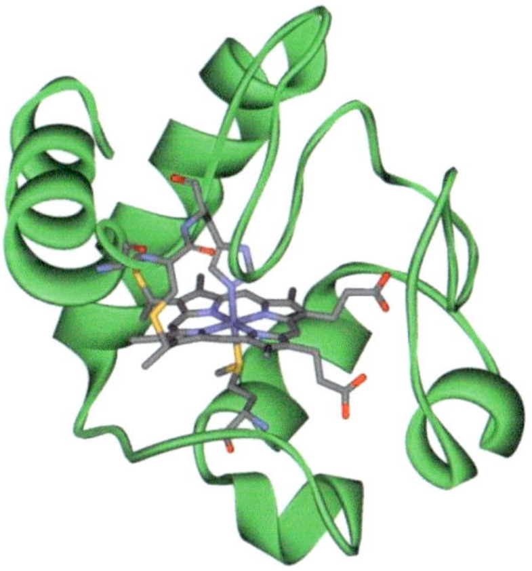

Fig. (5). Ribbon diagram (model) of Cytochrome c.

Cytochrome oxidase receives an electron from each of four cytochrome c molecules and transfers them to a single molecule of oxygen which is reduced completely to water. There are two pathways from the citric acid cycle to the cytochromes as shown previously (Fig. **4**). Hydrogen ions are shown being "pumped" into the inter membrane space. It is the inner-directed gradient of hydrogen ions that results in synthesis of ATP from ADP. ATP would not continue to be created if the electrons were not drawn out of the system by combining with oxygen. Oxygen must permeate the membranes to reach the matrix of the mitochondrion. Oxygen has just the precisely correct chemical properties including its diffusibility, its electronegativity, and the nature of its valence electrons to do this job. No other element suffices.

MEMBRANES AND ATP PRODUCTION

Biological membranes are complex in structure and function. After a brief overview, I shall focus on the mechanism by which membrane-enclosed structures of multicellular animals, with the use of oxygen, generate ATP. Biological membranes provide a variety of functions the simplest of which is to separate what is cell from the non-cell surroundings, or one cell from another cell. There are also membranes that create intracellular organelles (compartments) enclosing specific components with specific functions. There is a membrane around the cell nucleus and other intracellular organelles also have membranes. Our focus shall be on the double-layered membrane surrounding the mitochondrion.

Membranes contain lipids and the simplest structure, not quite a membrane, is the outer surface of a lipid droplet in water. Such droplets are created whenever a salad dressing is made by mixing oil and vinegar. The water and oil remain separate until well mixed to create small droplets of oil dispersed in the water-based vinegar. This results because oil is hydrophobic (water-hating) and water is hydrophilic (water-loving) and elegantly organized. Biological membranes have flexible, double-layered boundaries with intercalated proteins. The hydrophobic boundary excludes entry of molecules based on size, solubility and charge. The latter are organized to form pores in the membrane with gate keepers for some and preferred passes to other molecules. Phospholipids contain phosphate groups, combinations of oxygen and phosphorous, that ionize to leave a negative charge. Phospholipids thus have an uncharged portion (a chain containing carbon) that is hydrophobic and that associates with the same part of other phospholipid molecules. The charged phosphate groups, usually at one end of the molecule, are hydrophilic and orient to associate with water. Thus, in the water environment of a cell, phospholipids will naturally assemble to form a bi-layer with the "tails" of phospholipids together and the "heads" (phosphate groups toward the polar exteriors of the bi-layer). This creates an enclosed structure, with water inside and outside. Mitochondria have two such membranes; the inner and outer membranes with a space in between that is water-based. It is the inner membrane that is modified to contain a complex assemblage of chemical structures that are arranged with a specific architecture. This includes the cytochromes and enzymes that function in production of ATP (Fig. **6**). The cytochromes are proteins that contain heme groups and they function in the transport of electrons from lower energy levels creating a gradient as the electrons are passed from cytochrome to cytochrome. The heme group contains iron which readily can be oxidized and reduced with the aid of enzymes and the iron carries the electrons in the electron transport process.

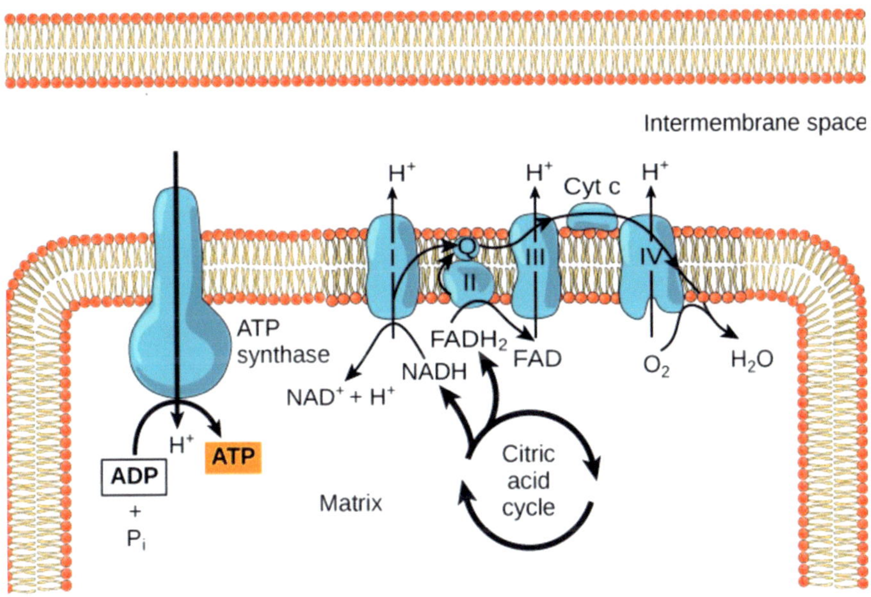

Fig. (6). Electrons and oxygen participate in the inner mitochondrial membrane to generate ATP.

Cytochromes c reductase (in Complex III) and also cytochromes oxidase (in complex IV) are necessary for this oxidation and reduction (Fig. **6**). These and other enzymes and cytochromes are arranged sequentially in the inner mitochondrial membrane (Fig. **6**). Electron transfer by the chain of cytochromes results in hydrogen ions (H^+) on the outside of the membrane. This higher concentration provides a proton gradient from outside to inside the mitochondrial membrane to provide the source of energy to synthesize ATP from ADP.

The flow of electrons (Fig. **6**) moves through the sequence from complex I to II, to II, to IV. Cytochrome c reacts with oxygen inside the membrane and water is produced. It is the elevated concentration of protons (H^+) that create the gradient that, with ATP synthase, transfers energy to create the third phosphate group on ATP. It is this step that produces ATP. Without the presence and participation of oxygen as the terminal acceptor of hydrogen ions and electrons, the flow could not continue. However, the source of the hydrogen ions and electrons is originally from food. The intricate design of this system, from the specific chemical make-up of the proteins and the membranes together with their precise integration into the architecture of the cell, is required for the life process to work.

ESSENTIAL OXYGEN

Total Oxygen in the Human Body

I do not know a way to directly and precisely measure the total oxygen content of the body. Gaseous and dissolved oxygen obviously varies with the respiratory cycle and the state of oxygenation of blood and tissues. A comparatively constant amount exists as oxygen molecules and as oxygen atoms in body chemicals.

However the amount of oxygen in the human body can be estimated as one way of examining its essentiality. Oxygen is present in the lungs, blood, tissues, cells and many different molecules. The body is approximately 50% to 60% water which itself is 89% oxygen. Only a small amount of oxygen is dissolved in the blood; generally less than 5% of that carried by red blood cells. Approximately 5 to 6 liters of air is the average maximum capacity of the human lung and about 20% of this is oxygen. Myoglobin of muscle binds oxygen and it is primarily in cardiac muscle and in skeletal muscle. The quantity of myoglobin in major muscles can be raised by exercise training. The majority of oxygen is carried by hemoglobin which varies with gender and other factors. Many of the body's chemicals contain oxygen in various amounts but only six different elements comprise about 99% of the body's mass: oxygen, 65%, carbon 18.5%, hydrogen 9.5%, nitrogen 3.2% calcium 1.5%, and phosphorus 1.0% [18]. Trace elements and rounding errors account for the remaining 1%. This may be surprising as one might tend to think that carbon is the major element in life unless one thinks about the abundance of water which is the solvent within which all the biochemistry of the living cells occurs. This is in addition to water's participation in many essential biochemical reactions plus the fact that water participates as a product in the terminal cytochrome reaction. These facts are votes in favor of oxygen as essential and paramount in life.

Extreme Breath Holding

Holding one's breath quickly demonstrates the essentiality of oxygen. Based on my own observations and personal experience, breath-holding by untrained individuals is possible usually for less than two minutes. (This should not be confused with the semi-medical test that times the interval, when otherwise breathing "normally", to the first stimulus to take a breath). Indeed, intentional breath-holding does not represent the end of oxygen availability but the limit imposed by the accumulation if carbon dioxide which is the stimulation to take the next breath. This, rather than the lack of oxygen, limits duration of breath-holding by novices. Breath-holding under water is a competitive sport with various rules for different events. Goran Colak has the Guinness world record for the longest oxygen-aided breath hold of 22 minutes and 30 seconds [19], set in 2013. This is an amazing time even under conditions of minimal exertion. It is made possible by training and, most significantly, by breathing pure oxygen which fills the lungs and "saturates" the tissues.

A more natural type of breath-holding is exhibited by the Ama (sea women) who are Japanese pearl divers [20]. They do considerable work in diving up to about 25 meters seeking abalone and other shellfish, and sometimes pearls. Ama are said to remain submerged, working heavily for more than 2 minutes, and to do this up to 40 times per day. They have trained themselves to repeatedly dive while breath-holding, even while swimming vigorously in cold water.

Some athletes, however, have trained diligently to dive even deeper into the ocean, under special test conditions using a cable for descent/return and other safety features. Great women divers include the women's 2002 record holder, Tanya Streeter, who dived to a depth of 525 feet [21]. The recent world record for free diving, set in June 2012, is held by Herbert Nitsch with a depth of 831 feet [22]. Extreme times are possible with training, and by incorporating a period of rapid and deep breathing with pure oxygen. This purges the system of carbon dioxide, provides five-times the oxygen in the lung, and also maximally loads the hemoglobin, a minor effect, and myoglobin. Pure oxygen breathed briefly in this manner is not harmful. It also increases the amount of oxygen dissolved in blood and tissue fluids which is only 5% when breathing air. Minimizing effort also reduces oxygen utilization and resting heart rates far lower than the "normal" of approximately 70 beats per minute are achieved. The biggest factors for record times are increased lung capacity, breathing pure oxygen (to increase oxygen stored and also saturation of hemoglobin and myoglobin), purging of carbon dioxide, and minimizing metabolism during the test.

There are three basically different types of human muscle (smooth, cardiac, and skeletal) and they have quite different capacities to store oxygen as well as other differences [23]. We previously described the requirement for rapid recycling of ATP (which requires oxygen to be efficient). It has been calculated [24] that a single cell (the report does not identify the cell type; we may assume the average for the body's total cells) uses about 10 million ATP molecules per second; thus its ATP molecules are recycled every 20-30 seconds. Muscle function is ultimately limed by oxygen supply.

A person can live for a few days without water and several weeks without food, but only minutes without oxygen.

What Bird Metabolism Reveals about the Need for Oxygen

Birds, to fly, must be light, strong and efficient in energy expenditure. Aerobic respiration using oxygen is ideal and birds have special design features. Ducks, for example, are capable of rapid and long distance flight and they have flight muscles that are dark in color because of their high myoglobin content that stores oxygen. Turkeys bred for Thanksgiving dinner cannot fly at all, and have large breasts that are white and myoglobin deficient. Dark meat is rich in myoglobin which turns reddish in the presence of oxygen. This is the reason the cut surface of your steak at the meat market appears dark red. Sometimes, special lighting is used to make this appear brighter red which is more appetizing. Myoglobin stores and delivers more oxygen to hard-working flight muscles of ducks.

Birds have another adaptation that aids in oxygen delivery; they have air sacks in addition to lungs. These air sacks can even extend into hollow bones. Also, birds have a different mechanism of respiration than humans; they have muscles that directly expand the lung rather than a diaphragm that creates reduced pressure as with the human lung. An excellent website reference [24] has detailed wonderful illustrations including animated depictions of the differences in the ways humans, birds, and insects "breathe". Ultimately, inspired oxygen must pass a blood-to-gas barrier to enter cells. Comparisons of the thickness of this barrier reveal this barrier is thinner in birds than in mammals, and also is thinner in smaller birds than in larger birds. For example the blood-to-gas barrier is 0.099 micrometers in the small, violet-eared hummingbird, but 0.56 micrometers in an ostrich, and even thicker in large land mammals [24].

Indeed, the humming bird is a most interesting case. These tiny birds have very high wing-beat rates and R. K. Suares and colleagues reported that the humming bird has a high density of mitochondria in wing muscles that may be near theoretical limits and also has a higher rates of mitochondrial respiration than is found in human muscle. Humming bird lungs are capable of almost 10-times higher oxygen diffusing capacities than the lungs of similar-sized mammals and they can have heart rates as high as approximately 1400 beats per minute with hearts twice as large as expected based on scaling considerations (information cited by Suarez, *et al.* [25]).

The tiny canary also takes in oxygen very efficiently. This may explain why canaries are more susceptible to toxicity from gases such as carbon monoxide, and this has been used by miners to alert them to the presence of deadly carbon monoxide underground in mines. The mechanisms underlying the canary's susceptibility are debated, but it seems plausible that, for its size, it respires more and takes in more of the toxic carbon monoxide.

All of the factors we have described point to oxygen as limiting for performance. Life on earth is designed or has adapted to using oxygen most efficiently.

What Can Insects Tell Us

Insects today are limited in size although fossil evidence indicates that in the past they attained much greater size. It is apparent that the delivery of oxygen within the insect body is a limiting factor for size. Paleontologists generally subscribe to the theory that in the past when there were giant insects, the atmospheric oxygen concentration was much higher than the steady approximately 21% oxygen in the Earth's atmosphere today. Instead of lungs with oxygen delivered by a complex system of heart, blood, lungs, arteries, veins and capillaries, insects have a simple system of spiracles (openings) and internal tubes and sacs [26]. Air that supplies the necessary oxygen generally enters by simple diffusion, but some insects have a rudimentary system to pump air in; and the aquatic, larval stages of some insects actually have gill-like structures. Perhaps the largest insect is the tropical Goliath beetle which is only about 15 cm long and its size is probably constrained by the ability of this insect's oxygen delivery system, in total. Humans and other large mammals have elaborate systems that deliver oxygen to all body cells to make large size possible.

Some insects (particularly fireflies) glow by a process called luminescence which requires oxygen. You may have memories that as a child you caught fireflies on a summer's eve. I have seen my grandchildren at first be concerned that fireflies might be hot to touch, and be relieved that they were not. The firefly's light is a cold, enzymatically-produced light and the process is called bioluminescence to indicate light of biological origin. It is an example of the use of enzymes in a controlled "burning" that we have previously discussed. The firefly uses the flashing light as a signal to help find a mate and, as children, we reaped a pleasant benefit [27].

Oxygen, the element we are focused on, is the source of the electrons that light the firefly's lantern. The system of structures and enzyme that permit and control this process are complex and unique [26 - 29]. Luciferin is a small molecule that is the substrate for the enzyme luciferase (Fig. **7**) shown in vastly different scales. Luciferase uses oxygen from the atmosphere in a reaction that generates an excited-state biochemical intermediate that then decays to the ground-state with emission of light [30]. Luciferin and luciferase have the root Latin word *Lucifer* that means "light-bringer".

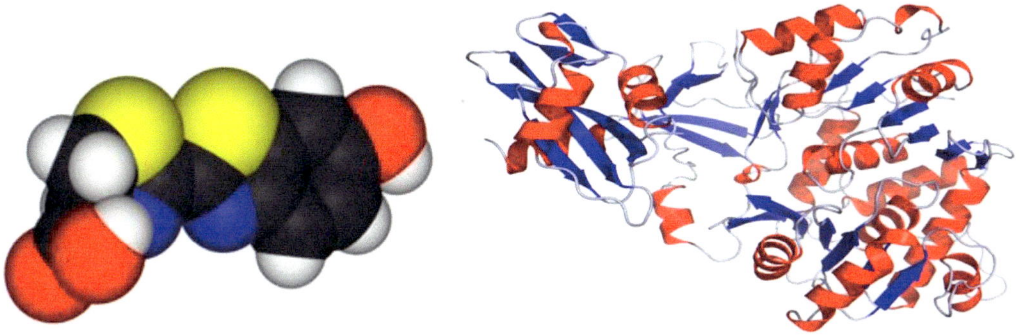

Fig. (7). Space-filling model of firefly luciferin (left) and, which is the substrate, and firefly luciferase, the enzyme (right), shown as a ribbon diagram of the protein structure.

For many years the basic components in firefly luminescence has been known [31]. Credit goes to Bruce Branchiti and colleagues for providing a more complete picture of the chemistry of the process [29]. It was known that oxygen and luciferin were involved; however, based on known chemistry, they were unlikely to react to produce light. Branchiti showed that the superoxide radical form of oxygen, with its unpaired electron was what caused light to be emitted.

The firefly's lantern only lights up when oxygen is available at the enzyme site; when the oxygen is used up, the light goes off [32]. This on-and-off reaction is rapid; too rapid to be done by the slow speed of muscles that control the flow of oxygen. Nitric oxide gas (the same gas your dentist may have offered you) is present in the light-emitting structure and helps to resolve the mystery. When the firefly light is off, no nitric oxide is being produced and oxygen is bound to the mitochondria and not available to the light organ. When nitric oxide is generated, oxygen is available at the light-emitting site. Nitric oxide forms very quickly and breaks down very quickly; sufficiently quickly to account for the rapid on-and-off flashing of the firefly lantern. The flash rate is carefully regulated, is species specific, and aids fireflies in identifying mates. The reaction is approximately 80% to 90% efficient in transforming chemical energy into light. By comparison, the ordinary incandescent bulb (now being replaced by act of Congress) is only about 10% efficient at converting electrical energy into light. Most of the energy of the incandescent light bulb is released as heat, and a typical LED (light emitting diode) while more efficient, only converts about 20% of the electrical energy into visible light [33].

What Can the Apparent Limitation on Cell Size Tell Us

The rate of diffusion of oxygen into cells is a limiting factor for size (volume) attained by cells. Observation tells us that as a cell becomes larger, the cell volume grows more rapidly than the cell surface area. (The formula for the area of a sphere is $4 \pi r^2$ and the formula for the volume of a sphere is $4/3 \pi r^3$). We can do the math, or we can cheat and, by inspection, note that if we divide a sphere into smaller spheres, the volume does not change but the surface area increases. Thus, the amount of surface membrane available for each unit of volume gets bigger. The converse is also true, and cells with large volumes eventually will have too little surface area to permit sufficient nutrients and/or oxygen to supply their needs.

Cells that comprise the human body differ considerably in size and shape. An "average" cell has a diameter of approximately 10 microns (a micron is a millionth of a meter) with considerable variation in shape. The largest human cell is the unfertilized ovum which approaches the limits of human vision (approximately 100 microns). A typical liver cell (hepatocyte) is cuboidal (primarily because close packing forces cells into this shape) rather than spherical. Axons of nerve cells can be very long in comparison to diameter of a cross section, and can stretch from the lower spine to the toe. If we go beyond human cells, there is controversy but the unfertilized egg of the ostrich can be half-a-foot long and weigh about 3 pounds. Most of this is stored food and the region which, when fertilized, divides to produce a new ostrich is much smaller. The giant squid has nerves that are said to be up to more than 3 yards in length and the giraffe's neck nerves can be a several feet long. The size of cells is certainly limited by the ability of nutrients including oxygen to enter through the cell membrane and the oxygen concentration within cells is described as a gradient. Obviously, being long and thin increases the surface area compared to a spherical cell. Oxygen, with all its special features, still becomes limiting for cell size.

What can Cyanide Poisoning and Time to Anoxic Brain Death Tell Us

In our search for evidence of the limitations placed on life by availability of oxygen, we can consider the length of time required for the cellular poison cyanide to kill. Cyanide kills by blocking the terminal step in oxygen mitochondrial oxygen uptake (a step previously described). Death by this poison thus might seem to give an estimate of the time we can remain alive without oxygen. It does not, however, because cyanide, although deadly, requires a rather large dose, compared to the deadliest of poisons. Indeed, cyanide was considered and even tested as a war gas in WWI. It was very ineffective, primarily because it was difficult to deliver lethal doses to soldiers through the respiratory route. Thus, the time required for death is related primarily to the time required for a sufficient dose of cyanide to be delivered and this makes it not very precise for determining how long life continues without oxygen.

However, with confidence we can say that the death of brain cells is the factor that limits life of multi-organ animals including humans without oxygen. The brain primarily uses glucose to derive its ATP and consumes about 20% of the total oxygen consumption when a person is at rest. The brain is only about 2% of the weight of the body but it receives approximately 15% of the heart output of blood and uses about 25% of the glucose consumed at rest. The brain cannot perform its essential functions without ATP and without oxygen ATP is used up and not recharged. It is generally known that a time of approximately 4 minutes or slightly more of anoxia usually results in brain death. Again, however, this does not prove that the lack of oxygen itself directly produces the limiting time we seek to know. The survivability of other body cells is known to vary during periods of anoxia and heart muscle is more sensitive than most. Protection of cells from anoxia involves complex relationships during re-oxygenation and this will be described further in subsequent chapters in relation to oxygen-radical formation. Although proof is lacking, it seems reasonable that availability of oxygen ultimately constrains size, activity, and even survivability of aerobic life forms.

The Near Tragedy of Apollo 13

Two days after the historic launch on April 11, 1970, the mission of Apollo 13 had to be aborted. Apollo 13 was commanded by James Lovell (1928-) with crew members John L. Swigert (1931-1982), and Fred W. Haise (1928-). It was the 7th manned Moon launch and it was intended to be the 3rd to land men on the Moon. This was not to be. An explosion erupted in the No. 2 oxygen tank in the service module (Fig. **8**) which was separate from the command module that contained the astronauts. The tank was ruptured and the fittings to the No. 1 oxygen tank were damaged and all the oxygen was lost from both tanks in the service module within about 3 hours, plus loss of water, electrical power, and use of the propulsion system [33 - 35]. Fortunately, there was oxygen in the command module and lack of oxygen was not the dominant concern. The lunar module was designed with consumables to sustain two people for 1.5 days, not three people for four days. The most serious problem concerned carbon dioxide produced by the astronauts themselves. The lunar module was equipped with lithium hydroxide canisters but they were not sufficient for the entire crew for the time required, and spares were beyond the crew's reach in the descent-stage module. The command module had an oxygen supply but it was mechanically incompatible [33, 34]. Engineers on the ground improvised a plan so the astronauts on board could attach the cube-shaped, command-module's canisters to the lunar module's cylindrical-shaped canisters. The astronauts named their rigged contraption "the mailbox". There were other problems which were also overcome and the astronauts returned safely [34].

Fig. (8). Apollo 13 Service Module damaged by explosion of oxygen tank number two; somewhat out-of-focus photo taken by crew.

The oxygen lost by the explosion supplied the fuel cell that provided power, potable water, and heat for the mission. There were provisions for back-up oxygen for the crew; however, the critical factor was removal of carbon dioxide, the by-product of the metabolism of the astronauts. A movie was made of the mission in 1995 with Tom Hanks as Lovell. The real-time, aborted Apollo 13 mission captivated the public. Because of the low power, communications during the return were limited and the public was in considerable suspense until the safe landing was reported in the South Pacific. In 2016, Lovell gave an interview with Jim Clash [36]. Responding to a question about the fearful time when the oxygen tank exploded, Lovell said:

> "Well, we were all apprehensive. But you have to have a positive attitude, number one. And number two, we were all from test pilot backgrounds, so naturally this was an adventure... Okay, here's the problem. What do we have to work with? Is it immediate that we have to do something? ... It turned out that the explosion had crippled the Command Module. So we still had the Lunar Module, which we used as a lifeboat to get home."

CONCLUDING REMARKS

In this chapter, we have encouraged reflection on the essential chemical and biochemical characteristics of oxygen for life on Earth. Thoughtful people can disagree about what constitutes life. As far as I can establish, life exists only on Earth. I have used the words "life as we know it", but maybe this is superfluous. Indeed, there are known forms of life that do not require oxygen, the anaerobes. They are the exceptions: bacterial in nature, and single cells, not multicellular organisms.

If we focus on "higher" multicellular forms of life, we can trace a clear path for the essentiality of oxygen. We deduce, at the cellular level, life must have a means of replicating its components with great (essentially complete) fidelity. For replication, life has DNA (deoxyribonucleic acid) arranged in a double helix that allows enzymes to unwind it to expose its code and make exact copies. Enzymes have the ability to serve as catalysts to speed up reactions and direct and control specific arrangements of atoms into specific molecules. This requires storage and retrieval of information. The entire process requires energy and ultimately depends on oxygen. Transcribing the information of DNA into RNA requires special proteins called enzymes. The machinery for protein synthesis is complex and uses the information from DNA; specific, transfer RNA molecules; and intracellular structures called ribosomes that act, by analogy, like factories to make proteins. There is a requirement for membranes to separate what is the cell from its exterior milieu and to separate functions like those within the nucleus and mitochondrion. For these functions, there is a necessity for a "fungible" source of energy which is ATP. Therefore, the essential characteristics of life surely include fidelity in preserving information, fidelity in using information, fidelity in ability to separate domains with membranes, fidelity in storing and transferring energy to pull coupled reactions, and (with some noted exceptions) it requires oxygen.

Although oxygen is a part of the molecules that make up the DNA, RNA, proteins, and membranes, its role as an acceptor of electrons in biological oxygen is most striking. Thus, without oxygen's ability to induce ATP to be generated in plentiful supply with water and carbon dioxide as readily eliminated by-products, complex multicellular plants and animals would not be possible. The lack of any one of these processes would prevent the development and functioning of life. All require ATP and ultimately rely on oxygen.

NOTES

[1] Lawrence Krauss wrote: "The average male human uses about 420 pounds of ATP each day of his life to power his activities. Considering that we all contain less than about 50 grams of ATP in our bodies at any one time, that involves a lot of recycling. Specifically, each molecule of ATP must be reenergized at least 4,000 times each day. The respiration of oxygen, even though it is tied to the very last stage of energy production in our bodies, increases the energy output obtained from a sugar molecule from 2 molecules of ATP purely *via* the process of fermentation, to 38 molecules of ATP when respiration is added".

[2] Chosen values are: 7.2 billion people as the earth's population (in may be 7.4); 30% under 15 years of age and average body weight of 22 kg; 70% over 15 years

of age with average body weight of 70 kg for an average weight for all of 56 kg; oxygen uptake of approximately 9.5 grams oxygen /kg-day, from default value, US Environmental Protection Agency.

[3] The newspaper continues to be published today and it is archived online.

REFERENCES

[1] Krauss LM. Atom, an Odyssey from the Big Bang to Life on EarthAnd Beyond. 1st ed., Little, Brown and Company 2001.

[2] Welp LR, Keeling RF, Meijer HAJ, *et al.* Interannual variability in the oxygen isotopes of atmospheric CO2 driven by El Niño. Nature 2011; 477(7366): 579-82. Nature Publishing Group, a division of Macmillan Publishers Limited. All Rights Reserved. [http://dx.doi.org/10.1038/nature10421]

[3] Anon. Hans Krebs - Biographical. Available from: http://www.nobelprize.org/nobel_prizes/medicine/laureates/1953/krebs-bio.html 1953.

[4] Anon. Fritz Lipmann - Facts. Available from: http://www.nobelprize.org/nobel_prizes/medicine/laureates/1953/lipmann-facts.html 1953.

[5] Anon. Otto Warburg - Facts. Available from: http://www.nobelprize.org/nobel_prizes/medicine/laureates/1931/warburg-facts.html 1931.

[6] Ede A. Hans Adolf Krebs. Available from: http://www.chemistryexplained.com/Hy-Kr/Krebs-Hans-Adolf.html 2014.

[7] Anon. Current Comments. To Remember Sir Hans Krebs: Nobelest, Friend, and Advisor. Essays of an Informational Scientist. Curr Contents 1982; 5: 627-33.

[8] Krebs HA, Johnson WA. The role of citric acid in intermediary metabolism in animal tissue. Enzymologia 1937; 4: 148-56.

[9] Anon. Monument to a Giant. Nature 1981; 296: 294.

[10] Evans EA, Burton K. The 1953 Nobel prize awards. Science (80-) 1953; 118: 711-2.

[11] Anon. Nobel prize shared. Sci News Lett 1953; 64(275)

[12] Blaschko H. Hans Krebs: nineteen nineteen and after. FEBS Lett 1980; 117 (Suppl.): K11-5. [http://dx.doi.org/10.1016/0014-5793(80)80565-5] [PMID: 6998728]

[13] Krebs H. New Horizons of human knowledge - Biographical and medical aspects of juvenile delinquency. Horizons of human knowledge. New York: Unesco Press 1981; pp. 79-88.

[14] Krebs H. Reminiscences and reflections. Oxford: Oxford University Press 1981.

[15] Sackin M. The Nobel Prize Winner from Hildesheim. Association of Jewish Refugees in Great Britain (AJR) in London. 1953; VIII.(12)

[16] Clark J. Definitions of oxidation and reduction (redox). Available from: http://www.chemguide.co.uk/inorganic/redox/definitions.html 2013.

[17] Bergman J. The perfect energy currency for the cell. Available from: http://www.trueorigin.org/atp.asp 2014.

[18] Darling D. Biological abundance of elements. Encyclopedia of sciene. Available from: http://www.daviddarling.info/encyclopedia/E/elbio.html 2016.

[19] Liang J. Croatian freediver goran colak breaks guinness world record for longest O2-aided breath hold. Deeper Blue. Available from: https://www.deeperblue.com/croatian-freediver-goran-colak-br-aks-guinness-world-record-longest-o2-aided-breath-hold/ 2013.

[20] Sepper M. Pearl Jewellery. Available from: http://mariliissepper.com/blogs/news/14279253-ama-the-japanese-pearl-divers 2016.

[21] Anon. Tanya streeter breaks both the men's and women's variable ballast freediving world records. Available from: http://www.scubadiving.com/article/news/tanya-streeter-breaks-both-mens-and-wom-ens-variable-ballast-freediving-world-records 2014.

[22] Nitsch H. The deepest man on earth. Available from: http://www.herbertnitsch.com/ 2016.

[23] Anon. Human Physiology/The Muscular System. Wikibooks. Available from: https://en.wiki-books.org/wiki/Human_Physiology/The_Muscular_System 2016.

[24] Anon. Avian Respiration. Ornithology. Available from: http://people.eku.edu/ritchisong/birdrespirat-ion.html 2016.

[25] Suarez RK, Lighton JR, Brown GS, Mathieu-Costello O. Mitochondrial respiration in hummingbird flight muscles. Proc Natl Acad Sci USA 1991; 88(11): 4870-3.
[http://dx.doi.org/10.1073/pnas.88.11.4870] [PMID: 2052568]

[26] Anon. The open door web site : Biology : The tracheal breathing system of insects. Ornithology. Available from: http://www.saburchill.com/chapters/chap0021.html 2014.

[27] Buck J, Case J. Flash control and female dialog repertory in the firefly *Photinus greeni*. Biol Bull 1986; 170: 176-97.
[http://dx.doi.org/10.2307/1541802]

[28] Stryer L, Berg JM, Tymoczko JL, Stryer L. Biochemistry. The calvin cycle and the pentose phosphate pathway. San Francisco: W.H. Freeman. Available from: http://www.ncbi.nlm.nih.gov/books/bv.fcgi?-rid=stryer 2002.

[29] Branchini B. Chemistry of firefly bioluminescence. Available from: http://photobiology.info.branchi-ni2.html 2016.

[30] Anon. Luciferase. FireflyScience.org. Available from: http://fireflyscience.org/luciferase.html 2016.

[31] Bittel J. How do fireflies glow? Mystery solved after 60 years. National Geographic. Available from: http://news.nationalgeographic.com/2015/07/150724-fireflies-glow-bugs-summer-nation-science/ 2015.

[32] Anon. How and why do fireflies light up? Scientific American. Available from: http://www.scientificamerican.com/article/how-and-why-do-fireflies/ 2005.

[33] Wilson E. What's That Stuff? Chem Eng News 1999; 77(3): 65.
[http://dx.doi.org/10.1021/cen-v077n003.p065]

[34] Anon. Mission operations report apollo 13 Available from: http://history.nasa.gov/alsj/a13/A13_MissionOpReport.pdf

[35] Williams DR. The apollo 13 accident. Available from: http://nssdc.gsfc.nasa.gov/planetary/lunar/ap13-acc.html 2016.

[36] Clash J. Astronaut Jim Lovell tells how he dealt with fear aboard apollo 13. Available from: http://www.forbes.com/sites/jimclash/2016/06/25/astronaut-jim-lovell-tells-how-he-dealt-with-fear-aboard-apollo-13/#48ffc1e24136 2016.

CHAPTER 3

Discovery and History of Oxygen

Abstract: Oxygen was first isolated by Carl Scheele, officially discovered by Joseph Priestly (because he published first), and named by Antoine Lavoisier. Air had historically been known to be required for both animal respiration and for combustion. With the careful laboratory work of Lavoisier, involving weighing and measuring, air was discovered to not be a single element but was found to be composed primarily of two gases, one of which was oxygen in the proportion of about 20%. Priestley published that in a sealed chamber this "pure form of air" (later named oxygen by Lavoisier), was what kept the mouse alive in his experiment. The work of Priestley and Lavoisier disproved and overthrew the Phlogiston Theory which speculated that a substance "phlogiston" was present in air which after combustion became dephlogisticated. The discovery of oxygen as an element and its roles in animal respiration and in combustion, were seminal events in the period of the late 1700s when chemistry transitioned from a largely "black art" (alchemy) to science. Thus, in addition to being essential to life (certain bacteria are exceptions), oxygen had a significant role in the development of modern chemistry.

Keywords: Air, Albert Szent-Gyorgyi, Benjamin Franklin, Bomb calorimeter, Burning, Calorie, Carbon dioxide, Carbonation, Combustion, Dephlogisticated air, Father of chemistry, Guinea pig experiment, Isolation-purification, Lavoisier, Linus Pauling, Marie-Anne Pierrette Paulze, Mouse experiment, Oxidation, Oxygen, Phlogiston, Priestley, Pure oxygen, Rubber eraser, Scheele, Scurvy, Second Law, Unitarian minister, Vitamin C.

DISCOVERY OF OXYGEN

Introduction

It is commonly written by historians of science [1, 2] that oxygen was first isolated (separated from air) by Carl Scheele (1742-1786), officially was disco-

Olen R. Brown

vered by Joseph Priestley (1733-1804) and was named by Antoine Lavoisier (1743-1794). However, the history of the discovery of oxygen is controversial because of how and when these scientists made their discoveries and when they published their results. There is evidence that Scheele was somewhat introverted, was very meticulous, delayed to report, and never aggressively sought credit during his lifetime. The evidence is, however, clear that Scheele was the first to isolate oxygen and this occurred in 1771-1772, but he did not publish until 1777 in his *Treatise on Fire and Air.* Priestley, however, published his work dealing with the isolation and some experiments with the gas in 1774.

The research by these men, especially Priestley, was foundational for setting precedents for chemical research that involved meticulous record keeping and careful observation. A confusing and incorrect theory of gases, called the phlogiston theory, dominated chemistry in the 1770s [3]. Phlogiston was incorrectly thought of as the "stuff" that was released from burning of a substance. More specifically, the phlogiston theory was an ancient idea that a "fire-like" substance called phlogiston was present in all combustible substances, and was released by combustion. Georg Ernst Stahl (1660-1735) is usually credited as the chief proponent and developer of the phlogiston theory which dominated Chemistry in Europe during his scientific lifetime and up until near the end of the 18th Century [4]. A primary proponent was Joachim Becher (1635-1682) who was a German chemist, physician and adventurer who thought that matter was made of three "earths": the vitrifiable, the mercurial, and the combustible. He experimented with changing sand from the Danube into gold and his extraordinary claims (which apparently were believed by some, at least for a time) led to his disgrace and more strikingly, forced him to flee [5].

The phlogiston theory was an unsuccessful attempt to explain combustion and rusting of metals (we now call the latter oxidation and know oxygen is involved). The phlogiston theory was overturned largely because of the work of Lavoisier (subsequently to be detailed in this chapter) that greatly advanced chemistry by proving that combustion required oxygen. This Lavoisier documented by carefully weighing the masses of substances before and products after reaction in closed containers. This proof of the role of oxygen in combustion solved a mystery and overthrew a complicated and ugly theory but also contributed greatly to what became known as “the chemical revolution” that stirred the scientific world beginning near the end of the eighteenth century.

Oxygen is a small, diatomic molecule. A classic, but now little used, depiction of diatomic oxygen shows the inter-lapping of the orbitals of shared electrons that create the double bond between atoms (Fig. **1**). Two atoms of oxygen are shown each with their outer orbitals completed by 8 electrons with 2 shared. The double bond results in four electrons that can be removed, one at a time, and this has consequences that will be later discussed. Oxygen is a colorless gas at ordinary temperatures near the earth’s surface. It becomes liquid at -182.96 °C, and is referred to as Lox. Oxygen freezes at -361.82 degrees, and it has a very large expansion ratio at ordinary pressure and temperatures; therefore it can be compressed as a good storage form of breathable oxygen.

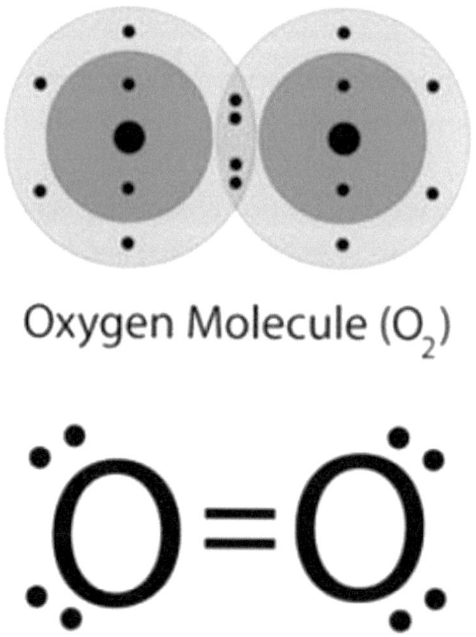

Fig. (1). Classical representation of diatomic oxygen.

Joseph Priestley

A thorough and authoritative account of Priestley’s life, research and discoveries concerning oxygen was assembled by Carmen Giuta of LeMonde College [6].

This on-line document has many references, including many to original publications. It provides insights into and quotations from the works of Priestley and his contemporaries who contributed to the transition of chemistry from ancient, mystical speculations into what became the foundations of modern, laboratory chemistry. In this document, under "Notes", Giuta records that Priestley often stated that he stumbled upon many of his discoveries by chance. Not being guided by an hypothesis, he was often surprised. It is interesting that Priestley emphasized the importance of having the right equipment to perform laboratory experiments. This has become even more important over the years and the quaint laboratory instruments of that day have evolved into complex machines that perform precise measurements and even chemical synthesis and robotic manipulations. Even pipetting of solutions is now done automatically. The cost of such equipment often makes it difficult for Research Deans to separate need from desire to keep up with one's neighbors down the hall.

Credit for the discovery of oxygen is almost universally given to Joseph Priestley (1733-1804) (Fig. **2**) for research done in 1774 and first published in 1775 [7]. Priestley, at the time of the discovery, was living and working at Boswood House in Wiltshire which was owned by Lord Shelburne (1737-1805) who was Priestley's benefactor. It was a common practice in that day for a wealthy person to support scientists. Priestley experimented by collecting pure oxygen gas by focusing the sun's rays on mercuric oxide, a red powder that was commonly known to produce gases when heated. He tested the gas on himself and on a mouse. Schofield [7] recorded that Priestley discovered that his isolated gas was: "… five or six times better than common air for the purpose of respiration… and I believe for every use of common atmospheric air". Priestley also found that plants produce oxygen, a discovery that he did not pursue but one which we shall.

Priestly was the first to isolate oxygen in a form that was essentially pure oxygen. He noted that a candle burned much brighter in pure oxygen than in what he called "ordinary" air. This caused him to wonder (in his written musings) whether pure oxygen would be therapeutic or unhealthy for humans. Indeed, in subsequent years, oxygen research has resulted in useful therapies, breathing gas mixtures for outer space and deep under the sea, and high-pressure oxygen chambers to treat human diseases. However, as I shall describe, the toxicity of oxygen limits its therapeutic uses just as Priestley had mused.

Fig. (2). Joseph Priestley.

It is historically interesting that more than 240 years ago, Priestley wrote a description of an experiment with oxygen that is intriguing today [6]. The air of which Priestley speaks was derived from "mercurius calcinatus" (an oxide of mercury). Priestley's account[1] is novel:

> "On the 8th of this month I procured a mouse and put it into a glass vessel, containing two ounce-measures of the air of mercurius calcinatus. Had it been common air, a full-grown mouse, as this was, would have lived in it about a quarter of an hour. In this air, however, my mouse lived a full hour; and though it was taken out seemingly dead, it appeared to have been only exceedingly chilled; for, upon being held to the fire, it presently revived, and appeared not to have received any harm from the experiment."

These quaint words of Joseph Priestley describe important biological properties of oxygen still studied today. Oxygen is required for human life; it becomes toxic in pure form; and it can be life-saving as a therapeutic in some circumstances. Thus, oxygen truly is both boon and bane (see Chapter 7).

When Priestley was conducting his experiments, chemistry was developing rapidly out of alchemy (an early form of chemistry practiced in the Middle Ages). Alchemists sometimes sought to change base metals into gold and hoped to discover an elixir that would keep people alive and forever young. Mankind has an innate and pervasive desire to extend life, and to find deep meaning including a desire for eternal life. The latter is clearly supernatural and not reasonably a part of the experimental sciences today. Priestley like many scientists of his time was interested and involved in many things besides science, and this included religion. Indeed, Schofield [7] described Priestley as learned in an unusual variety of fields: theology, grammar, education, aesthetics, politics, and natural philosophy. Unfortunately, not many scientists in today's world are interested (or versed) in more than one of these today. He was well known for his Unitarian religious beliefs which included anti-Catholic and anti-Protestant sentiments. Expressing his very liberal (for the time and place) theology got him into trouble and his church and home were burned. This persecution caused Priestly to flee England in 1794 for the United States, which was greatly enriched by his presence.

Priestley also investigated scientific curiosities other than oxygen. One such was India rubber, a substance brought back to England from the New World discovered by Columbus. The prolific author Isaac Asimov (1919-1992) [8], credits Priestley with naming the eraser that today is on the common pencil, a "rubber" because it would "rub-out" the marks made by a graphite pencil.

Other details of Priestley's life are noteworthy. It was at about the time that the American Colonies gained independence from England, that Priestly discovered oxygen. As stated, he subsequently came to America, and there he met Benjamin Franklin. Some consider Franklin to have been the first great American scientist and his dangerous experiment (which he fortunately survived) with a kite, a key, and a thunderstorm is legendary. Franklin, better known today as a statesman, raised funds that built a hall in Philadelphia which was much later reported by

Time magazine [9] to have been "*...expressly for the use of any preacher of any religious persuasion who might desire to say something*". Priestly learned a great deal from Franklin about life in the American Colonies. It would be interesting to know what they shared about science. It is not recorded whether he used Franklin's hall.

Priestley was, however, deeply religious. As a young man he had studied to become a Unitarian minister, while pursuing science only as a hobby [7]. In later life, next door to his church in Leeds, England was a brewery and he became curious about the foam produced during the brewing of beer and ale. This foam results from carbon dioxide released during fermentation by yeast. Priestly found that carbon dioxide dissolved in water to produce a fizzy, pleasant, tart taste– was this the first soda pop?

Priestley's work with oxygen was foundational for today's current interests in the basic biology and chemistry of oxygen, and the medical interest in using anti-oxidants to thwart the harmful effects of oxygen at the cellular level in the body. His interests in oxygen involved also an indirect connection to Priestley's studies of carbon dioxide, to a supposed cure for scurvy, and eventually to the discovery of vitamin C.

Vitamin C cannot be made, at least not adequately, by humans; thus, it must be part of the diet. Without vitamin C, a disease called scurvy results and it can be fatal. The French Ministry of Finances asked Lavoisier (about whom more will be written later in this Chapter) to investigate a treatment for scurvy that Priestley reportedly had discovered– that "fixed air" (actually carbon dioxide) when dissolved in water made the carbonated drink that Priestley had fooled with. This "fizzy water" was, widely and erroneously, reported to prevent scurvy. Scurvy was a serious illness of sailors on long voyages. Scurvy limited exploration by sea. England, with its large navy and dependence on ships for trade and for defense, was deeply in need of a preventative for scurvy. Carbonated water was a disheartening, false lead and not a preventative or cure.

However, the juice of the lime (which as it turned out, has large amounts of vitamin C) accidentally was found to prevent scurvy. This lead to limes being taken on long sea voyages by the English, and to English sailors being called "limies". Many years later, Albert Szent-Gyorgi (1893-1986) who won a Nobel Prize for other work, discovered vitamin C in fruits such as the lime and in paprika, which he obtained from his wife's kitchen and serendipitously found was a rich source for isolating the vitamin. Szent-Gyorgi went on to show why scurvy results from a deficiency of this vitamin. In addition to protection against scurvy, vitamin C is a strong antioxidant that protects cells against the damaging, oxidizing effects of oxygen free radicals; thus, there is another connection to our oxygen story. However, there is little firm scientific data that large doses of Vitamin C are helpful. Anyone who is today taking a supplement containing a large amount of vitamin C does so, aware or not, largely because of the work of another scientist, Linus Pauling (1901-1994), the only person to win two unshared Nobel Prizes (Nobel Chemistry and Nobel Peace). (I shall write specifically about Pauling and his universally acknowledged great scientific work in Chapter 7). Indeed, there is continuing controversy about the amount of vitamin C needed by humans. Linus Pauling has promoted the idea that vitamin C is required in large amounts and is not just a vitamin needed in trace amounts. Many scientists disagree. The use of high doses of vitamin C as a water-soluble anti-oxidant to destroy oxygen free radicals for the prevention of heart disease and cataracts, and prevention and treatment of the common cold are not approved by the U.S. Food and Drug Administration.

Carl Scheele

Carl Wilhelm Scheele (1742-1786) was born in Pomerania, then a part of Sweden [10]. It is said that he had little formal education and his training in science began when he was apprenticed at an apothecary in Gothenberg where he developed knowledge and extraordinary laboratory experimental dexterity while working with primitive laboratory equipment [11 - 13]. There is some controversy about credit to Priestley for the discovery of oxygen because Carl Wilhelm Scheele (in 1771 and 1772) discovered that heated manganese oxide evolved a gas that he called "fire air". He studied the properties of the gas but did not publish his findings until 1777. In a consensus resolution of this difficulty, the Royal Society of Chemistry and the American Chemical Society joined in honoring Priestley (not Scheele) on August 7, 2000 by designating the site at Bowood House in

Calne, Wiltshire, England as "An International Historic Chemical Landmark" for the discovery of oxygen on August 1, 1774. It also recognized the significance of Priestley's research for the development of the founding principles of modern chemistry [14].

Because of his role in the discovery of oxygen, Scheele later was recognized for his dedicated work and in 1775 he was elected into the Stockholm Academy of Sciences [10]. Perhaps because of his economic circumstances he delayed marrying the widow of a friend until two days before his death. It is reported that he worked long hours often in cold and drafty conditions [10, 13]. He experimented with prussic acid, which generates cyanide gas, and described its odor and taste not knowing of its poisonous nature. He also worked with the poisonous metal arsenic. It has been speculated that these factors may have contributed to his death at the age of only 43; however, his cause of death remains officially unknown. He published late and little about oxygen although many credit him with its discovery in spite of this honor officially going to Priestley. He discovered chlorine and manganese, and made other scientific contributions. However, he was misled by the popular phlogiston theory and he referred to the oxygen he had discovered as "fire air". Scheele is listed among the 40,000 or more people in the Notable Names Database (NNDB) where biographical facts are posted [13].

Lavoisier

The life and work of Antoine Lavoisier (1743-1794) is fascinating for what it reveals about the chemistry, politics and social history of the times in which he lived. His contributions relative to the chemistry of oxygen make his story essential. He had the unusual, for its time, assistance of his wife Marie-Anne Pierrette Paulze (they married when she was 14 and he was 28) [15 - 17].

Lavoisier helped establish a new formal language to use in describing chemical reactions. Although he did not originate the idea of the "Law of Conservation of Matter" in chemical reactions, his careful work clearly established this universal principle. This justifies Lavoisier being called by many, the "Father of Chemistry" (Fig. **3**). Thus, he was deeply involved in the revolution that overtook all of

Chemistry in the 1700s and 1800s and his studies with oxygen were a part of this. He also was involved in the political French Revolution and it also greatly affected his life. Tragically, Lavoisier was beheaded in 1794, when he was only fifty-one years old, on false charges and because he had helped collect taxes for the government, the losing side, during the French Revolution [16, 17].

Fig. (3). Lavoisier with laboratory equipment used in investigating oxygen.

Essentially all historians whose works I have read agree in proclaiming Lavoisier as the "Father of Chemistry". This was primarily for his use of, and insistence on, carefully weighing and measuring reactants and products of chemical reactions. With this principle he led in the conversion of chemistry from a qualitative art (sometimes "black art") to a quantitative science. This careful approach helped establish that the mass of reactants in a chemical reaction did not change but could be altered in form. This concept was shared by others, but he was its dominant proponent.

He was the son of a wealthy attorney and inherited a large fortune at the age of 5 on the death of his mother. He studied law but never practiced law. In his mid-twenties he bought a share in the *Ferme General* which had the right to collect taxes. This later was one source of charges that lead to his execution. His wife, Marie-Anne was the daughter of a senior member of the *Ferme General*. She edited and published his memoirs, translated articles from English for him (Fig. **4**), and assisted in the laboratory (Fig. **5**) [16].

Fig. (4). Reproduction of a portrait of Antoine Laurent Lavoisier and his wife Marie-Anne Pierrette Paulze, by the French painter Jacques-Louis David in 1788, and now in the New York Metropolitan Museum of Art.

Lavoisier also did extensive experiments on combustion and erroneously concluded that all acids contained oxygen, one of his few recorded errors in deduction made from his laboratory experiments.

He met Joseph Priestley and they discussed their work on oxygen. He, correctly, was a strong opponent of the phlogiston theory. His opposition primarily was responsible for laying this theory to rest, and he published important research in: "*Reflections on Phlogiston*" (English translation of title). In collaboration with Pierre Simon de Laplace (1749-1827), he synthesized water by combustion and proved water was not an element. Among other contributions, he helped reform chemical nomenclature, and wrote what is considered to be the first textbook on chemistry: "*Elementary Treatise on Chemistry*", published in 1789. Oxygen chemistry was prominently featured in this text [17].

Fig. (5). Lavoisier in center and his wife, Madame Lavoisier, assisting at right with others, one of whom is having his basal metabolic rate measured.

Lavoisier's research in respiratory physiology grew out of his interest in combustion where similarities had long been noted by many scientists. He constructed the first laboratory device (known today as a bomb calorimeter) for measuring the energy content (as heat) of a substance. In Lavoisier's experiment, the heat produced from a confined, live guinea pig was measured. He reported on this in a memoir: "*On Heat*". Interesting details from this work were also included in a 2004 paper: "*Is a Calorie a Calorie?*" by Andrea C. Buchholz and Dale A. Schoeller [18]. A quote from this paper is instructive for insight into the times:

> "Because humans and animals are warm and animal heat is the essence of being alive, that question, as variously phrased, was, what is the innate fire, the vital force, animal heat... it was somehow related to food."

These authors described that Lavoisier's laboratory device had an outer shell that was packed with snow which melted to maintain a constant zero degrees, around an inner shell, holding the guinea pig. As the snow melted from the guinea pig's heat the water was collected and weighed. From this, the released heat was determined. In 10 hours the guinea pig produced 29.6 large calories of heat, allowing Lavoisier to conclude: "... *la respirationest done une combustion"*. Which translates roughly to: respiratory gas exchange is a combustion. Indeed, it can be compared to a candle burning with regard to heat production. It was many years before the work of Krebs and others (reported in Chapter 2) resolved the mysteries that primarily involve the magic of enzyme catalysts and the ultimate participation of oxygen as the final acceptor of hydrogen ions and electrons in the (relatively) "cold burning" of food where the excess heat based on the Second Law of Thermodynamics, disclosed the secret. We may recall that the Second Law (also called the Law of Increased Entropy) requires that the usable energy of a closed system decreases over time. Usable energy is required for life processes.

John Dalton

John Dalton (1766-1844) (Fig. **6**) contributed to the chemical history of oxygen, although somewhat indirectly. He is most famous, of course, for his "Atomic Theory"; however, this work relied on observations made with gases, including oxygen. An interesting quote is reported by Fred Senese [19], from a publication by Dalton[2]:

> "We might as well attempt to introduce a new planet into the solar system, or annihilate one already in existence, as to create or destroy a particle of hydrogen."

The idea of atoms is said to have arisen in his mind as a physical concept developed from his intense study of the physical properties of the air and other gases [19]. This site also gives biographical details. He was born in England into a Quaker family. He attended a Quaker school where he was appointed an instructor when he was only 12 years old, and he was made principal as a teenager. Throughout his life he avidly pursued an interest in meteorology.

Fig. (6). John Dalton.

Dalton also is credited with discovering that red-green color blindness is hereditary [20]. There is an interesting connection here. A modest Quaker, he is reported to have refused many honors but did accept an honorary doctorate of science degree from Oxford University. Being awarded this honor ordinarily necessitated that one wear a red graduation gown; however, as a Quaker he ordinarily was required to wear black. However, being color blind, red was a color Dalton could not see. This provided the excuse he needed (used) to disregard the Quaker rule against wearing red [21][3]:

> "I believe that asceticism is in a very deep sense contrary to the real Quaker spirit, which desires in all things to abstain from any interference 'in the will of man' with Divine discipline... with the same aversion as it entertains for pre-arranged forms of worship. Friends... often believed themselves required to submit to the adoption of plain dress... but the plainness was not devised for that purpose, but chosen... because of its inherent suitableness and rightness... subordinating everything to principle. Its chief

significance is that of a protest against bondage to passing fashions… our very dress should show forth that inward quietness of spirit which does not naturally tend towards outward adornment, and the Friends' recognized dress is therefore one of extreme sobriety in colour and simplicity in form."

HISTORY OF OXYGEN

Introduction

In the 2nd Century BC, a Greek philosopher named Philo of Byzantium (ca. 280 BC- ca. 220 BC) wrote concerning his ideas about the gases in air[4]. In these writings, Philo described inverting a glass vessel over a burning candle with the neck of the vessel submerged in a saucer of water. Philo theorized that some part of air was converted into fire (then thought to be one of the fundamental elements: fire, earth, air, water) to account for the fact that the water rose in the neck of the vessel as the candle burned. We now know, of course, that burning of the candle consumed oxygen from the air and this caused the reduced pressure that drew the water into the space now containing less gas. This reinforces the truth that the "experiment is always right", it is just our interpretations that may be wrong.

Much later, Leonardo da Vinci (1452-1519) observed that some part of air was consumed during combustion. This was insightful, even remarkable, although the observation was not linked to the actual discovery of oxygen which did not occur until more than 250 years after Leonardo's death.

Brief History of Oxygen Formation on Earth

In this chapter to this point, some of the early historical period of the discovery of oxygen from the standpoint of the chemists involved has been described. In Chapter 1 the creation of oxygen in the universe, as currently proposed, was reviewed including its formation by solar reactions involving physics, not chemistry. In Chapter 4, the focus will be on that part of oxygen's history that relates to the continued formation of oxygen by photosynthesis. There is another history of oxygen: the story of its formation in the early atmosphere of the earth.

Malcolm Dole [22] has told this story well and much more completely than I shall attempt here. It is logical that the primitive earth retained only those gases that could be retained by chemical fixation: the inert gases escaped. Oxygen, as we have stated, exists on Earth in combinations as water, silicates, and other compounds, and as molecular oxygen (dioxygen). Dole [22] states that the early history of the chemistry of atmospheric oxygen began when the earth's temperature favored a photochemical process for its formation. Above an altitude of approximately 10 kilometers, in the troposphere, oxygen formed and diffused downward under the influence of gravity where it rapidly was mixed in the troposphere by wind and temperature-induced turbulence. Most of this oxygen was removed by reactions (oxidations) with elements in the earth's crust. Calculations show that the abundance of the accumulated, atmospheric oxygen was small and insufficient to provide much shielding from the ultraviolet rays of the sun.

However, algae subsequently accelerated production of oxygen *via* photosynthesis. Indeed, it is estimated that all the present atmospheric oxygen could have been created within the past 2,000 years by photosynthesis alone. In studying this, Berkner and Marshall [23] made complex comparisons of ratios of oxygen isotopes from different sources, including in mixed air and in oxygen newly produced *via* photosynthesis. By examinations of the isotope O^{16}/O^{18} ratios from these sources, they concluded that there is an approximate balance of oxygen consumption and generation by photosynthesis. Thus, oxygen concentrations appear to have remained approximately in a steady state for long periods into the past.

Ozone, a Special Form of Oxygen

Discovery and Significance

Ozone is a colorless to pale blue gas with a pungent smell and it can be detected by the sensitive noses of some people at a concentration as low as 10 parts per billion. For comparison, ozone is 1,000 times as "smelly" as ordinary alcohol (ethanol). Indeed, the name ozone is derived from the Greek word *ozein* meaning smell and it is sometimes detected around electric motors. The odor of ozone was associated with electrical storms before it was chemically identified in 1939 by Christian Frederich Schonbein (1799-1868) (Fig. **7**). These, and many other facts about ozone are provided by Mordecai B. Rubin in a detailed history of ozone [24].

Fig. (7). Christian Frederich Schonbein.

The molecular formula of ozone (**O_3**) was identified (Fig. **8**) by Jacques-Louis Soret (1827-1890) in 1865. The Soret peak (a strong absorption band seen in hemoglobin by spectroscopy) was named for him. This intense peak absorption at the blue end of the visible spectrum of wavelengths is also seen in the cytochromes. Ozone is too low in concentration to have any effect on sky color.

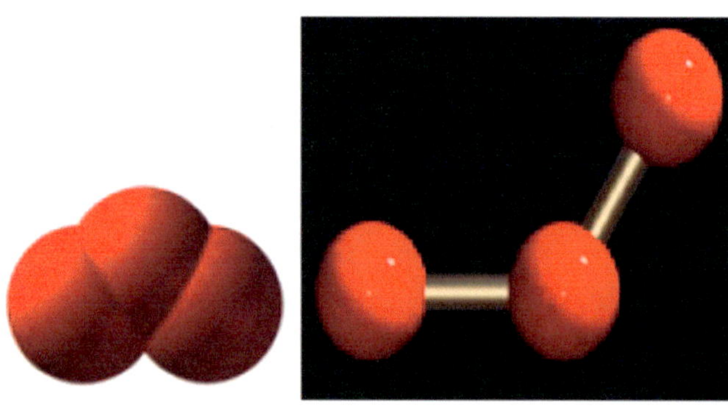

Fig. (8). Tri-atomic space-filling model (left), and ball and stick model (right) of ozone.

Ozone is found at ground level as a part of smog generated by chemical pollution and can reach levels that are harmful to humans. Thus, ground-level ozone is considered undesirable ("bad") because it is associated with various upper respiratory complaints. The U. S. Environmental Protection Agency (EPA) regulates it as a hazardous chemical and efforts are made to reduce ozone generation. The permissible exposure level for humans was lowered in 2008 to 75 parts per billion [25].

Ozone also is found in a "layer" in the upper atmosphere (Fig. **9**) where it blocks some UV light which otherwise would reach harmful levels on earth. This ozone is described as "good" ozone in contrast to "bad" ozone, previously described to accumulate primarily from human activity near the earth's surface in the breathing zones of humans.

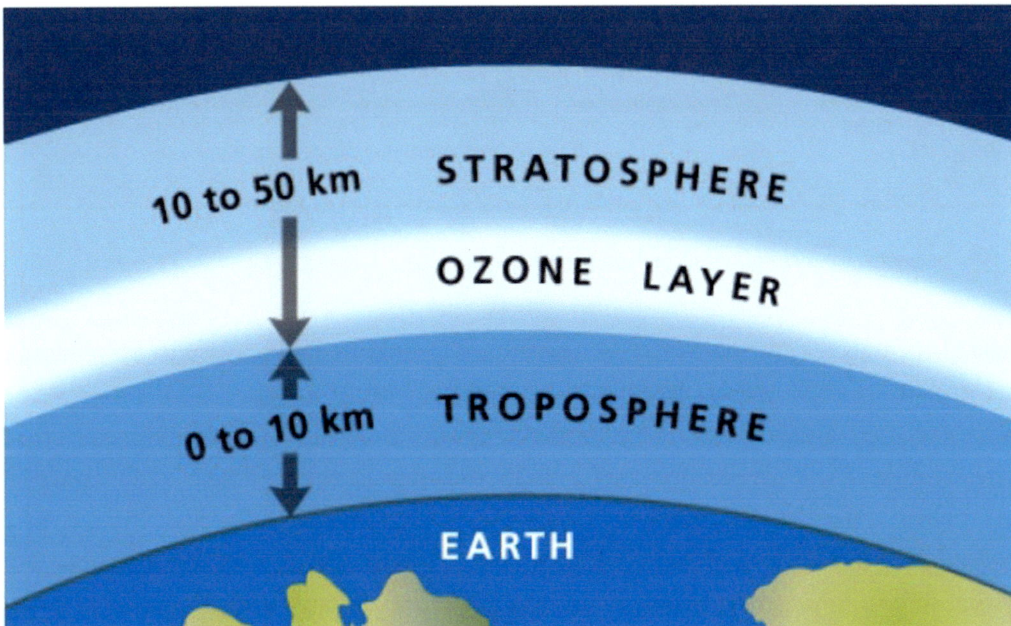

Fig. (9). The ozone layer in the stratosphere absorbs harmful UV rays from the Sun.

Ozone has become known to the general public for two reasons. There were news reports, almost daily, that ozone had a beneficial role in the upper atmosphere, but these reports also said that it was being depleted by man-made chemicals that were escaping into the air. These chemicals included certain chloroflurocarbons (CFCs) that were used in air conditioners in our homes, commercial buildings, and automobiles. Chlorine can be liberated by UV to attack and destroy ozone with chlorine released to recycle (Fig. **10**).

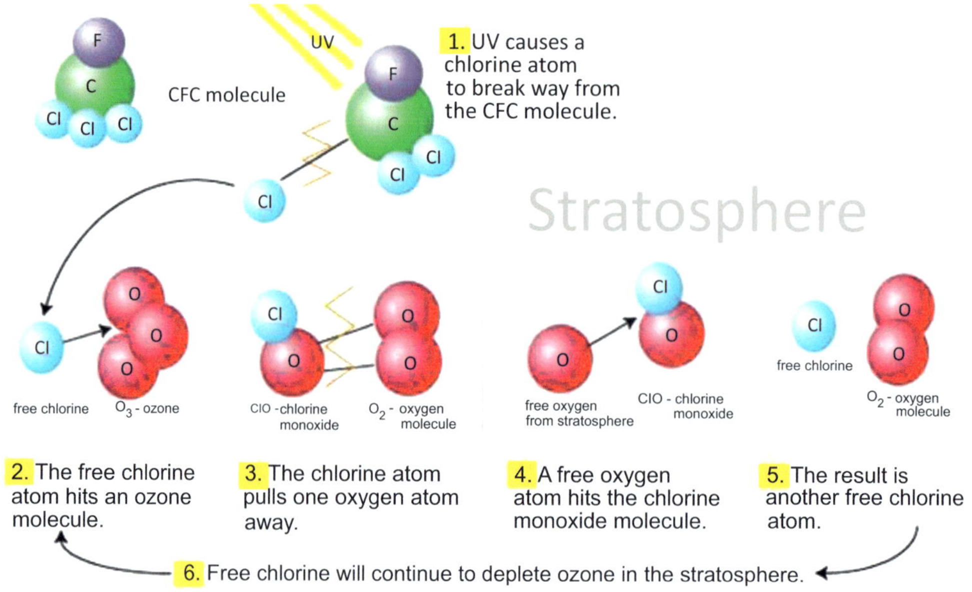

Fig. (10). Interactions of chlorofluorine molecules with UV light in the stratosphere.

These chlorinated hydrocarbons, by agreements made in 1989, were to be phased out by 2000 as reported by Robert W. Pease [26] in an article[5]: *Ozone Chicken Littles Are At It Again.* There was controversy but the point of view expressed by Pease in this *Wall Street Journal* article did not prevail. Pease [26] credits atmospheric chemist F. Sherwood Rowland, UC-Irvine, with the theory he developed in the early 1970s that CFCs deplete ozone in the upper atmosphere and that this is detrimental because the loss of ozone allows too much harmful UV light from the sun to reach the surface of the earth. Some of the arguments against the banning expressed by Pease included the following: Roland's theory ignores that there is an equilibrium concentration of ozone because ozone molecules are created and destroyed continually by natural processes (Fig. **10**). The amount of ozone varies during the year in a somewhat cyclic fashion. He further argued that the narrow band of UV rays that reacts to release chlorine, which actually destroys ozone, also can chemically react with oxygen to create ozone (Fig. **11**). Thus, the determination of whether there is a net destruction of ozone is a function of the concentrations of oxygen and CFC molecules in the ozone layer. He also stated that calculations "based on high-altitude CFC samplings and data supplied by the

National Oceanic and Atmospheric Administration show 60,000 ozone molecules are created for every chlorine atom released from a CFC molecule".

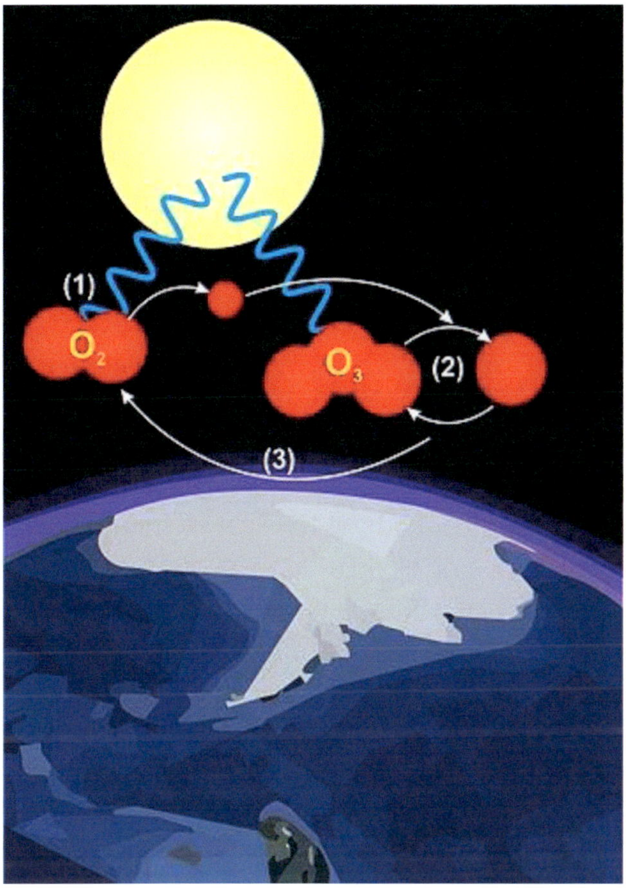

$O_2 + UV \rightarrow 2\,O$
$O + O_2 \rightarrow O_3$ (ozone production)
$O_3 + UV \rightarrow O + O_2$ (ozone destruction)
$O + O_2 \rightarrow O_3$ (ozone production)
$O_3 + O \rightarrow O_2 + O_2$ (ozone destruction)

Fig. (11). Reactions of oxygen species with ultraviolet light that create and destroy ozone.

It was during this debate that the use of the term "Ozone Hole" came into prominence, primarily in the press. This hole occurs over Antarctica and is not really a hole but a reduction of the concentration of ozone which fluctuates in size and position. It is interesting to note that currently the "Ozone Hole" is diminishing; proponents of banning say it is because of the banning.

Banning these chemicals resulted in the use of less efficient and more costly chemicals. Near the same time; however, media were reporting that ozone was harmful and present in smog, primarily a problem in the air of large cities. Note that scientists were worried about ozone at ground level while trying to promote its presence high in the atmosphere. This requires some explanation which will follow.

Ozone is a gas composed of 3 atoms of oxygen. It is formed in the stratosphere by the action of UV light from the Sun, and is highest in concentration there but also is present, as noted, in traces throughout the atmosphere (Fig. **9**). The reaction involves oxygen and the energy of sunlight, and is made complicated by the kinetics of the reaction which are significantly affected by the low concentrations of the reactant molecules:

O_2 + sunlight (UV) = O- + O-

O- + $O_2 = O_3$

Oxygen absorbs certain wavelengths of ultraviolet light (present in sunlight) and there is sufficient energy to split molecules of oxygen into oxygen into atomic oxygen which then reacts in the presence of a third substance (in this case the abundant nitrogen present in the air) to produce ozone (other reactions are also possible).

O_3 is more reactive than O_2 and reacts readily with molecules containing hydrogen, nitrogen, chlorine and bromine. NASA has calculated that the total mass of ozone is approximately 3 billion metric tons (1 metric ton = 1,000 kilograms). This is only approximately 0.00006% of the atmospheric gases [27].

Ozone is present at low levels in the atmosphere (in the troposphere, including at ground level) where its concentration is increased by the presence of components of smog. Near the earth's surface the average concentration varies but is reported to be approximately 10 ppb (ppb is parts per billion parts of air by volume). This is 0.000001%. The US Environmental Protection Agency in its final rule, Dec. 28, 2015 [28] has determined that greater than 70 ppb exposure for more than 8 hours is deemed to be "unhealthy" and can irritate the throat and lungs of some people and can aggravate asthma and emphysema.

Ozone also accumulates in the stratosphere (a region of the atmosphere 10 to 50 kilometers in altitude) which contains 90% of the ozone (Fig. **9**). It performs the beneficial function of shielding us from the effects of UV light from the sun. Ozone actually absorbs most of the UV from sunlight and it is a common conclusion that without ozone, the sun's intense UV radiation could make the surface of the earth uninhabitable. The peak concentration of ozone occurs at about 32 kilometers altitude where, although it varies, it can be as high as 15 ppm (0.0015%) by volume of the atmosphere. Ozone concentration is sometimes measured in Dobson Units (DUs) which are a bit complicated, but derived practically. One DU is 2.69×10^{16} ozone molecules per square centimeter. It can be thought of (at standard temperature and pressure (STP) which is 0 degrees Celsius and 1 atmosphere) as equivalent to a layer of pure ozone 0.01 mm thick above a point on the Earth's surface. The ozone layer has an average thickness of about 300 Dobson Units. This would make a layer only 3 millimeters thick if it were pure ozone at STP [22]. There are different types of UV light with different energies. It is commonly stated that ozone screens out all of the most energetic ultraviolet (UV-c), most of the UV-b but only about ½ of the UV-a (much less energetic). Excessive UV-a and UV-b can cause harmful effects including sunburn.

NOTES

[1] Stated by Joseph Priestley in *Experiments and Observations on Different Kinds of Air*, Vol. II, (London, 1775).

[2] John Dalton, a *New System of Chemical Philosophy*. Kessinger Legacy reprints, parts 1-2 (1808).

[3] Caroline Stephen, *Quaker Strongholds,* printed 1890, published by London Kegan, Paul Truener and Co., LLC; historical reproduction of the work by BiblioLife, LLC.

[4] In "*Pneumatic",* as related in the Story of Human Error, Joseph Jastrow Ed., London, D. Appleton-Century Co. 1936.

[5] Appeared in *The Wall Street Journal, 23 March 1989, p. A24:3*

REFERENCES

[1] Parrington JR. A Short History of Chemistry. 3rd ed., Dover Publications 2011.

[2] Brown WH. The Norton History of Chemistry. 1st American Ed., NY: W. W. Norton.

[3] Bowler PJ. Making Modern Science: A Historical Survey. Chicago: University of Chicago Press 2005. [http://dx.doi.org/10.7208/chicago/9780226068626.001.0001]

[4] Chang KK. Georg Ernest Stahl. Encyclopaedia Britannical. Available from: http://www.britannica.com/biography/Georg-Ernst-Stahl 2016.

[5] Editors. Johann Joachim Becher. Encyclopaedia Britannical. Available from: http://www.britannica.com/biography/Johann-Joachim-Becher 2016.

[6] Giuta C. Elements and Atoms: Chapter 4, Oxygen, An Element in Air: Joseph Priestley. Available from: https://web.lemoyne.edu/giunta/ea/PRIESTLEYann.HTML 2016.

[7] Schofield RE. The enlightenment of Joseph Priestley A study of his life and work from 1733 to 1773. Penn State University Press 2004.

[8] Asimov I. The intelligent man's guide to the biological sciences. NY: Pockert Books, Inc. 1964.

[9] Isaacson W. The amazing virtues of Ben Franklin (cover article). Time Magazine. Available from: http://content.time.com/time/magazine/article/0,9171,1005149,00.html 2003.

[10] Anon. Carl Wilhelm Scheele, 1742-1786. Available from: http://mattson.creighton.edu/History_Gas_Chemistry/Scheele.html 2014.

[11] Scheele CW. Chemical treatise on air and fire: A selection (illustrated edition) Republishe. Gloucester, UK: Dodo Press 2009.

[12] Anon. Carl Wilhelm Scheele. Available from: http://www.nndb.com/people/492/000095207/ 2016.

[13] Anon. Carl Wilhelm Scheele. Notable Names Database. Available from: http://www.nndb.com/people/492/000095207/ 2016.

[14] Joseph Priestley. Discoverer of Oxygen National Historic Chemical Landmark - American Chemical Society. Available from: http://www.acs.org/content/acs/en/education/whatischemistry/landmarks/josephpriestleyoxygen.html 2016.

[15] Boytim J. Antoine-Laurent Lavoisier. Chemical Heritage Foundation. Available from: http://www.chemheritage.org/discover/online-resources/chemistry-in-history/themes/early-chemistry-and-gases/lavoisier.aspx

[16] McKenna C. Antoine-Laurent Lavoisier. The Catholic Encyclopedia. Available from: http://www.newadvent.org/cathen/09052a.htm 2016.

[17] Anon. Antoine Lavoisier. Antoine Lavoisier (En.) Browsebiography.com. Available from: http://www.browsebiography.com/print_p.php?url=antoine_lavoisier_en 2016.

[18] Buchholz AC, Schoeller DA. Is a calorie a calorie? Am J Clin Nutr 2004; 79(5) (Suppl.): 899S-906S. [PMID: 15113737]

[19] Senese F. Dalton's atomic theory. General Chemistry Outline! Available from: http://antoine.frostburg.edu/chem/senese/101/atoms/dalton.shtml 2016.

[20] John Dalton Biography. The biography.com website. Available from: http://www.biography.com/people/john-dalton-9265201 2016.

[21] Books on Google Play. Available from: http://books.google.com/books?id=UU1FAAAAIAAJ&pg=PP9&source=gbs_selected_pages&cad=2#v=onepage&q&f=false 2016.

[22] Dole M. The natural history of oxygen. J Gen Physiol 1965; 49(1): 5-27. [http://dx.doi.org/10.1085/jgp.49.1.5] [PMID: 5859927]

[23] Berkner L, Marshall L. The history of oxygenic concentration in the earth's atmosphere. Discuss Faraday Soc 1964; 37: 122. [http://dx.doi.org/10.1039/df9643700122]

[24] Rubin M. The History of Ozone. The Schonbein Period, 1839-1868. Bull Hist Chem 2001; 26(1): 40-56.

[25] 2008 National Ambient Air Quality Standard (NAAQS) for Ozone. Federal Registry. Available from: https://www.epa.gov/ozone-pollution/2008-national-ambient-air-quality-standards-naaqs-ozone 2016.

[26] Pease R. Ozone Chicken Littles Are at It Again. Available from: http://www.ourcivilisation.com/ozone/ 2016.

[27] Ozone Hole Watch: Facts about ozone. Available from: http://ozonewatch.gsfc.nasa.gov/facts/ozone.html 2016.

[28] National Ambient Air Quality Standards for Ozone; Final Rule. Fed Regist 2015; 80(206) Available from: https://www.gpo.gov/fdsys/pkg/FR-2015-10-26/pdf/2015-26594.pdf.

CHAPTER 4

Oxygen Generation by Photosynthesis

Abstract: Evidence supports that all of the molecular oxygen present in Earth's atmosphere today could have been produced in the last 2,000 years. Photosynthesis by green plants, algae and related phytoplankton are the source of this atmospheric molecular oxygen. Green plants are the "sugar factories of the world" and have intricate microscopic systems within leaves that use sunlight, water, and carbon dioxide to produce carbohydrates and oxygen. Capture of photons from sunlight requires a complex array of pigments that include the green molecule chlorophyll. The photons are captured *via* absorption by pigments that are organized sequentially in a specific order within leaf structures called chloroplasts. Leaves have a complex, organized macroscopic structure and an even more complex microscopic structure. This organization begins with the shape and structure of the leaf itself. Leaves have an aerodynamic shape that allows them to survive in wind and rain and are oriented to capture sunlight. They have an outer waxy coating which retards water loss but with provisions (stomata) that regulate entrance and efflux of carbon dioxide, oxygen, and water vapor. The biosynthesis of carbohydrate from carbon dioxide requires energy input, and the leaf has intricate mechanisms for capturing sunlight to make chemical bonds in ATP. Great political turmoil recently has arisen over claims that carbon dioxide in the atmosphere is the determining factor in increasing the global temperature by amplifying the "greenhouse" effect. Based on speculation and questionable computer models, man-made (anthropogenic) carbon dioxide is said to be the cause of what initially was called "global warming", but more recently has been changed to "climate change". The ability to scientifically measure photosynthesis globally has become relevant for making intelligent decisions about climate change and atmospheric CO_2 concentration. There is reason to believe that natural (and/or future artificial) photosynthesis can maintain the balance of atmospheric CO_2.

Keywords: Aerobic life, Algae, Antennae, ATP, C3 plants, C4 plants, Calvin Cycle, CAM plants, Carbon dioxide, Chlorophyll, Climate change, Desert plant,

Olen R. Brown

Enzyme, Leaf, Lysozyme, Melvin Calvin, Oxygen, Peroxisomes, Petagram, Photosynthesis, Photosystem I, Photosystem II, Phytoplankton, RuBisCO, Stomata, Sunlight, Thylakoid membrane, Transpiration, Water vapor.

BIOLOGICAL PRODUCTION OF OXYGEN

Introduction and Overview

Molecular oxygen is significantly reactive even though this reactivity is reduced by the arrangement of its valence electrons as addressed in Chapter 2. Were it not for the production of oxygen by photosynthesis (of green plants, algae, and phytoplankton) essentially all of the oxygen on Earth would now be combined with other elements and likely only traces would be present in air. Thus, life as we know it would not exist on Earth. Photosynthesis also provides an important "sink" that captures CO_2 by "fixing" it into the complex carbohydrates of life.

The examination of photosynthesis in this Chapter, even though focused and limited in scope, will be designed to provide the basic information necessary for understanding two processes: oxygen generation and carbon dioxide removal with fixation into carbohydrates by green plants. Oxygen, the breath of life, is liberated by green plants (that also use oxygen for their metabolic respiration) and carbon dioxide (a simple form of carbon, somewhere between inorganic and organic carbon) is fixed fully into organic carbon. The carbon is initially fixed into a three-carbon sugar phosphate and eventually into sucrose that serves as a starting material for structural and energy requirements for the plants which then support animal life. A point that may be under-appreciated is the fact that both oxygen and carbon dioxide are gases and their presence in air makes them available as starting materials and products that readily enter and exit cells. Oxygen and carbon dioxide are present in air at nearly constant concentrations of approximately 21% and 0.04% (on a volume basis) in Earth's atmosphere today. Thus, the ratio of oxygen to carbon dioxide is approximately 500-to-1. Biological systems that capture and release these gases are efficient over this large range, and in photosynthesis 1 molecule of carbon dioxide is taken up for each 1 molecule of oxygen released.

I shall describe these complex photosynthetic systems in what may appear to be considerable detail. However, these descriptions are only an outline of the complexity present at ever deeper levels within plant cells. Photosynthesis efficiently manages the recycling of these two gases that have such different abundances with exquisite balance. Indeed, later in this chapter we shall explore how the large enzyme (RuBisCO) of the Calvin Cycle is able to catalyze two competing reactions involving carbon dioxide and oxygen as substrates [1]. Carboxylation of a substrate, common to both enzyme actions, results in carbon assimilation in one case and a competing reaction that reduces photosynthetic efficiency in the other case. Indeed, RuBisCO binds carbon dioxide stronger than oxygen and a region of the enzyme structure is a temporary reservoir for CO_2.

The energy that drives photosynthesis comes from photons of light delivered from the sun over approximately 93 million miles of cold, nearly empty space– about the right distance to render a nuclear fusion reactor safe for life. To use these photons, plants have unique antennae as part of an exquisite system that guides and delivers the energy so that it can be fixed in the bonds of biosynthesized organic molecules within the leaf. Recent studies using bacteria by researchers at Lawrence Berkeley National Laboratory and UC Berkeley [2] revealed that the mechanism of this biological energy-capture requires utilization of quantum mechanics for photosynthesis in certain bacteria (and probably in all systems). They found evidence of "quantum entanglements" that persist over picoseconds in this complex biological system, helping to account for its efficiency. If particles, such as two electrons in this case, are "entangled" any change in one is instantly reflected in the other. Their results were interpreted to indicate that photons exhibit this property in the bacterial light-harvesting complex. They suggested that this has implications for future attempts to design quantum-based technologies for computing. It is currently being used to create hybrid photosynthetic machines.

Thus, carbon dioxide, a single atom of carbon oxidized by two atoms of oxygen from the atmosphere where it is present at a concentration only two thousands that of oxygen, *via* the miracle of photosynthesis, creates the carbon backbone underpinning life on earth.

Measuring Global Photosynthesis

Assessing the global rate of photosynthesis is important but complex and technically difficult. A leap forward in ability to assess global photosynthesis was made possible by satellites. Understanding this involves exploring how photosynthesis works.

To thrive and grow all plants photosynthesize using chlorophyll, a complex molecule that gives the green color to plant life. The word photosynthesis implies conversion of the energy of sunlight into the chemical energy in the bonds of organic compounds that are usable by plant cells (Fig. **1**). Chlorophyll absorbs light from most of the visible spectrum, except for the green wavelengths which are reflected and this accounts for our perception that leaves are green.

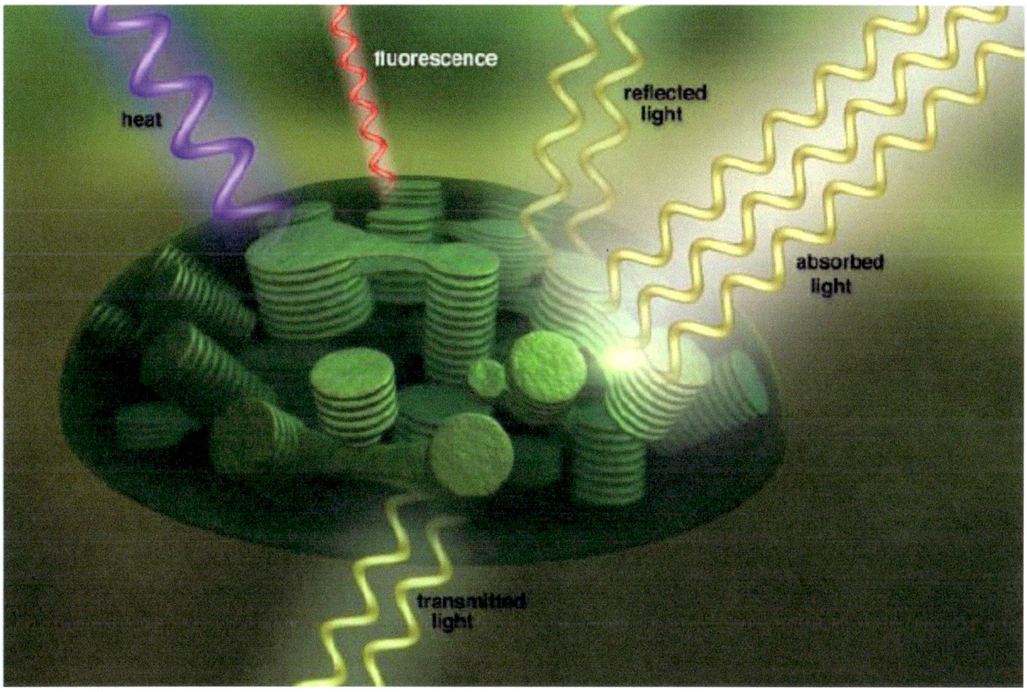

Fig. (1). Measuring photosynthesis from space using sensors on satellites. Image credit: NASA Goddard's Conceptual Image Lab/T. Chase.

Measuring the extent of photosynthesis is necessary for understanding atmospheric carbon dioxide concentrations but it also is a surrogate for assessing the degree to which plants are thriving. Plants that are healthy and which have adequate nutrients and water photosynthesize abundantly by absorbing sunlight and releasing some energy as fluorescence. Thus, the extent of fluorescence indicates the extent of photosynthesis. Fig. (**1**) diagrammatically shows that light is absorbed by chlorophyll-containing green structures within a plant and some is re-emitted as fluorescence. Healthy plants exhibit high fluorescence while low or no fluorescence is indicative of plants that are stressed by temperature, lack of water, disease or inadequate sunlight, as examples. This fluorescence, associated with events occurring inside plants at a cellular level, has been detected and measured from a great distance away, over large geographic areas, by using satellite imaging [3]. The fluorescence is invisible to the human eye, but is detectable by instruments on satellites orbiting hundreds of miles above Earth. NASA scientists have developed a way to convert this satellite data into global maps that allow photosynthesis over a broad land area to be estimated in considerable detail [3]. The satellite maps permit assessment of plant health and also allow estimation of photosynthesis rates by regions and also globally [3].

Maps were produced in 2013 by Joanna Joiner and colleagues of NASA's Goddard Space Flight Center that increased resolution of photosynthesis detection by16-fold for resolution in space and 3-fold for temporal resolution beyond the proof-of-concept maps released in 2011 [3]. Improved global measurements have helped farmers identify early indications of crop stress and have greatly increased the ability of ecologists to better measure global photosynthesis and carbon cycle processes. Quotations from Joiner [3] provide the significance of these advances:

> "For the first time, we are able to globally map changes in fluorescence over the course of a single month... This lets us use fluorescence to observe, for example, variation in the length of the growing season."

The dynamics of changes in vegetation, including the northward regional progress of plant growth during spring in the northern hemisphere was being observed by this improved technology by 2013. It also provided new data for comparison with traditional observations of "greenness" as a measure of plant growth and activity.

Thus, fluorescence measurements complement other data by providing direct and immediate information about plant productivity. For example, the technology permitted researchers to observe when plants started to "shut down" in the fall even before their leaves began to change to fall colors. The initiation of spring plant growth was also detected early in the spring of 2012 [3].

Detailed activity of plant photosynthesis thus can be mapped by this technology that identifies the faint fluorescence signal collected by the Global Ozone Monitoring Instrument 2 (GOME-2) instrument on Metop-A, a European meteorological satellite. The technology overcame the difficulty presented by the mixing of the fluorescence signal and the signal from sunlight reflected from Earth's surface and from clouds, and the absorption of sunlight by gases in the atmosphere [3].

To use fluorescence for quantification of photosynthesis, Joanna Joiner and colleagues had to separate multiple light signals each of which has a spectral signature analogous to a fingerprint [3]. As indicated, these signals came from fluorescence, Earth's surface, and the atmosphere. The fingerprint associated with fluorescence had to be separated from the other data. Disentangling the signals from atmospheric influences was not possible in the early results in 2011 when Joiner and colleagues first attempted to measure global land plant fluorescence from space. That early study used data from a spectrometer aboard a Japanese satellite called the Greenhouse Gases Observing Satellite (GOSAT). At that time, researchers analyzed a dark region of the infrared portion of the solar spectrum where there was only minor interference by background light. It was possible to distinguish the faint fluorescence signal from plant photosynthesis activity. The current, new technology is more complex but it permits enhanced resolution utilizing more frequent measurements. Earlier measurements with GOSAT averaged data over areas of 200 square kilometers every month. GOME-2 permits averaging data over areas as small as 50 square kilometers over periods as short as approximately 10 days. This more frequent sampling over smaller regions led to the conclusion that agricultural areas in the U.S. Midwest are among the most productive lands on Earth. The satellite data also was compared with data from older technology (tower-based observations of carbon dioxide taken up by plants) to help establish its reliability.

Photosynthesis in More Detail: Roles of Oxygen and Carbon Dioxide

Photosynthesis occurs in green plants and in phytoplankton (and in a few other types of organisms) [4]. Green plants are the sugar factories of the world. They use the energy of sunlight, carbon dioxide and water to synthesize a phosphorylated three-carbon compound that is eventually converted to sucrose. Oxygen is released in the process. This is the source of practically all the oxygen present in the air today. Therefore, photosynthesis is most relevant to our understanding of the linkage between the concentration of oxygen in the atmosphere and its function as the source of respiration for aerobic life on earth.

A simple depiction of photosynthesis is: $6\ CO_2 + 6\ H_2O = C_6H_{12}O_6 + 6\ O_2$

In words, six molecule of carbon dioxide and six water molecules produce one molecule of a six-carbon compound and liberate 6 molecules of oxygen into the atmosphere. The more complicated explanation of photosynthesis that follows is required to provide an overview of the process and as an introduction to a more detailed explanation of the mechanisms involved.

Photosynthesis liberates molecular oxygen (O_2) and is the source of all molecular oxygen in the air we breathe. In this process, carbon dioxide and water from the air enter a leaf through open stomata. Photons of light (not shown in the equation depicting photosynthesis) of a certain range of energies from sunlight contribute energy. Some of this energy ends up in the molecular bonds of one molecule of the 6-carbon compound (glucose) which can be further metabolized and used in biological oxidation (which consumes oxygen) to store energy in ATP. The rate of this process is controlled by enzymes, temperature, intensity of sunlight, water vapor and carbon dioxide concentrations, the type of plant, and availability of minerals and other factors.

Overview and Production of Oxygen by the Light Reaction of Photosynthesis

Because photosynthesis is so complex, I choose to describe its mechanisms system-by-system which introduces some overlap. Chlorophyll is intimately involved in photosynthesis [4] where it serves two types of functions: it absorbs light and transfers energy to a specific chlorophyll molecule in a cellular region called the reaction center. A space-filling molecular model of chlorophyll is shown in Fig. (**2**). Chlorophyll is present inside chloroplasts which are organelles in plant cells. Chlorophyll is thus part of the plant's "photosystem". The photosystem is embedded in a special type of membrane called the thylakoid membrane that is present inside chloroplasts in leaves. Light absorbing molecules of chlorophyll are exquisitely arranged at the submicroscopic scale so that energy from photons of light can be efficiently transferred to specific sites.

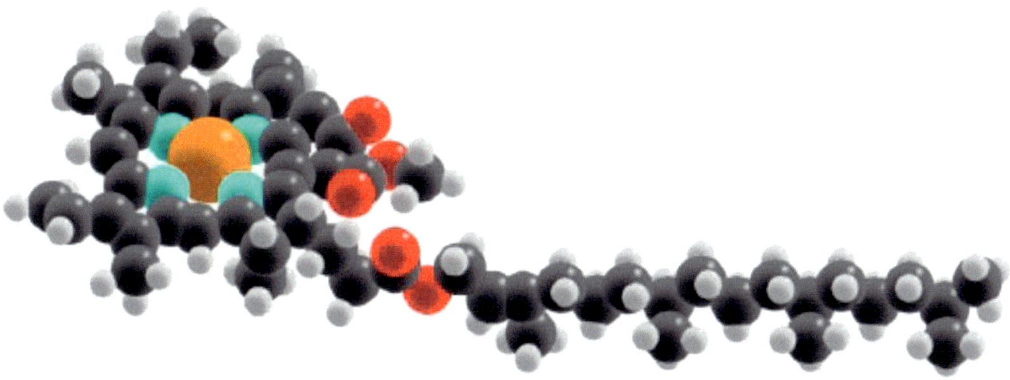

Fig. (2). Space-filling model of chlorophyll a, the most abundant chlorophyll in green plants on earth.

Further details of photosynthesis will be reviewed as it occurs in the leaves of green plants, although the process is similar in the other photosynthetic life forms.

A leaf is complex and exquisitely designed for its function. It is located on branches that give maximum exposure to sunlight. Some plants, including the sunflower, *Helianthus annuus* (ever present in my environment when growing up in Oklahoma, although Kansas claims it as the State flower) turns and adjusts the flower head to face the sun throughout the day. The leaf shape seems designed to minimize damage from wind and the leaf structure contains vessels to transport water and nutrients. Openings (stomata), too small to be observed, regulate water, carbon dioxide and oxygen into and out of the leaf. The leaf also has a waxy barrier to reduce water loss. Indeed, both the presence of the waxy leaf coating and the opening and closing of leaf stomata are tightly regulated by plants. Certain water plants do not have the coating, and certain desert plants regulate stomata opening and closing to coordinate with the night-and-day cycle (Fig. **3**).

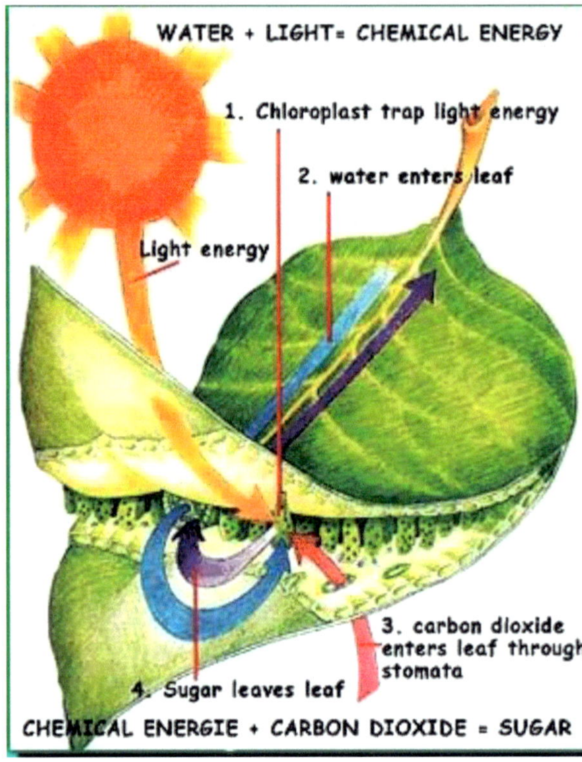

Fig. (3). Overview of photosynthesis.

Thylakoid Membrane and Chlorophyll

Within the tiny chloroplast there are even smaller membrane-bound organelles. These organelles contain the stroma and grana composed of thylakoid membranes (Fig. **4**). The inner membrane surrounds a space (the stroma) that is fluid-filled and contains various enzymes where light-dependent reactions occur within the thylakoids [4]. Imbedded within the thylakoid membranes are the antenna complexes that include proteins and light-absorbing pigments (which we will subsequently describe). Although all cells in green parts of green plants have chloroplasts, most of the light energy captured by plants occurs in chloroplasts in the leaves. Chloroplasts are miniscule and their abundance is amazing: Cells in the interior tissues of a leaf can contain from 450,000 to 800,000 chloroplasts per square millimeter [4].

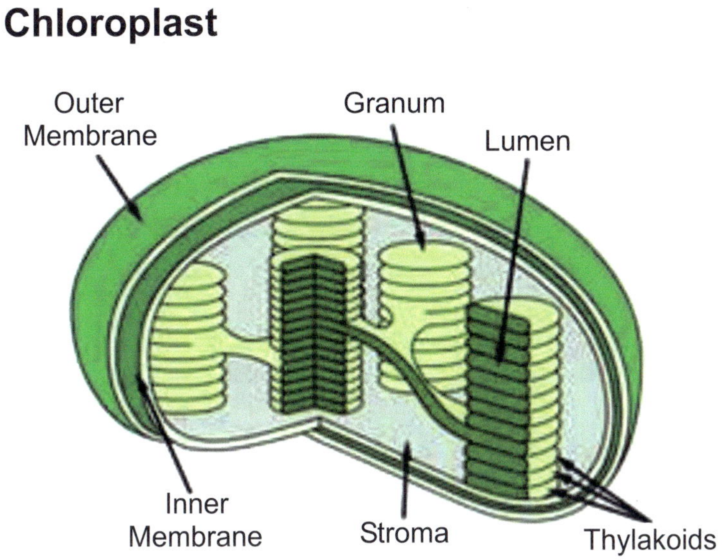

Fig. (4). Chloroplasts are complex, pigment-containing organelles within the leaves of green plants that have complex pigment-containing organelles.

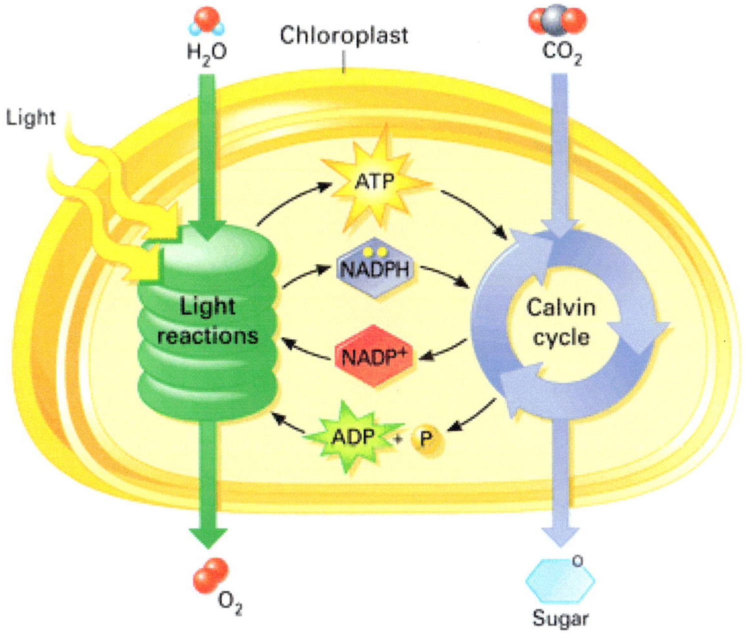

Fig. (5). Depiction of the flow of energy from photons through the photosynthetic reaction center.

Reaction Centers and Antennae

Microscopically, inside the leaf, the real mysteries are hidden from our eyes. A diagrammatic overview shows the relationship among important contributors to photosynthesis (Fig. **5**). Photons of sunlight energy are absorbed by leaf pigment molecules that function like antenna. They are specifically designed and located for this transfer of energy. After a journey whose path is directed by laws of energy transfer, and dependent on precision in location and precision in chemical details, the photons arrive at the reaction center.

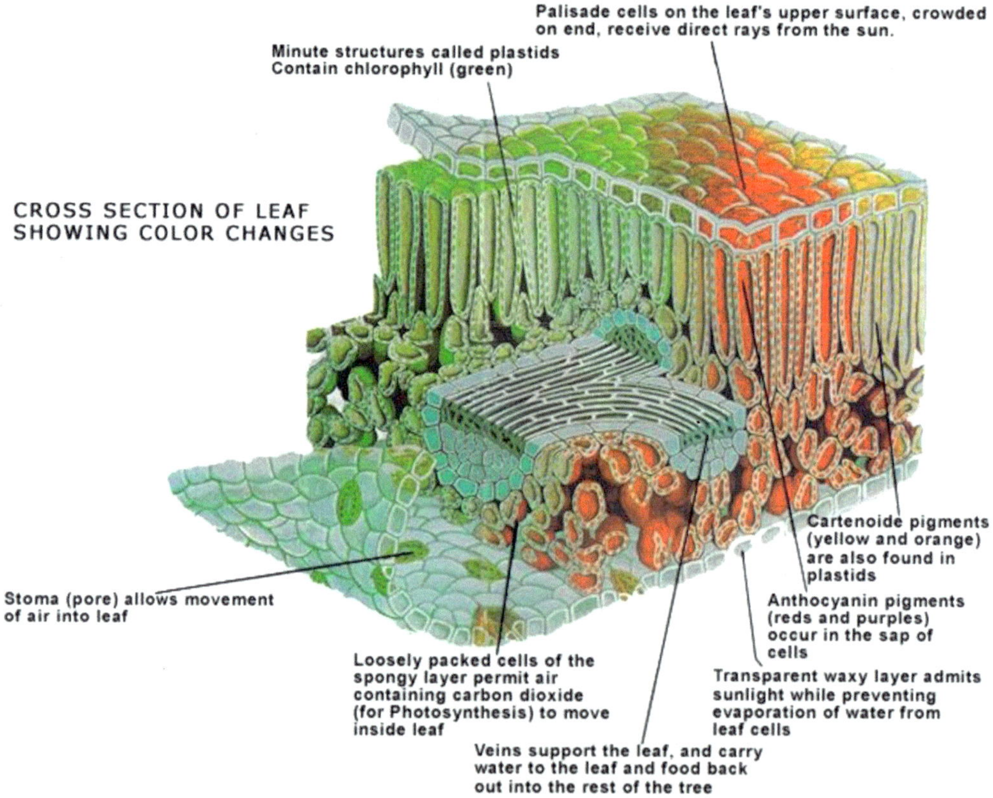

Fig. (6). Integrated picture of the complex systems that coordinate to make sugar and oxygen.

The closer we focus, the greater is the observed precision and complexity (Fig. **6**). The light reactions involve Photosystems I and II that interact with the Calvin Cycle (to be described subsequently). Ultimately, water and light generate oxygen and electrons to make ATP and a reduced form of a coenzyme called NADP. Carbon dioxide is "fixed", initially into a 3-carbon phosphorylated sugar which is condenses to a 6-carbon sugar and ultimately made into sucrose.

Atmospheric carbon dioxide, over a wide range of low concentrations [4] in the atmosphere, strongly regulates the rate of photosynthesis (Fig. **7**). The current atmospheric concentration of carbon dioxide is approximately 400 ppm. Obviously, by increasing photosynthesis rate, plants are capable of contributing more to the process of decreasing carbon dioxide and generating oxygen into the atmosphere. Considering only carbon dioxide, it appears that the rate of photosynthesis can increase significantly up to a carbon dioxide concentration approaching 900 ppm.

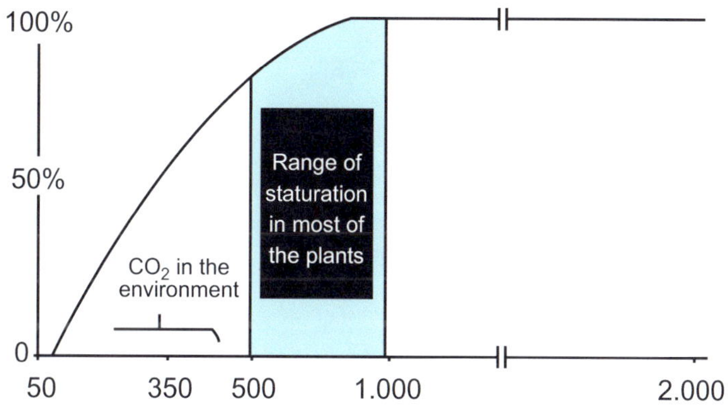

Fig. (7). Effect of carbon dioxide concentration on rate of photosynthesis. The photosynthesis rate as percent (Y-axis) is plotted *versus* the carbon dioxide concentration in parts per million in the atmosphere (X-axis). The colored zone is the region of saturation for most plants.

The Calvin Cycle

The rate-limiting enzyme in photosynthesis is ribulose-1,5-bisphosphate carboxylate/oxygenase (RuBisCO). It is reported to be the most abundant protein on earth [4, 5]. It catalyzes the initial step in a series of reactions known as the Calvin Cycle (Fig. **8**). The cycle was named for Melvin Calvin (1911-1997) who was awarded the Nobel Prize in Chemistry in 1961 "... for his research on the carbon dioxide assimilation in plants" [6]. The Calvin cycle (Fig. **8**) is shown in relation to the light-catalyzed reactions of photosynthesis. The cycle fixes carbon from carbon dioxide in the air ultimately into a sugar whose bonds are rich in available energy. The oxygen in carbon dioxide is released as diatomic oxygen to account for all the dioxygen in Earth's atmosphere today. The Calvin Cycle (Fig. **9**) is the primary way inorganic carbon enters into the organic chemistry of life processes.

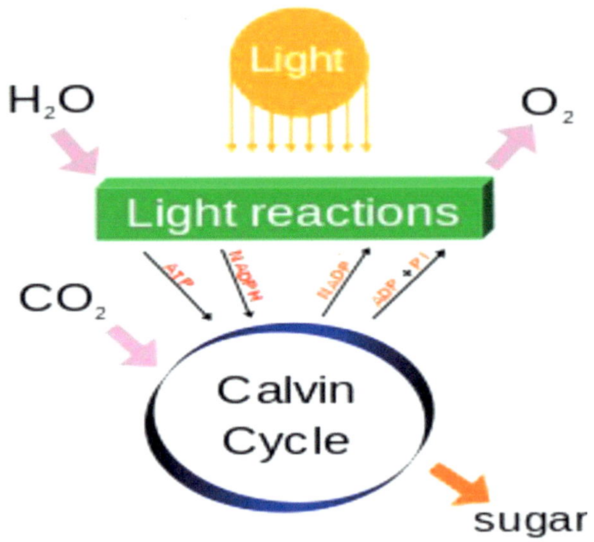

Fig. (8). Interactions of the Calvin cycle and sunlight with generation of molecular oxygen.

In the Calvin Cycle, ATP is required at two steps. It refers to the light-independent part of photosynthesis, after light has been absorbed and done its work with creation of ATP and release of oxygen. The Calvin cycle has 3 stages: carbon fixation, reduction of the carbon, and regeneration. The regeneration stage requires further explanation. In order to remove reduced carbohydrate (glucose) from the cycle a 5-carbon compound must be regenerated as the molecule to which a one-carbon unit is added. Only one CO_2 molecule is incorporated at a time (one turn of the cycle); therefore, the cycle must be completed three times to produce the equivalent of one three-carbon molecule and six times to produce all the carbons needed for a single molecule of glucose (which has 6 carbons).

Note the prominent role of RuBisCO which catalyzes the first step in a dedicated pathway that accepts carbon dioxide [5]. It is generally true that for a chain of enzyme-catalyzed steps the first step is usually rate-limiting and this is true for RuBisCO. When rate-regulation is required, the first step provides an efficient (non-wasteful) point at which to control the rate of the operation of the cycle.

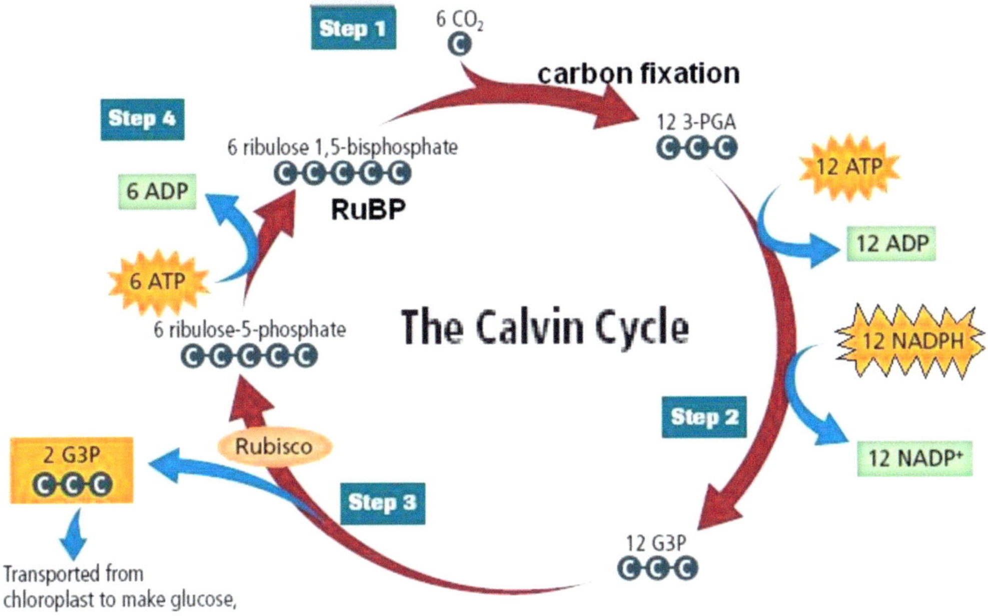

Fig. (9). The Calvin cycle in more detail.

Photosynthesis is very complex and it proceeds in three significantly different ways in different types of plants. These plants are called C3, C4, and CAM. As we have noted already, a key component in photosynthesis is the enzyme called ribulose-1,5-biphosphate carboxylase/oxygenase (RuBisCO) [7]. Enzyme names contain a lot of information. Its name is explained when we examine the reactions it catalyzes (Fig. **10**). In this case, we see that the substrate for the enzyme is the 5-carbon sugar containing 2 phosphate groups. The "oxygenase" part of the enzyme name indicates that it adds oxygen (shown in green) to an unstable intermediate. The "carboxylase" part of the name indicates that it adds a carboxyl group (in this case two such groups) from cleavage of the preceding intermediate.

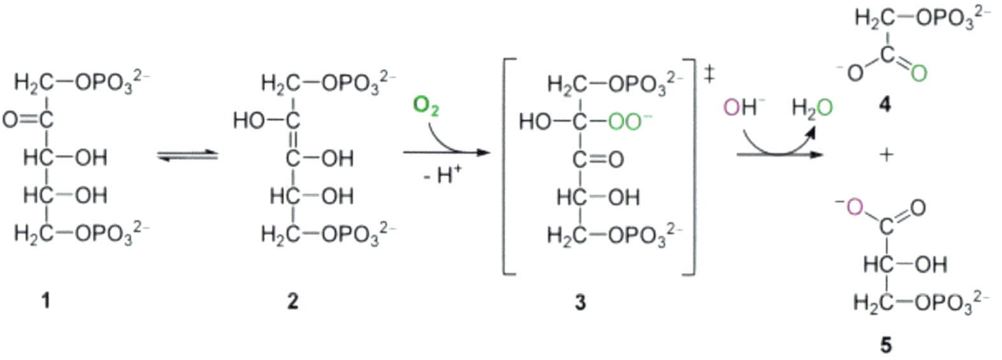

Fig. (10). Reactions catalyzed by RuBisCO.

RuBisCO is found in all photosynthetic species including cells of green plants, algae, cyanobacteria, and phototropic and chemoautotropic proteobacteria. Magnesium is required for the enzyme to function. Both pH and magnesium concentrations regulate the activity of this enzyme. RuBisCO catalyzes the reaction of carbon dioxide with ribulose-1,5-bisphosphate, and it also catalyzes the reaction of ribulose-1,5-bisphosphate with molecular oxygen [5, 7]. A space-filling model of RuBisCO provides additional insight (Fig. **11**). It is composed of long protein chains (dimers, shown in white and grey) and short protein chains (shown in blue and orange). Like all enzymes (almost without exception), it is a protein with a large molecular mass in comparison to the 5-carbon sugar and carbon dioxide which are its substrates. Using crystallized protein, x-ray diffraction and computers, the positions in space for the atoms of the enzyme have been determined. The first enzyme whose structure was determined in this detail was lysozyme which is found in tears and it also is produced by certain bacteria which can break down the cell walls of some types of bacteria and they subsequently lyse (burst). The x-ray diffraction technique allows determination of the shape and size of the enzyme's reactive site and its relationship to particular amino acids and binding sites. The large size of enzymes and variations in amino acid sequences permit the existence of specific regions where the shape and charge on the protein is conducive to binding of the substrate. Of course, the two-dimensional structure of the protein is specified by the genetic code, and folding of the protein creates the three-dimensional shape including the reaction site. It turns out that, among enzymes, RuBisCO is very inefficient at its catalytic function which is binding and catalyzing the fixation of carbon dioxide [7]. This is not apparent from its structure even with its three-dimensional structure determined.

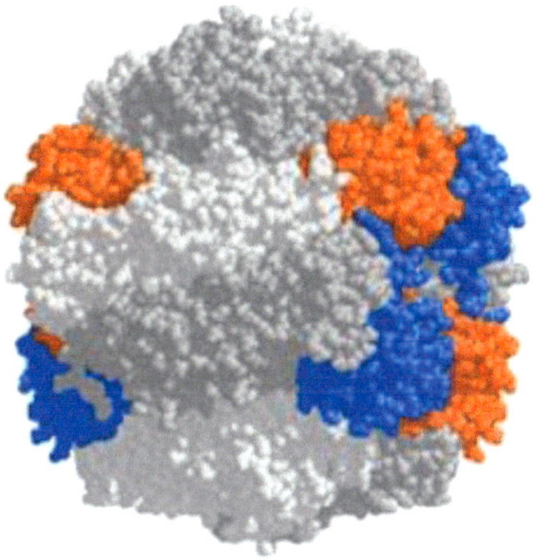

Fig. (11). Space filling model of the enzyme RuBisCO (a large molecular weight protein).

Both CO_2 and O_2 are substrates for RuBisCO. When carbon dioxide is the substrate, the first stable products are two molecules of glycerate-3-phosphate which are products of carbon fixation. When molecular oxygen is the substrate, the products are phosphoglycolate and 3-phosphoglycerate. The former is recycled by photorespiration (in mitochondria *via* cytochromes) which generates CO_2. At the concentrations of oxygen and carbon dioxide found in air, the ratio of the two reactions is about 4 to 1 in favor of the reaction that fixes carbon dioxide.

RuBisCO alone accounts for approximately 50% of the soluble proteins in leaves of plants classified as C3 plants which are said to make up approximately 95% of plant species on Earth [5]. C3 plants use RuBisCO to make a 3-carbon compound as the first stable product of carbon fixation. In these plants, RuBisCO creates two different reaction products which results in some loss of carbohydrate production during a process called photorespiration [5, 7]. C4 plants have a mechanism that involves incorporation of the carbon of carbon dioxide in a 4-carbon compound that subsequently can be metabolized to carbon dioxide which can then be used to produce carbohydrate by the conventional C3 pathway. In C4 plants, RuBisCO makes up approximately 30% of soluble leaf protein. Only about 3% to 5% of the world's plants are C4. There also is a minor group of interesting plants that have developed a slightly different photosynthetic process that is especially beneficial in arid environments. These plants photosynthesize by a pathway called CAM for Crassulacean acid metabolism. CAM derives from the family name of these plants (*Crassulaceae*), and the AM stands for acid metabolism. The acids are 4 carbon compounds including malic acid which accumulates to the extent that in the early morning hours leaves are reported to be quite sour to the taste. These plants include: pineapple, agave, cacti, and orchids. Their stomata (leaf pores) remain closed during the day to reduce water loss but open at night when carbon dioxide is collected. Since carbon dioxide is a gas there is no direct mechanism for significant storage in the plant. The plant has a way around this; it converts the carbon dioxide to a 4-carbon compound that is used during the day in photosynthesis.

The tremendous importance of the enzyme RuBisCO and the Calvin cycle can be appreciated by considering the words of Melvin Calvin in his acceptance speech at the Nobel Prize award ceremony for research on the cycle named for him [8]. In the opening part of the speech he said:

> "It becomes obvious that photosynthesis not only provides an explanation for the existence of the autotrophic organisms but also furnishes food for man and animals. In other words, photosynthesis is the absolute prerequisite for all life on earth and the most fundamental of all biochemical reactions. It has been estimated that plants and microorganisms on earth transform about 6,000 tons of carbon from carbon dioxide to carbohydrate per second, with at least four-fifths of this amount contributed by organisms in the oceans."

Simple calculations show that the 6,000 tons of carbon per second is equal to approximately 190,000 teaspoons (approximately 1 ton) of sugar per day for each person on earth. This is equivalent to 172 petagrams of carbon incorporated from the atmosphere into carbon dioxide by photosynthesis each year, based on Calvin's figures.

RuBisCO, compared to many enzymes, is considered to be slow in promoting its catalytic function. One molecule of RuBisCO enzyme catalyzes the fixation of approximately 3 to 10 molecules of carbon dioxide per second [9]. Catalase for example (an enzyme found inside peroxisomes which are membrane-bound organelles in cells) is among the fastest enzymes known and can convert 40 million molecules of hydrogen peroxide into water and oxygen per second) [10]. It has been claimed widely that catalase is diffusion-limited, meaning that the rate of catalysis is limited by the rate at which the substrate (hydrogen peroxide) can diffuse into the active site of the enzyme. Thus, it has been called the "perfect" enzyme because no higher rate is possible and every collision of substrate with enzyme results in cleavage. You have undoubtedly observed catalase in action when you have treated a cut with a weak solution of hydrogen peroxide and observed the almost violent eruption of bubbles (the oxygen released by tissue catalase exposed by the injury). By comparison, RuBisCO is a slackard.

RuBisCO is most active during the day because its substrate ribulose1-5-bisphosphate is not produced in the dark. When light is not the limiting factor, the rate of catalytic action of RuBisCO is faster with increases in carbon dioxide concentration in the air. This has very important consequences for removal of carbon dioxide, which is a significant factor in preventing carbon dioxide build-up in the atmosphere.

There are limits on the potential increase in photosynthesis. Although increased temperature can stimulate photosynthesis (with an approximate doubling of rate with a 10-degree Centigrade rise in temperature), increased temperature (to very high levels) eventually denatures enzymes and photosynthesis fails. Limits also are created by the eventual limits imposed by other factors including availability of nutrients. However, photosynthesis acts to ameliorate the potentially harmful effects of increases in atmospheric carbon dioxide brought about by human activities. It is clear that the rate and global amount of photosynthesis will tend to counteract increases in carbon dioxide in the atmosphere and no distinction is made by nature on the source of this carbon dioxide including that from human activities. A full consideration of the role of increased carbon dioxide in the atmosphere and global warming/climate change is not appropriate; however, the effects of photosynthesis cannot logically be left out of this calculation.

Photosynthesis Removes Carbon Dioxide from the Air

A closer examination of photosynthesis is warranted because of carbon dioxide's predominate role in cellular respiration (carbon dioxide is the way cells eliminate carbon which has been derived from biological oxidation). Oxygen reacts in the marvelous sequence of previously described enzyme-catalyzed reactions that store energy taken from the bonds of glucose into the special molecule ATP. In the balance of nature, carbon dioxide arises from several sources. It derives from cellular respiration which may be properly viewed as a "cold" (body temperature) burning/oxidation. It is generated from decay of plant and animal matter (primarily by microscopic life forms). It comes from combustion at high temperature by burning of organic matter. The primary way carbon dioxide is removed from the atmosphere is through the photosynthetic activity of green plants, algae and phytoplankton. In examining the balance of carbon dioxide in nature, we must consider both the processes that "fix" carbon dioxide in plant material, and the processes (plant respiration; energetics) that add carbon dioxide back to the atmosphere. The latter process is necessary for plants to live and to photosynthesize. Plant respiration has been estimated to release annually about half of the amount of carbon dioxide fixed by photosynthesis. The global rate of photosynthesis recently has been reexamined by Dr. Lisa Welp of Scripps Institute of Oceanography together with colleagues who found photosynthesis to be higher than previously reported and to result in incorporation of approximately 550 to 640 (average = 595) petagrams of carbon dioxide annually [11]. A pentagram is 10^{15} grams (one million billion grams). For comparison, 1 petagram also is about 1.1 billion U.S. tons. If global carbon dioxide were to double (to approximately 0.08%). Photosynthesis would likely increase by at least 40%. Let us examine this in more detail.

Photosynthetic plants both contribute and remove carbon dioxide from the atmosphere. The net effect of photosynthesis on atmospheric carbon dioxide can be estimated. The units for these calculations are petagrams of carbon dioxide on a global basis annually. The amount of carbon dioxide that is removed from the atmosphere by photosynthesis is approximately 595 petagrams [11]. In the process of growing and living (respiration) while removing the carbon dioxide, photosynthetic organisms add to the atmosphere approximately 298 petagrams.

Eventually the living things that photosynthesize will die and decay which produces approximately 187 petagrams. The net result for the amount of carbon dioxide removed from the atmosphere by photosynthesis is: 595 removed and 298 + 187 added back = 595 – 485 = 110 petagrams. It is generally agreed that the amount of carbon dioxide currently added to the atmosphere by those activities described to emerge after the Industrial Age by "human activity" (land use changes due to farming, burning of fossil fuels, and cement production) annually, is approximately 35 petagrams [12, 13]. Therefore, the net of 110 pentagrams of carbon dioxide removed can be compared to the 35 petagrams of carbon dioxide that is said to be contributed by "Post-Industrial Age" human activity. Thus, there is an excess capacity of photosynthesis to remove carbon dioxide including that produced by human activity. Indeed, photosynthesis increases from both increased temperature and increased carbon dioxide concentration. How much photosynthesis may be increased further by the associated biological adaptation is not known but there is every reason to believe it will result in increased photosynthesis.

This is significant because it indicates the tremendous capacity of nature to remove carbon dioxide from the atmosphere as part of a balanced recycling of carbon and oxygen.

Carbon Dioxide, an Oxygen-Containing Molecule Essential for Life

Carbon dioxide has a dual role in our story of oxygen. It is essential to the primary process, photosynthesis, that generates the oxygen found in the atmosphere. It is important to repeat that the break-down products of cellular metabolism are eliminated primarily as carbon dioxide. Carbon dioxide is ideally suited for this. It is gaseous and easily diffuses out of cells. It contains carbon in a 1:2 molar ratio with oxygen and carbon dioxide is a "waste product" of most processes that degrade cell compounds and processes that "generate" cellular energy. Therefore, both for aerobic unicellular life forms and for complex multicellular life forms, including humans, carbon dioxide is an ideal molecule for disposal of "waste"– waste, meaning what is produced in excess during life processes and needs to be excreted or eliminated. Recall that carbon dioxide is generated specifically by metabolism of cells, including when cells "burn" food chemicals such as glucose to store energy in the form of ATP by coupled reactions. That this carbon is gaseous is ideal.

Carbon dioxide is composed of 1 atom of carbon and 2 atoms (1 molecule) of oxygen. Everywhere at Earth's temperatures and pressures it is a gas that is generally somewhat uniform in average concentration (there are exceptions) in the air up to several miles in altitude (the Troposphere). It currently approaches approximately 0.04% by volume of air when averaged over a reasonable time. Thus, carbon dioxide is a trace gas in air which is more than 99.96% other gases, including the oxygen we require to live which is present at approximately 20.95%. By analogy, if a flock of birds contain crows and ravens and if ravens represent, by analogy, oxygen, and crows represent carbon dioxide, there would be 537 ravens for each 1 crow. Carbon dioxide is important because it, like oxygen, is required for life on earth, including human life, as we know it.

CARBON DIOXIDE AS A GREENHOUSE GAS

Introduction

Carbon dioxide in the atmosphere has come into prominence because it is a "greenhouse" gas and because human activity has resulted in production of an increased amount that has recently increased annually. A greenhouse gas is a molecule that contributes to the warming of earth by absorbing terrestrial radiation that is emitted by the Earth's surface as a consequence of the warming effect of incoming solar radiation. In my opinion, the current "cries of wolf", whether described to be global warming from anthropomorphic, recent activities, or more lately re-designated "climate change", are not based on proven science, but rather are based largely on politics[2].

Earth's Security Blanket, the Warming Atmosphere

Earth's atmosphere is such that it allows warming sunlight energy, streaming to earth from our Sun through the vastness of cold space, to pass through it but then certain wavelengths of the sunlight energy that otherwise would be reflected back into space, are trapped (absorbed) by certain types of air molecules and this energy remains to warm the earth [14]. This retention of heat from the Sun ensures that the global mean temperature rises sufficiently to permit the life we know on earth– the greenhouse effect. There are several atmospheric gases that contribute to this warming effect: water vapor, carbon dioxide, methane, nitrous oxide (N_2O), ozone and traces of some other gases. Barrett [14] concluded that ozone is a "very minor" contributor in the troposphere, but a "very important" contributor in the stratosphere. N_2 and O_2 have zero dipole moment and cannot interact with (absorb) infrared radiation and they are not greenhouse gases although they account for 99.93% of the dry atmosphere [14]. Water vapor, is used only in computer simulation of the effects of anthromorphically increased carbon dioxide based on its "feedback" effect not its "forcing" effect [14] in the models of the United Nations' Intergovernmental Panel on Climate Change (IPCC) Fifth Assessment Report [15]. I shall discuss this subsequently. Water, however, is unquestionably the strongest greenhouse gas based on its infrared absorption characteristics and its atmospheric concentration [14 - 16]. Water has a dipole moment with three fundamental vibration modes and the bending vibration at 1595 cm^{-1} absorbs terrestrial radiation [14]. Carbon dioxide has zero dipole moment. It has three vibrational modes: symmetric stretching at 1388 cm^{-1}, asymmetric stretching at 2349 cm^{-1}, and bending at 677 cm^{-1}. The latter two cause changes in the dipole moment and are active radiatively (they cause absorption) [14]. N_2O has a dipole moment that is small and because it is present in such low concentration in the atmosphere, it is not of significance as a global warming gas. Methane has zero dipole moment but the fundamental vibration is in the 0 to 2500 cm^{-1} range at 1306 cm^{-1}. Ozone has strong absorbance at 1043 cm^{-1} but because of its low concentration, its contribution is minor [14]. Barrett [14] has made a detailed analysis of the absorption frequencies of water vapor in relation to contributing as a greenhouse gas in table 1 in his publication. Data are shown as contributions to the absorbance of the Earth's radiance by the first 100 meters of the atmosphere, as follows: absorption relative to water vapor: Water vapor = 1.0; CO_2 (calculated at 285 ppm) = 0.249, (calculated at 570 ppm) = 0.271; CH_4, 0.018; N_2O = 0.007. Barrett states [14] that at 285 ppm CO_2 (the pre-industrial CO_2 estimated level) this green house gas absorbs 72.9% of the available radiance, leaving 27.1% that is transmitted of which an amount equivalent to 22.5% of the total passes through the window and the other parts of the spectral range transmit only 4.6%". For the case with CO_2 doubled (570 ppm), the 4.6% decreases only to 4.1%".

Barrett [14] has also made calculations of the contributions of the greenhouse gases on the greenhouse effect (global warming). Barrett reported values in table 2 of his publication (as % contributions at 285 ppm CO_2 and 570 ppm CO_2, respectively): water vapor = 78.5, 77.1; CO_2 = 19.6, 20.9; methane = 1.4, 1.4; N_2O = 0.6, 0.6).

There is general agreement that there are important "greenhouse gases" other than carbon dioxide in Earth's atmosphere. It is reasonable and logical as you would surmise, that the greenhouse effect (the amount of warming) by these gases is complex, but that the amount of warming is dependent on the nature of the gas, its ability to transmit and reflect different light energies (wavelengths), and the amount that is present in air. Indeed, air contains several greenhouse gases whose potentials for trapping energy in the atmosphere have been compared by a factor termed the global warming potential (GWP) based on their abilities to absorb energy (their radiative efficiency) and how long they remain in the atmosphere. Of course, the concentration of the gas in the atmosphere is also a determining factor. By definition, carbon dioxide is assigned a GWP of 1.0 (on a molecular basis) and for comparisons, based on 100-year lifetime, methane has a GWP of 28-36, nitrous oxide 265-298, and several chloroflurocarbons are in the tens of thousands [15, 16]. The properties of these gases depend on the physical nature of the elements as molecules and are reliably known. The current concentrations of the gases in the atmosphere are somewhat less reliably known and are dependent on averaging measurements made at various places.

The global warming gases the panel listed [15] and their potentials (as specified by the panel), compared to carbon dioxide, are as follows (based on an arbitrary 100 year time "horizon"): Carbon dioxide 1; methane 21; nitrous oxide 310; and 14 gases containing fluorine that ranged from 140 to 23,900. Water vapor is not considered as a radiative forcing molecule. However, as previously stated, it is more potent than carbon dioxide and it is present in approximately 19-times the concentration in the atmosphere. Water vapor, however, varies greatly. Global warming potentials are not provided for several gases and aerosols because there is no agreed-upon method to estimate the contribution of gases that are short-lived in the atmosphere, are spatially variable, or have only indirect effects [15].

Thus, we can see that molecule-for-molecule methane is more than 20 times as effective as carbon dioxide at trapping heat in the atmosphere. It is generally agreed that over the last two hundred and fifty years, the concentration of methane in the atmosphere increased by more than 150%. Over the same time period, carbon dioxide has been estimated to have increased approximately 30%. Anthropogenic sources of methane include agricultural activities, landfills, coal mining, wastewater treatment, stationary and mobile combustion and all petroleum and gas uses, and certain industrial processes.

A Scientific "Consensus" Supported by Political Fiat

Let us avoid all the complication produced by the difficulty in measuring the amounts of atmospheric carbon dioxide. We can debate, or accept, the validity of historical values and the errors in such measurements/estimates of both carbon dioxide and of temperature over the ages (growth rates of trees as measured by tree ring data, measurement in gas bubbles from ice cores, and other means). We can also debate whether the nature of the cause-and-effect relationship supports that a rise in temperature reflects the increase in carbon dioxide or the other way around. One can debate whether the temperature measurements by different means over recorded history are equivalent. For example, there are proposed and measured heat island effects of the changes in locations (sites) of temperature measurement sites (with more currently in large cities). There are legitimate concerns about whether satellite measurements or surface measurements are more appropriate; or whether land, sea, or air temperatures are most valid; or whether historical records of warm periods, or geological evidence of sea levels and ice and glacier coverage are most important; or even debate about whether various reports have distorted the data, or omitted certain data, or whether small differences are reliable. It is not currently possible to have consensus other than by fiat. Indeed, the classic story of "The Boy Who Cried Wolf" (Fig. **12**), credited to Aesop (620 – 564 B.C.)[3] has a moral:

"Even when liars tell the truth, they are never believed. The liar will lie once, twice, and then perish when he tells the truth".

Fig. (12). The Boy who cried wolf; Franc is Barlow 1687.

A DIFFERENT APPROACH FOR YOUR CONSIDERATION

I propose a different approach. Let us focus on nature's photosynthetic ability to keep carbon dioxide in balance and on mankind's ability to improve on photosynthesis artificially and implement it technologically to remove rising carbon dioxide while producing useful energy from the trapped sunlight. I shall devote additional emphasis to this positive approach. However, before doing so, a critically important element is left. The IPCC repeatedly has stated that it does not include water vapor as a forcing agent in its assessment that carbon dioxide drives the temperature increases that they predict in their computer models. These models indicate that global warming is a global threat and that it arises from carbon dioxide produced into the atmosphere by human activities. They dismiss water vapor, they say, based on the fact that water vapor molecules are transient, and they are transient because water is compressible and its concentration in the atmosphere is temperature-dependent. The water molecules are not long-lasting; they are created by evaporation and fall from the air by condensation. In other words, unlike carbon dioxide, water condenses; it is a condensable gas. They choose to use 100-year residence times for carbon dioxide and completely omit water vapor from their computer models. As global temperatures rise, more water vapor is present in the atmosphere, and this enhances the effects of global warming by carbon dioxide; thus, by their terms, water vapor is not a forcing agent for global warming but is an amplifier.

There has been great controversy about the effects of human activity on the concentration of carbon dioxide and possible consequences. Currently, politicians have won the day. Further discussion of the politics of this is outside the scope of this book. However, carbon dioxide has essential roles in metabolism and in the carbon cycle of nature and those will be discussed. Simply stated, the carbon cycle refers to the sum total of effect of the processes that convert the element carbon into and out of its various forms. Carbon dioxide is a molecular form of carbon in which one atom of carbon has combined with two atoms of oxygen. Gasoline, coal, trees, our bodies, and carbonaceous rocks are examples of other forms of carbon combined with other elements and in other ratios.

It is logical and reasonable that if we wish to know the source of the carbon dioxide that is found in the air we must understand the relationships and concentrations of the various sources that can lead to the carbon dioxide in the air around us. If we intend to understand why there are changes in the air concentrations of carbon dioxide, we must measure or otherwise determine how changes in the various forms of carbon affect the air concentrations of carbon dioxide. This, obviously, is an enormous undertaking. It is reasonable that the outcome, in terms of the quantity of carbon dioxide in the air, is interdependent on the quantities of many different classes of molecules. As an example, a tree grows by virtue of photosynthesis taking carbon dioxide out of the air and fixing it into the trunk, leaves and other carbon-containing structures of the tree. When the tree dies (or even annually when deciduous trees shed their leaves) some carbon is converted by decay to carbon dioxide which returns to the air. This is clearly a cycle; and one may speak of various cycles including the net effects of all the cycles to increase or decrease the atmospheric carbon dioxide concentration. It is reasonable and logical that any cycle can be in balance at any given time or over any given time, or it may result in a net accumulation or decrease in a component of the cycle. Pertinent to our interest, the amount of carbon dioxide in the air over any given time may be increasing, decreasing or not changing. If increasing carbon dioxide is the cause of global warming, that carbon dioxide must be increasing in the atmosphere due to a net effect of all the components of the carbon cycle to create an increase in atmospheric carbon dioxide. This logic and reasoning is clear and not controversial and I expect most can agree on this.

There are other aspects relating to the carbon cycle about which all reasonable parties can likely agree. If the argument is that, by their activities, humans are causing an increase in atmospheric carbon dioxide that is sufficient to cause global warming, then the net amount of the carbon dioxide released into the atmosphere which is not negated by processes that "fix" carbon dioxide, *e. g.*, remove it from the atmosphere, must be positive (increased). It is obvious that the amount of carbon dioxide in the air, and the amount of carbon in various other states at a time before man-attributed global warming began, are relevant. We can, however, simplify this by recognizing that the net balance of carbon flowing into carbon dioxide in the air from human activities, must be increasing for the argument of human-caused global warming to be valid. There is no escaping this logic and reasoning[4].

Data and Some Calculations

At this point, is it necessary to consider the quantities of the various molecules in the carbon cycle that affect the net change that can occur in atmospheric carbon dioxide. Although there is some disagreement about quantities, careful work has been done and, like a financial bookkeeper, we can see if we have mostly accounted, or not accounted at all, for individual "transactions" that increase or decrease carbon dioxide.

In this regard, it is interesting to estimate how much of earth's oxygen is sequestered in the carbon of carbon dioxide in the atmosphere and in the oceans and seas, using figures from the scientific literature. This oxygen is potentially available to be recycled by photosynthesis occurring in green plants, in algae and in phytoplankton.

The total weight (mass) of the earth's atmosphere has been calculated by various means and there is general agreement that the amount is approximately 5.1480 x 10^{21} grams (5.1480 x 10^{6} petagrams) [17]. I shall, for consistency in comparisons, use the unit petagram which is 10^{15} grams.

The mass of carbon dioxide in all earth's atmosphere can be calculated from the percentage of carbon dioxide present in the atmosphere which is an average of many measurements and the assumption is that it is well-mixed throughout the atmosphere. Carbon dioxide concentration is usually reported in parts per million (ppm) which is not an amount of mass but of volume. This is valid for comparing gases in the atmosphere, because different gases have different molecular weights but one gram molecular weight of any gas occupies 22.4 liters (at standard temperature and pressure). If we want to calculate carbon dioxide starting with the known mass of the air and the air contains approximately 400 ppm of carbon dioxide, we must first convert the mass of the atmosphere into the corresponding number of moles of air. Therefore, 5.1480 x 10^{21} grams of air divided by 28.966 (the molar mass of the gases in the atmosphere [17], taking into account their percentages) = 1.7773 x 10^{20} moles of air. Since 400 ppm carbon dioxide = 400 divided by 1,000,000, the total in the air = 1.7773 x 10^{20} moles of air x 400/1,000,000 mole fraction that is carbon dioxide, which is 7.09 x 10^{16} moles of carbon dioxide in the atmosphere.

The molar mass of carbon dioxide is 44 grams; therefore: the total carbon dioxide in the atmosphere is approximately 3120 petagrams of carbon dioxide. Since carbon in the atmosphere is 12/44 of this amount, there are approximately 851 petagrams of carbon in the carbon dioxide in the atmosphere. This, with rounding, agrees reasonably with the amount reported by NASA, 800 petagrams of carbon in the atmosphere [18].

Let us calculate the total amount of oxygen in the atmosphere, and compare that to the amount of oxygen bound up in carbon dioxide in the air. The air is approximately 20.95% oxygen (209,460 ppm). We can calculate the mass of oxygen in the air using the same procedure we used for carbon dioxide. The answer is 1.19 x 10^{21} grams of oxygen in the atmosphere (as O_2). Thus, on a weight (mass) basis there are 1,190,000 petagrams of oxygen molecules but only 3,120 petagrams of carbon dioxide in the atmosphere. The mass of carbon dioxide in the atmosphere is only 0.262% of the mass of oxygen.

Since there are two atoms of oxygen for each atom of carbon in a molecule of carbon dioxide, there is 32/44 x as much oxygen. Thus, there are approximately 3120 x 32/44 = 2269 pentagrams of oxygen contained in the carbon dioxide in the air. There are 1,079,000 pentagrams of oxygen in the air. Thus, 2182 divided by 1,079,000 = 0.002022 which is approximately 0.00002%. This is the approximate amount of oxygen sequestered in carbon dioxide in the air. This is a tiny amount; therefore, most of the potential oxygen in air is free (as O_2) and it is not bound.

The amount of carbon dioxide in all the oceans is estimated to be approximately 73,700 petagrams and most of this is in the form of bicarbonate with only about 1487 petagrams as dissolved carbon dioxide gas. Thus, the oceans provide a huge storage site for carbon dioxide. The actual release of carbon dioxide from that stored as bicarbonate depends on several factors but the most significant (if we do not assume large changes in ocean pH) is wave action to stir the ocean and allow surface release as well. We can also estimate the amount of oxygen that is sequestered in carbon dioxide in the oceans and seas and the value is 0.0005%. This compares to the value of 0.00002% for the oxygen sequestered in carbon dioxide in the air. Thus, the oceans and seas contain approximately 25-times as much oxygen sequestered in carbon dioxide as is present in carbon dioxide in the atmosphere.

Under today's conditions, there is almost no difference in the rate of dissolution and rate of solution of carbon dioxide by the oceans. It is claimed by the IPCC [15] that in preindustrial times the "natural" rate of release of carbon dioxide to the atmosphere was balanced by the rate of gaseous carbon dioxide dissolving eventually into the oceans. It is the position of the IPCC that this balance kept carbon dioxide concentration in the atmosphere at approximately 0.0275% for "many thousands" of years prior to the Industrial Age. This compares to approximately 0.0395% carbon dioxide in the air today. Therefore, the increase in carbon dioxide in the air from preindustrial times until today is approximately 44%. However, photosynthesis removes carbon dioxide from the atmosphere. It has been reliably reported that photosynthesis can be increased by at least 40% by doubling today's atmospheric carbon dioxide concentrations (as in a greenhouse).

I believe that photosynthesis itself has been underestimated in quantity and potential. This significant and important fact was addressed in the previous section: "Photosynthesis removes carbon dioxide from the air". Considering all factors, the ocean exchange of carbon dioxide reasonably and logically can be ignored in our evaluation of its effect on changes in atmospheric carbon dioxide, relevant to this assessment.

However, one may speculate that additional dissolved carbon dioxide will decrease the pH (acidity) of the oceans sufficiently to damage corals and other sea creatures that are dependent on dissolved bicarbonate in seawater for their carbonate-containing structures. Evidence for the rebound of what was a decline in coral reefs, speculated to result from the increased atmospheric carbon dioxide, during a period when release of carbon dioxide into the atmosphere was increasing by human activities, discounts this. There are many speculated causes for both the decline and recent rebound of coral reefs. There is no logical and reasonable basis to invoke a carbon tax because of changes in coral reefs from unknown causes.

For Your Consideration

I believe scientific investigation, even at its best, only shines a light into the darkness to reveal a bit more with each new discovery, but never unveiling all. There is always room to learn more. The greater the consequences of an idea, thesis, or hypothesis, the more we should reserve some skepticism; the more we should examine our data. Winning an argument, even though it is about a scientific question, by fiat or by appeal to majority or to politics, and not based on facts and logic, is not winning the argument scientifically. If the facts, themselves, are in question it is folly to build a theory that exalts one result over all others. If a theory or model does not encompass all the variables, it can only be provisional. To take actions that incur great consequences based on an unproven theory is imprudent. Only when objections to one's theory are absurd, should one dismiss another's ideas without consideration. Calling an absurdity is a dangerous game that should rarely be used except when dealing with ideas proven or otherwise demonstrably false. Attempt to win other than on the evidence and its logical application is not scientific; and even worse, it is immoral or misguided because its ends are immoral or misguided.

I confess a bias toward the way Freeman Dyson thinks [19]:

> "My first heresy says that all the fuss about global warming is grossly exaggerated. Here I am opposing the holy brotherhood of climate model experts and the crowd of deluded citizens who believe the numbers predicted by the computer models... The models solve the equations of fluid dynamics, and they do a very good job of describing the fluid motions of the atmosphere and the oceans. They do a very poor job of describing the clouds, the dust, the chemistry and the biology of fields and farms and forests. They do not begin to describe the real world that we live in. The real world is muddy and messy and full of things that we do not yet understand. It is much easier for a scientist to sit in an air-conditioned building and run computer models, than to put on winter clothes and measure what is really happening outside in the swamps and the clouds. That is why the climate experts end up believing their own models".

I do not know if the Earth, globally, is currently warming significantly. If it is, I do not know if the increase is significant in terms of effects that have measureable consequences. If it is, I do not know if this is a trend that will continue in the future. If it is a continuing, I do not know if it is caused in a significant way by carbon dioxide; if it is, I do not know that it is caused by human activities. If it is, I do not know that any currently proposed actions will change the outcome. How can I make these statements?

It is safe to assume that some readers will take the opposite view on each of these; they will be convinced that man's actions, by the agency of carbon dioxide, are causing the Earth to significantly grow hotter and that this will continue to devastation, unless actions are taken, such as the carbon tax and stopping the use of coal and petroleum-based energy.

In this book, I cannot address all the data, and all the factors that create the computer model that has predicted the results described in the immediately preceding paragraph. Suppose all the data used to program the model are precisely correct, except for one. Suppose the effect of photosynthesis is not correct. What is the evidence for this? I have discussed the difficulties in actually measuring global photosynthesis. Thus, the problem is compounded by the difficulty in knowing the amount of carbon dioxide removed by global photosynthesis. It seems pretensious to claim to know this. But aside from this, the estimated amount, as you might suspect, has been changing. Without being critical, I simply say that Melvin Calvin (Nobel, 1961; previously described) stated that "... plants and microorganisms on earth transform about 6,000 tons of carbon from carbon dioxide to carbohydrate per second". Doing the math, this equates to 172 petagrams of carbon annually. Estimates have varied: a value was reported in 1990 of 120 petagrams [20]; in 1990, 120 petagrams [21], and 150 to 175 petagrams [21, 22]. The highest estimate (175 petagrams) was justified by the authors. I could not find the source on which Melvin Calvin based his estimate in 1961. The current, highest estimate will go higher if atmospheric carbon dioxide rises. Temperature and humidity will increase and will also work in the direction of increasing carbon sequestration by photosynthesis. The potential for photosynthesis to increase is estimated to be at least 40%. Doing the math, using the highest figure of 175, this equates to 245 petagrams of carbon sequestered annually by photosynthesis. This is 27-times what is attributed to mankind today. It cannot be proved that response would occur, but it is worth pondering.

There is further confusion in the interpretation of the effect of global photosynthesis. In a paper on-line, stated by the authors to have been taken from a book[5], in Part 8. Global Photosynthesis and the Atmosphere, it is stated:

> "The amount of CO_2 removed from the atmosphere each year by oxygenic photosynthetic organisms is massive. It is estimated that photosynthetic organisms remove 100 x 1015 grams of carbon (C)/year (ref. cited). This is equivalent to 4 x 1018 kJ of free energy stored in reduced carbon, which is roughly 0.1% of the incident visible radiant energy incident on the earth/year. Each year the photosynthetically reduced carbon is oxidized, either by living

organisms for their survival, or by combustion. **The result is that more CO_2 is released into the atmosphere from the biota than is taken up by photosynthesis.** The amount of carbon released by the biota is estimated to be 1-2 x 1015 grams of carbon/year (emphases added)."

I assume that the formatting did not allow subscripts or superscripts and that this accounts for the way CO_2 is written and that 1015 should be interpreted as 10^{15}. The data presented show the opposite of the conclusion. My objective here is not to criticize the authors, but to show that all interpretations, including mine, should be examined.

Another way of expressing the huge impact of global photosynthesis, is that it is estimated [23] to capture "... approximately 130 terawatts, which is about 6-times larger than the current power consumption of all human civilization".

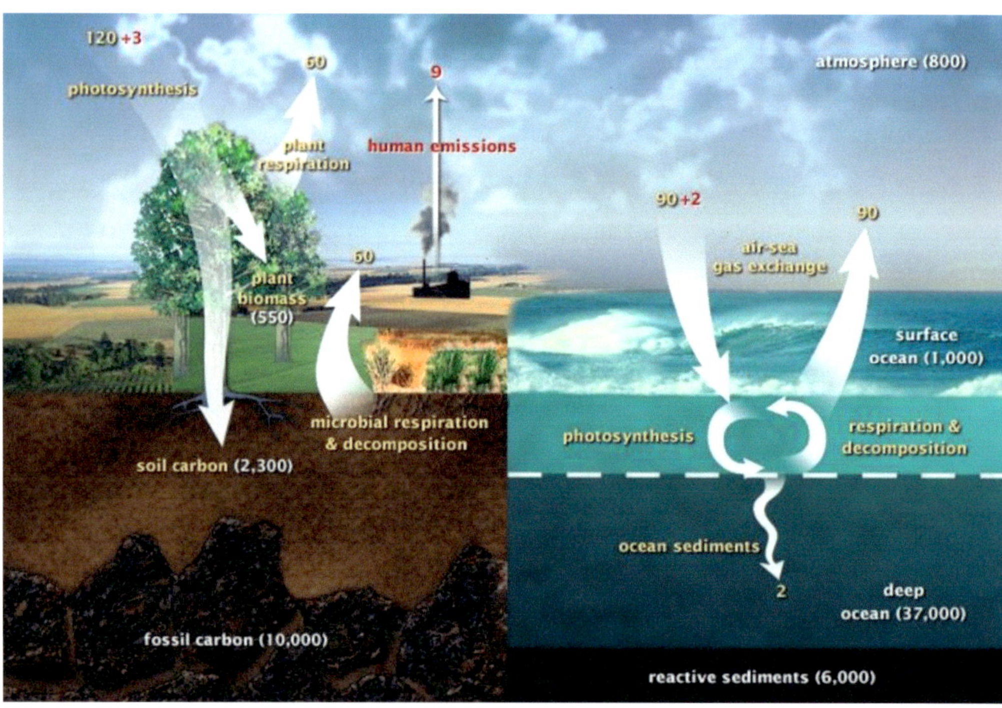

Fig. (13). The fast carbon cycle, showing the fluxes of carbon in gigatons per year.

This book is not about the subject of global warming, a subject so vast and controversial that it would require an entire book to address. Photosynthesis is relevant as the source of oxygen in the atmosphere, and as a significant mechanism that removes carbon dioxide from the atmosphere. The bottom line summary can be easily comprehended by inspection of the information in Fig. (**13**). This figure comes from an on-line document [18] from the National Aeronautics and Space Administration (NASA) with the date of June 16, 2011 which I accessed on 08-12-2016. NASA describes this figure as follows: "This diagram of the fast carbon cycle show the movement of carbon between land, atmosphere, and oceans. Yellow numbers are natural fluxes, and red are human contributions in gigatons of carbon per year. White numbers indicate stored carbon. (Diagram adapted from U.S. DOE, Biological and Environmental Research Information System.)".

Plant a Tree

The relevant point for this book is as follows. Carbon emissions and removals from the atmosphere, measured in gigatons of carbon annually, for the following categories are: human emissions, contributes 9; land photosynthesis removes 120, but contributes 60 for a net of 60. Thus, total human activity is 9/60 = 15% of what plants remove. Would it not be more reasonable just to plant a tree? Or, perhaps we could plant more food crops (of the efficient C3 photosynthetic type) and reap the benefits worldwide of the increased atmospheric carbon dioxide that is proven to stimulate photosynthesis and sugar production [24].

The "plant a tree" statement was written in jest. Let us do some analysis. I will list some data and their sources. From this I shall make some calculations and sprinkle them into the mix, and I will encourage you to make calculations of your own.

Assume that the current, consensus data is correct that global carbon dioxide in the atmosphere is increasing at the rate of approximately 2 ppm annually [25]. Assume that a tree, by photosynthesis, removes 3.78×10^{-11} petagrams of carbon dioxide annually [26]; this number equates to 1.03×10^{-11} petagrams of carbon annually. Of course, we know this number is only an estimate, and there is a large potential for variation. It is based on a mature *Pinus raduata,* a pine tree. NASA uses the figure of 800 petagrams of carbon (in the fast cycle) for total atmospheric carbon (approximately 2.0 x 10^{18} grams of carbon dioxide). If carbon dioxide is increasing by 2 ppm annually, then the increase is 14.5 petagrams of carbon dioxide. Note: this says that the annual increase in carbon dioxide (2 ppm) in the atmosphere results from an additional 14.5 petagrams of carbon dioxide.

Let us calculate the number of trees we would need to plant world-wide, to prevent this increase. The required calculation is: 14.5 x 10^{15} petagrams must be removed by photosynthesis, divided by 3.78 x 10^{-11} petagrams of carbon dioxide per tree per year. The answer is approximately 52 trees. This is the number each person on earth would need to plant, based on these data! Another way of figuring this is that we would have to plant a number of trees equivalent to 12% of the existing trees.

However, human (anthropomorphic) carbon dioxide production is said to be 9 petagrams of carbon which is 33 petagrams of carbon dioxide [18]. Dividing 14.5 by 33 results in 0.44 which is 44%, the amount the world would have to cut carbon dioxide emissions from human-related sources in one year to stay at the current carbon dioxide concentration of 400 ppm! Is this a realistic, achievable goal?

There is another potential salvation, the trees themselves. We have previously documented that photosynthetic sequestration of carbon dioxide from the atmosphere is stimulated by increased carbon dioxide. I have searched out some data that will be useful. If you prefer, look up your own numbers. I found an estimate of 3 trillion (3 x 10^{12}) for the total number of trees, globally [27]. The global world human population is calculated/estimated to be approximately 7.4 billion [28]. An "average" American generates approximately 2.3 tons of carbon dioxide annually; the carbon footprints of 18 average Americans can be neutralized by one acre of hardwood trees; a single tree can absorb 48 pounds of carbon dioxide annually; "Planting trees remains one of the cheapest, most effective means of drawing excess CO_2 from the atmosphere" [29]. There are approximately 750 billion metric tons (750 gigatons, or 750 petagrams) of carbon in the atmosphere (800 by NASA's calculations [18]); "... human activities-particularly the extraction and burning of fossil fuels and the depletion of forests- are primary sources of carbon emissions, totaling about 6.6 billion metric tons per year"; "A fifty-year-old forest on average absorbs 0.8 metric tons of carbon per acre per year and a sixty-five-year-old forest 1.6 metric tons per acre per year" [30]. Forests with contiguous canopies remove approximately 0.5 to 2 metric tons of carbon per year per hectare and grasslands have similar rates annually but the long-term storage is less; Australia has 149 x 10^6 hectares of forest and 440 x 10^6 hectares of grassland; "... Australian forests store about 10.5 billion metric tons of carbon (excluding soil carbon)... has accumulated over an assumed life of 100 years for native eucalypt re-growth. This translates to our forests storing an amount of carbon equivalent to almost 38.5 billion tones of gaseous carbon dioxide from the atmosphere, about 70 times Australia's annual net greenhouse gas emission" [31]. The global distribution of photosynthesis has been plotted by satellite-acquired data (Fig. **14**).

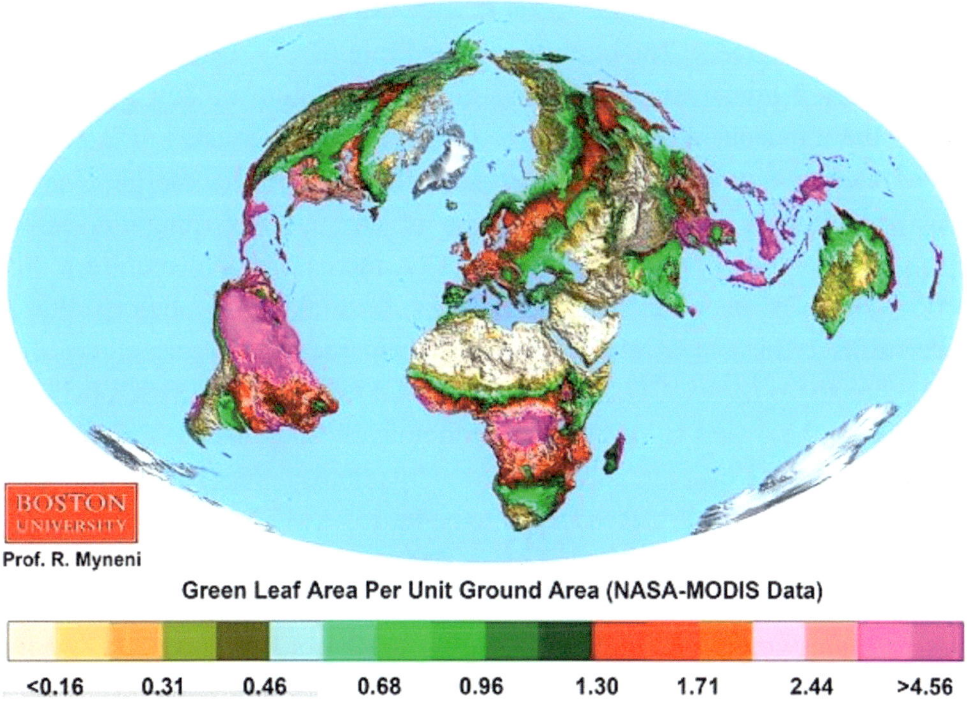

Fig. (14). Global distribution of green leaf area which correlates with areas of plant photosynthesis.

The Mauna Loa Keeling data show that atmospheric carbon dioxide has been increasing at the rate of approximately 2 ppm over the past several years and is currently approximately 400 ppm at that site, 3400 feet of elevation on the volcano Mauna Loa. It is stated that this site is suited to detect changes in the northern hemisphere from increased human endeavors that produce carbon dioxide. This is the stated purpose. It is only recently that global satellite measurement methodology has been applied to begin collection of data globally about carbon dioxide globally in the atmosphere (Fig. **15**). This data clearly show large variability by region and by season. To deny that the Mauna Loa data show an increasing CO_2 and a current high would, in my opinion, equate to "denial". To not accept that it tells us what is happening globally is scientific prudence. By design and stated intent it tells something we can know (or determine) from many direct data sources: that anthropomorphic CO_2 production has increased in the northern hemisphere. It is known that carbon sequestration happens. As the global satellite data accumulate we will know if and to what extent carbon sequestration is able to compensate. The broader point is what can humans do to change it? Let us look at some numbers. The current accepted rate of CO_2 increase is 1.9 ppm (approximately 2 ppm) annually. Using the NASA figure of 800 gigatons of carbon in the atmosphere (a gigatons is 10^{15} grams) this equates to 2.9 x 10^{18} grams of carbon dioxide (or 2900 petagrams). The anthropomorphic annual amount according to NASA is 9 gigatons of carbon (33 petagrams carbon dioxide). As previously stated, to stop the CO_2 increase would require a 44% reduction annually in what man contributes. I think it is obvious this is inconceivable. I think it is even more inconceivable that a carbon tax does anything directly to reduce CO_2. The hope is that it will change people's behavior either because they want to or because economically they cannot do otherwise (they are forced to). However, nature removes by photosynthesis an amount most recently measured by satellite technology to be up 150 to 175 petagrams of carbon [22]. Most importantly, data show that increasing CO_2 can increase photosynthesis by as much as 40% with existing biota. Calculations show that only a 14% increase in photosynthesis calculated only on the basis of what trees remove, would offset the annual increase in CO_2. It is known that trees only account for part of the total CO_2 sequestered photosynthetically and that my 14% does not take into consideration all these other sources. Indeed, it is believed that algae may sequester an amount equal to that of trees and the amount by grasses and field crops is significant but very difficult currently to assess.

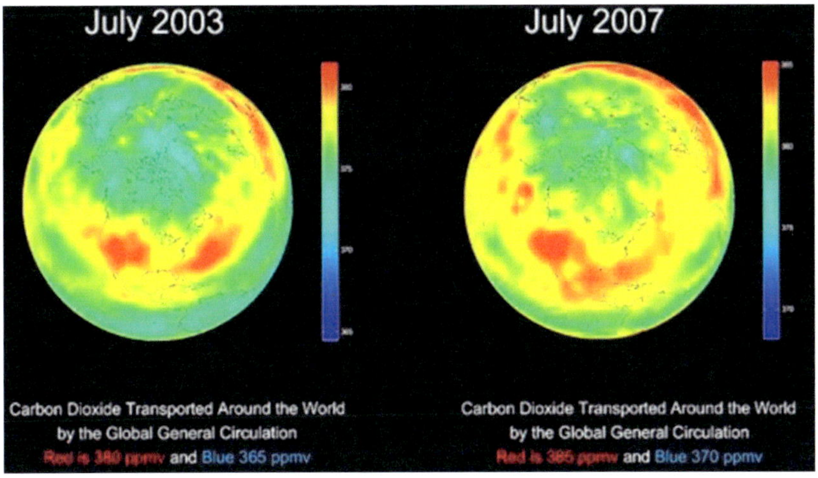

Fig. (15). Global distribution of carbon dioxide that reveals variable distribution.

Carbon Dioxide Added to the Air by Humans in the Modern Era

The primary contributors of additional carbon dioxide released into the atmosphere after the Industrial Age activities of humans, reasonably and logically, must result from the burning of fossil fuels, the production of cement, and the clearing of forest land and/or other changes in land use that release carbon dioxide to the atmosphere. There also are the possible secondary effects of any associated global temperature increase (by the carbon dioxide "greenhouse" effect *via* trapping heat from sunlight) on the various processes that take-up or release carbon dioxide into the atmosphere. Release of carbon dioxide into the atmosphere from human production of cement has been estimated to currently to be approximately 1.8 petagrams annually. This is approximately 5.1% of the total carbon dioxide projected to be released annually by human activity, and can reasonably be ignored. The ability of photosynthesis to compensate was explained previously in the section: "Photosynthesis removes carbon dioxide from the air". Burning of fossil fuels in the United States has been calculated to release into the atmosphere approximately 5.3 petagrams of carbon dioxide in 2011. This sequesters annually only approximately 0.0005% of the atmospheric oxygen as gaseous, oxidized carbon.

Impact of Plant Growth

The positive impact of increased atmospheric carbon dioxide on plant growth and photosynthesis is well known. A review published in 1993 [32], of research based on studies using greenhouses, reported that a doubling of CO_2 concentration stimulated the growth of 156 different species of plants by an average of approximately 37% with 41% for C3 plants and 22% for C4 plants. Herbaceous crop species increased most strongly (58%).

A 23 page report [33] that evaluated the net effect for carbon sequestration of several "perennial energy crops" were defined as fast growing species that can be grown in marginal soils- crops the authors say are likely to increase as countries seek to decrease dependence on fossil fuels and reduce carbon dioxide emissions. The authors concluded (in part): "... the annual net SOC (soil organic carbon) storage change exceeds the minimum mitigation requirement (0.25 milligrams of carbon per hectare per year) for herbaceous and woody perennials by far (1.14 to 1.88 and 0.63 to 0.72 milligrams carbon per year per hectare, respectively)".

A 2004 report [34] addresses the question: What have we learned from 15 years of free-air CO_2 enrichment experiments relative to photosynthesis, canopy properties, and plant production? The authors stated: Free-air CO_2 enhancement (FACE) experiments allow study of the effects of elevated carbon dioxide on plants and ecosystems grown under natural conditions without enclosures. Information from 120 peer-reviewed articles describing the physiology and production in 12 large-scale experiments with carbon dioxide at from 475 to 600 ppm were subjected to meta-analysis. Some of the results from previous chamber experiments were confirmed: for example, light-saturated carbon uptake, diurnal carbon assimilation growth and above-ground production increased. Trees were more responsive than herbaceous species to elevated carbon dioxide. Grain crop yields increased far less than anticipated from prior enclosure studies. The authors stated: "The broad direction of change in photosynthesis and production in elevated CO_2 may be similar in FACE and enclosure studies, but there are major quantitative differences: trees were more responsive than other functional types; C_4species showed little response; and the reduction in plant nitrogen was small; and largely accounted for by decreased Rubisco (RuBisCO)".

The reasonable conclusion is that Earth's temperature surely has varied in the past over both short intervals and over long intervals of time. It will surely vary in the future. Predicting the effects on global temperature (or climate) of one source (human activity) of one gas (carbon dioxide) based on computer models is fraught with uncertainty. It is most evident that the design of the computer climate model and the data that is entered into the model will determine the predicted global temperature changes. At this time, it is my opinion that no reliable model is available; therefore, no prediction is reliable. As an example, consider the words of Wolfgang Knorr in 2009 [35]: "Is the airborne fraction of anthropogenic CO_2 emissions increasing?" Knorr, a member of the Department of Earth Sciences, University of Bristol, UK) also wrote:

> "This study re-examines the available atmospheric CO_2 and emission data including their uncertainties. It is shown that with these uncertainties, the trend in the airborne fraction since 1850 has been $0.7 \pm 1.4\%$ per decade, i. e. close to and not significantly different from zero. The analysis further shows that the statistical model of a constant airborne fraction agrees best with the available data if emissions from land use change are scaled down to 82% or less of their original estimates. Despite the predictions of coupled climate-carbon cycle models, no trend in the airborne fraction can be found."

As one prime example of the unreliability of current models is the difficulty in reliably factoring in the effect of atmospheric water vapor. I understand the argument for leaving water out because it is a compressible gas, it cycles over short time intervals, and it changes in concentration with temperature changes. After all, the term relative humidity affirms the role of temperature. To see the enormous difficulty in incorporating water vapor into the model, consider the effects of clouds. As temperature rises, the sky will be cloudier. Clouds affect sunlight's ability to heat the Earth in ways that are too complex to model, even if we had sufficient input data, which we also lack. If the climate modelers are right, then ignoring water vapor could be justified. If they are wrong, the predictions of their models are no better than crude guesses.

Think about the last time you were in the Colorado Rocky Mountains (or another place with low humidity and frequent changes in cloud cover). Even a few moments of shade from clouds affects your sense of temperature, and you reach for a jacket. I recognize that these are short-term effects; however, long-term effects are logically made-up of complied short-term effects. Also, as an example in logic, consider the short-term effect of humidity on local, transient temperatures, and the effects of cloud cover on day and night temperature cycles. Also, consider the Sun itself. We have not even discussed the variability of output of the Sun's energy and its approximately 11-year cycles.

Consider some statements made by David Biello that include statements attributed to "world leaders" who met in Copenhagen in 2009. These were published in *Scientific American* on January 1, 2010 [36]. Biello wrote:

> "Scientists do not know what 'number'- in terms of temperature or concentrations of greenhouse gases in the atmosphere- constitutes a danger... When it comes to defining the climate's sensitivity to forcings such as rising atmospheric carbon dioxide levels, 'we don't know much more than we did in 1975' (attributed to Stephen Schneider of Stanford University, who first defined the term "climate sensitivity" in the 1970s). (Biello further wrote): "... but what eludes precise definition is how much other factors play a role- the response of clouds to warming, the cooling role of aerosols, the heat and gas absorbed by oceans, human transformation of the landscape, even the natural variability of solar strength". (Biello then quoted Gavin Schmidt of NASA Goddard Institute for Space Studies to say): "We may have to wait 20 or 30 years before the data set in the 21st century is good enough to pin down sensitivity". (Biello then wrote): "But how much heating and added CO_2 are safe for human civilization remains a judgment call. European politicians have agreed that global average temperatures should not rise more than two degrees C above preindustrial levels by 2100, which equals greenhouse gas concentrations of roughly 450 ppm". (Biello cites geochemist Wallace Broecker of Columbia University as saying): "We're at 387

> now, and we're going up at 2 ppm per year. That means that 450 is only 30 years away. We'd be lucky if we could stop at 550". (Biello references Goddard's James Hansen as arguing that): "... atmospheric concentrations must be brought back to 350 ppm or lower-quickly". (Biello quotes James Hansen as saying): "Two degrees Celsius (of warming) is a guaranteed disaster". (Biello cites Physicist Myles Allen of the University of Oxford as saying): "To solve these problems, we need to eliminate net emissions of CO_2 entirely. Emissions need to fall by 2 to 2.5 percent per year from now on". (Biello states): "Keeping atmospheric concentrations of greenhouse gases below 550 ppm, let alone going back to 350 ppm or less, will require not only a massive shift in society- from industry to diet- but, most likely, new technologies, such as capturing CO_2 directly from the air". (Biello then quotes physicist Klaus Lackner of Columbia University as saying): "Air capture can close the gap."

And that makes my point: science is warning of us of a threat, that may or may not be imminent (we know that some level of global warming is threatening); capturing CO_2 is a solution for which photosynthesis is proven to be remarkably efficient and adaptive at accomplishing; politicians taxing carbon is not.

TOXICITY OF CARBON DIOXIDE

Carbon dioxide is essential for both plant and animal life, including humans. It is "naturally" present in the atmosphere in a small amount. It is inhaled with each breath (approximately 0.04% in air) and exhaled in a higher concentration (approximately 4.2%). It stimulates our breathing mechanism to take a breath. It is produced as a byproduct of the cellular mechanisms that allow our bodies to grow, our muscles to move, our heart to beat, our lungs to function, and our brain to think.

Recently, carbon dioxide was declared to be a toxic substance and an air pollutant by the US EPA[6]. However, Merriam Webster's dictionary, on-line edition, defines a pollutant as follows: noun- a substance that makes land, water, air, *etc.*, dirty and not safe or suitable for use: something that causes pollution. It further states:

the action of polluting, especially by environmental contamination with man-made waste; also: the condition of being polluted. This leads us to consider contaminant and contamination. Synonyms for contamination are: adulterant, contaminant, impurity, defilement, pollutant, and related words are: blot, blotch, spot, stain, taint, dirt, filth, grime, muck, scum, sludge, smut, soil, blemish, defect, disfigurement, fault, flaw, abnormality, imperfection, irregularity.

One definition of the scientific discipline known as toxicology is: "the study of poisons". A correct saying in toxicology is: "The dose makes the poison". One must consider the concentration and dose to determine if a substance acts with no effect, has a beneficial effect, or is toxic.

To evaluate further the biological effects of carbon dioxide on life, we can turn to the literature of toxicology, medicine, botany, microbiology, chemistry and biochemistry. The air we inhale has the following composition: Oxygen 20.9%; carbon dioxide 0.04%; water vapor a small, variable amount, perhaps 0.75%; nitrogen 78.4%. The air we exhale: oxygen 15.3% (this varies with activity); carbon dioxide 4.2%; water vapor near saturation; nitrogen 74.3%. Thus, the air we exhale with each breath has about 105 times the carbon dioxide found in the air we inhale.

Other relevant conclusions about the safety of carbon dioxide are provided by the OSHA (Occupational Health and Safety Administration). After reviewing the evidence, the OSHA concluded that limiting carbon dioxide in workplace air to 1% (10,000 ppm) for 8 hours per day and to 3% for a 15 minute exposure will protect employees from the adverse effects associated with excessive exposures to carbon dioxide. It can be noted that 1% is 25 times the amount of carbon dioxide currently in the atmosphere, and 3% is 75 times this amount. Thus, even the maximum atmospheric increase in carbon dioxide in the future by any global warming scenario would not be toxic. If not toxic, then various speculated changes in the earth have to be added on *ad hoc*, to provide any possible, however unreasonable and illogical, bases for regulating carbon dioxide air concentration. If there were no costs, economic or otherwise, this regulation is not justified scientifically. Because of the cost and likely ineffectiveness, this solution seems to be a non-solution.

Therefore, on balance, atmospheric carbon dioxide at current concentrations is best characterized scientifically as a healthy and even essential partner in the global recycling that is an integral part of the relationships among oxygen, organic carbon in its various forms, carbon dioxide, and life on earth.

NOTES

[1] Blaise Pascal (1623-1662). *Pensées,* 580, 1670, tr. William F. Trotter, 1931.

[2] Men, it has been well said, think in herds; it will be seen that they go mad in herds, while they only recover their senses slowly, and one by one. Charles Mackay (1812 or 1814-1889) *Extraordinary Popular Delusions and the Madness of Crowds.*

[3] By Francis Barlow, 1687; called by him: *"De pastoris puerro et agricus"* The shepherd boy 119; Aesop's Fables Number 210, Perry's numbering system.

[4] Science is not the exclusive property of scientists; it is a common human enterprise, a heritage of every thinking person on the planet. Frank S. Hsu (1943-) *The Physical Universe, an introduction to Astronomy.* University Science Books, 1992.

[5] Concepts in Photobiology: Photosynthesis and Photomorphogenesis. Edited by G.S. Singhal, G. Renger, S.K. Sopory, K.-D. Irrgang and Guvindjee, Narosa Publishers/NewDeli; and Kluwer Academic/Dordrecht, pp. 11-51.

[6] Science is the search for truth- it is not a game in which one tries to best his opponent; to do harm to others. Linus Pauling (1901-1994). *A Nobel scientist speaks.* Liberation (New York) 2, No. 11, February 1958.

REFERENCES

[1] Calvin Cycle. New World Encyclopedia. Available from: http://www.newworldencyclopedia.org/entry/Calvin_cycle 2016.

[2] Sarovar M, Ishizaki A, Fleming GR, Whaley KB. Quantum entanglement in photosynthetic light-harvesting complexes. Nat Phys 2010; 6(6): 462-7. [http://dx.doi.org/10.1038/nphys1652]

[3] Hansen K. Seeing photosynthesis from space: NASA scientists use satellites to measure plant health. Available from: http://www.nasa.gov/content/goddard/seeing-photosynthesis-from-space-nasa-scienti-

sts-use-satellites-to-measure-plant-health

[4] Photosynthesis. New World Encyclopedia. Available from: http://www.newworldencyclopedia.org/p/index.php? 2015.

[5] Carmo-Silva E, Scales JC, Madgwick PJ, Parry MA. Optimizing Rubisco and its regulation for greater resource use efficiency. Plant Cell Environ 2015; 38(9): 1817-32. [http://dx.doi.org/10.1111/pce.12425] [PMID: 25123951]

[6] Melvin Calvin - Biographical. Available from: http://www.nobelprize.org/nobel_prizes/chemistry/laureates/1961/calvin-bio.html 2014.

[7] Parry MA, Andralojc PJ, Scales JC, *et al.* Rubisco activity and regulation as targets for crop improvement. J Exp Bot 2013; 64(3): 717-30. [http://dx.doi.org/10.1093/jxb/ers336] [PMID: 23162118]

[8] Calvin M. The Nobel Prize in Chemistry 1961. Melvin Calvin. Award Ceremony Speech. Available from: http://www.nobelprize.org/nobel_prizes/chemistry/laureates/1961/press.html 2016.

[9] Goodsell DS. Rubisco. RCSB Protein Data Bank. Melvin Calvin. Award Ceremony Speech. Available from: http://pdb101.rcsb.org/motm/11 2000.

[10] Su Meiyi Li TR. Catalase Kinetics http://webmitedu/chrissu/Public/5310lab3pdf

[11] Welp LR, Keeling RF, Meijer HAJ, *et al.* Atmospheric CO_2 driven by El Nin. Nature 2011; 477(7366): 579-82.

[12] Joes *vs.* the Volcano. Available from: http://www.snopes.com/volcano-carbon-emissions/ 2016.

[13] Global Carbon Project (GCP). Global Carbon Project (GCP). www.globalcarbonproject.org/

[14] Barrett J. Greenhouse molecules, their spectra and function in the atmosphere. Energy Environ 2005; 16(6): 1037-45. [http://dx.doi.org/10.1260/095830505775221542]

[15] IPCC Fifth Assessment Report. Available from: http://www.c2es.org/science-impacts/ipcc-summaries 2016.

[16] US EPA Climate Change Division. Understanding Global Warming Potentials. Available from: https://www3.epa.gov/climatechange/ghgemissions/gwps.html 2016.

[17] Petagrams of carbon. Available from: http://how-it-looks.blogspot.com/2011/07/petagrams-of-carbon.html 2016.

[18] Riebeek H. The Carbon Cycle : Feature Articles. NASA Earth Observatory 2011.

[19] Heretical thoughts about science and society. Available from: https://www.edge.org/conversation/freeman_dyson-heretical-thoughts-about-science-and-society

[20] Woodwell GM, Hobbie JE, Houghton RA, *et al.* Global deforestation: contribution to atmospheric carbon dioxide Science 1983; 222(4628): 1081-6. [http://dx.doi.org/10.1126/science.222.4628.1081]

[21] Novel Method for Measuring Global Photosynthesis Rate. Scripps Institution of Oceanography, UC San Diego. Available from: https://scripps.ucsd.edu/news/1895

[22] Salleh A. Global photosynthesis faster than thought. Available from: http://www.abc.net.au/science/articles/2011/09/29/3327059.htm 2011.

[23] Global Forestry Carbon. Available from: http://globalforestrycarbon.com/photosynthesis.html

[24] Ridley M. Benefits of global greening. Available from: http://www.rationaloptimist.com/blog/the-benefits-of-carbon-dioxide/ 2016.

[25] US Department of Commerce NESRL. ESRL Global Monitoring Division - Global Greenhouse Gas Reference Network.

[26] Trees, YOU and CO2 - Plant Talking Points, Science Education and Outreach. Available from: http://botany.org/PlantTalkingPoints/CO2andTrees.php#check

[27] Tree Counter Is Astonished By How Many Trees There Are : Goats and Soda : NPR. Available from: http://www.npr.org/sections/goatsandsoda/2015/09/02/436919052/tree-counter-is-astonished-by-how-many-trees-there-are

[28] World Population Clock: 7.4 Billion People. Worldometers. Available from: http://www.worldometers.info/world-population/ 2016.

[29] Carbon and Tree Facts. Available from: http://www.arborenvironmentalalliance.com/carbon-tree-facts.asp

[30] Carbon Sequestration Facts. Available from: http://www.forestecologynetwork.org/climate_change/sequestration_facts.html 2016.

[31] Which plants store more carbon in Australia: forests or grasses? Available from: http://www.chiefscientist.gov.au/2009/12/which-plants-store-more-carbon-in-australia-forests-or-grasses/ 2016.

[32] Poorter H. Interspecific variation in the growth response of plants to an elevated ambient CO_2 concentration. Vegetatio 1993; 104105(2): 77-97.
[http://dx.doi.org/10.1007/BF00048146]

[33] Agostini F, Gregory AS, Richter GM. Carbon Sequestration by Perennial Energy Crops: Is the Jury Still Out? BioEnergy Res 2015; 8(3): 1057-80. Available from: http://link.springer.com/10.1007/s12155-014-9571-0

[34] Ainsworth EA, Long SP. What have we learned from 15 years of free-air CO_2 enrichment (FACE)? A meta-analytic review of the responses of photosynthesis, canopy properties and plant production to rising CO_2. New Phytol 2005; 165(2): 351-71.
[http://dx.doi.org/10.1111/j.1469-8137.2004.01224.x] [PMID: 15720649]

[35] Knorr W. Is the airborne fraction of anthropogenic CO_2 emissions increasing? 2009; 36(21)

[36] Biello D. Climate Numerology. Sci Am 2010; 302(1): 14-5. Available from: http://www.nature.com/doifinder/10.1038/scientificamerican0110-14.
[http://dx.doi.org/10.1038/scientificamerican0110-14]

CHAPTER 5

Oxygen, Vital Element in Water

Abstract: Water is the most abundant molecule in which atomic oxygen is found on earth. Water is composed of two atoms of hydrogen (the simplest of the elements) and one atom of oxygen covalently bonded (four valence electrons are shared). Water covers about 71% of the land mass on earth and by weight water is approximately 0.33% of the atmosphere. It is the only element that is present in the earth-like temperature range as liquid, solid and gas. There is an earth-water cycle, and it nourishes, cleans and sustains the land with transpiration to clouds and condensation as rain, snow, sleet and hail. Our bodies vary in oxygen content, but the average is around 50%; slightly less in women than in men, and more in the skinny than in the obese. Oxygen has physical and chemical properties that are unusual and not predicted, compared to other small, simple compounds. It has maximum density at about 4 degrees Centigrade which causes lakes and ponds to freeze from the top down with interesting consequences. It is a good solvent for many chemicals and it forms the milieu of the cells of our bodies. It forms hydrogen bonds with itself which introduces unexpected structure in collected water molecules. Without it, life as we know it (and what other kind of life might there be?) would not be possible– we even search for it in space and on other worlds as a sign there might be life there.

Keywords: Atmosphere, Aquifer, Boiling point, Covalent bond, Diatomic, Dipole, Freezing point, Heat of vaporization, Hydrogen bond, Hydrophilic, Hydrophobic, Ocean, pH, Sea, Surface tension, Thales, Transpiration, Universal solvent, Water, Water cycle.

INTRODUCTION

Water covers approximately 71% of the earth's surface [1] and only 2.5% of this is fresh water and most of that is in ice with less than 0.3% in rivers, lakes and the

Olen R. Brown

atmosphere [2]. Amazingly, according to published data [2], only 0.003% of earth's fresh water is found in "biological bodies and manufactured products". It is the only natural substance that exists in all three states: solid, liquid, and gas at the usual temperatures and pressures found on earth's surface [3]. Most scientists agree that without water life cannot exist.

Water is a small molecule, composed of two atoms of hydrogen and one atom of oxygen with shared electrons forming the oxygen to hydrogen bonds. Two valence electrons come from the two hydrogen atoms, and six valence electrons come from the one oxygen filling the orbital with eight electrons. The mass of water is 18 grams per molecule, and water is approximately 89% oxygen by mass. Water has a number of unusual physical properties. Compared to common liquids, water has the highest: surface tension, heat conductance, heat capacity, and latent heats of fusion and vaporization. High surface tension dictates size of drops in rainfall and controls biological properties within cells. Also, small bugs take advantage of this and some can walk and skate on water. Surface tension pulls water together when it is in small droplets (water is attracted to itself and tends to form a ball shape). In a small tube, water will form a meniscus which rises up the sides of the tube, and the meniscus is elevated higher in proportion to the smallness of the tube. Water will rise in the vasculature (small tubes) of plants, and will rise to the top of even tall trees, due to these forces plus the osmotic forces in the roots and transpiration in leaves that overcome the pull of gravity. This unexpected behavior of water results also from the propensity of water molecules to stick together and to stick to other surfaces. The latent heats of fusion and vaporization work to modulate heat near the earth's surface, and the heat of vaporization is in control of the regulation of transfer of heat between ocean, seas, and the atmosphere.

WATER, A SMALL BUT UNUSUAL MOLECULE

Thus, water is a small, simple molecule but it has unusual and unexpected properties. One atom of oxygen and each of two atoms of hydrogen are held together by coordinate, but unequal, sharing of electrons (Fig. **1**). The valence (bonding) electrons are held more tightly (closely) to the oxygen atom which creates a slightly negative charge on one "side" of the molecule. The opposite occurs in the region around the two hydrogen atoms which are slightly and positively charged. Thus, the molecule is a dipole. The hydrogen atoms of the water molecule bond at an internal angle of close to 105 degrees and the water molecule looks something like Mickey mouse's head (Fig. **1**).

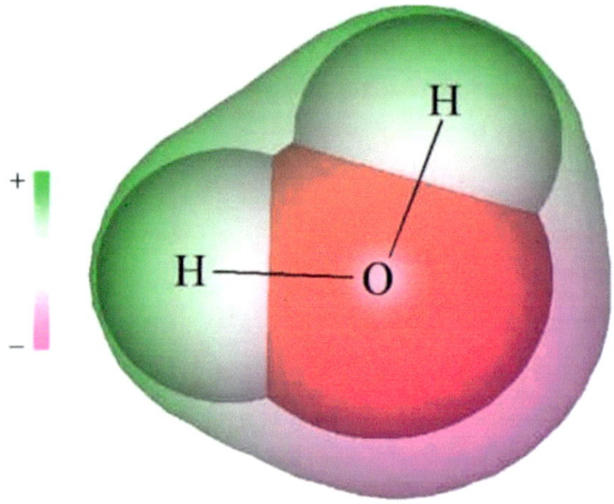

Fig. (1). Space-filling model of oxygen; one atom of oxygen (red) with two atoms of hydrogen attached by covalent bonds at an angle of 105 degrees, resulting in concentration of positive charge on one side and negative on the other.

In a collection of water molecules, there also is considerable orientation and organization of individual molecules. This results because water can form 4 intermolecular hydrogen bonds– an unusual number of such bonds for such a small, simple molecule (Fig. **2**). Although the hydrogen bonding is comparatively weak (only about 0.1 as strong as the typical covalent bond between atoms in a molecule) it is responsible for most of the unusual properties of water including: high freezing and boiling points, high heat capacity, high heats of fusion and evaporation, high surface tension, bipolar nature, and great ability to be a solvent for many chemicals. The high surface tension of water, as just discussed, causes it to rise in capillaries including the vascular systems of plants, even tall trees [4]. If water did not have (for its size and simplicity) the unusual, complex extremes in these physical characteristics, it could not function as it does in myriad ways, both inside cells and in nature's water cycle to clean, feed and restore the earth.

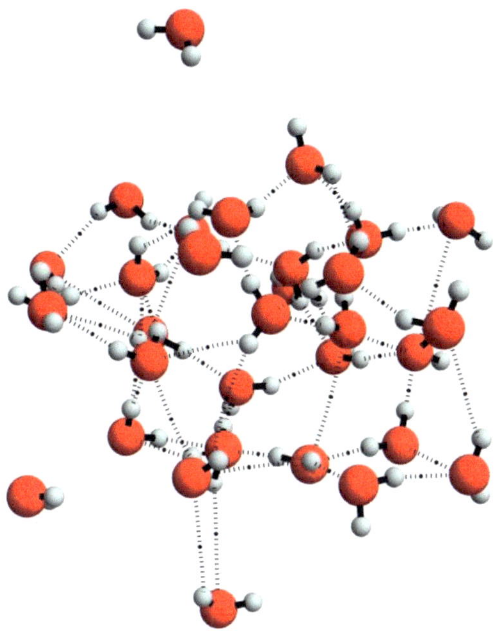

Fig. (2). Hydrogen bonding of water molecules; one molecule can bond with a maximum of four other water molecules (two water molecules outside the bonding range are also shown).

THE BLUE (AND RED) SKY

Water molecules (actually their OH bonds) strongly absorb infrared light, but very little light in the visible part of the light spectrum; thus, even humid air is transparent until water droplets form (first as micro-droplets). Also, the presence of water vapor at the concentrations found, in the absence of condensation with cloud formation, does not interfere with the light needed for photosynthesis and the generation of oxygen. Air is not completely transparent, and we also enjoy the blue color of deep water where sunlight eventually is filtered out and where photosynthesis is marginal (except as noted in Chapter 11– there are certain marine bacteria that can get along with very low light).

Why is the sky blue? This seems like a question that a child might ask, but the answer intrigued serious scientists including Sir Isaac Newton, John Tyndall and Lord Rayleigh, several ancient Greeks, and Leonardo da Vinci was interested in the problem (as stated by Sadi-Carnot [5] who cited Pedo Liliiefield [6]). The account by Leonardo da Vinci was recorded in his book *Codex Leicester*, circa 1508[2]. Leonardo wrote [7], in his famous backward but clear script, under the item "Water 17. Why is the sky blue?":

> "I say the blue appearance of the atmosphere is not its own color but is caused by the heated moisture having evaporated into minute, imperceptible particles; these particles are struck by the sun's rays and then become bright beneath the infinite darkness of the region of fire that lies upon the atmosphere from outside like a cover."

The reason the sky is blue also accounts for why sunsets are red, and it involves water droplets, water vapor, oxygen and nitrogen gases, dust particles in the atmosphere, and of course, the Sun. Molecules of oxygen and nitrogen predominate and they absorb, reflect and transmit sunlight in different ways, including to scatter the blue light from the spectrum of sunlight (Fig. **3**), and we "see" the scattered blue light.

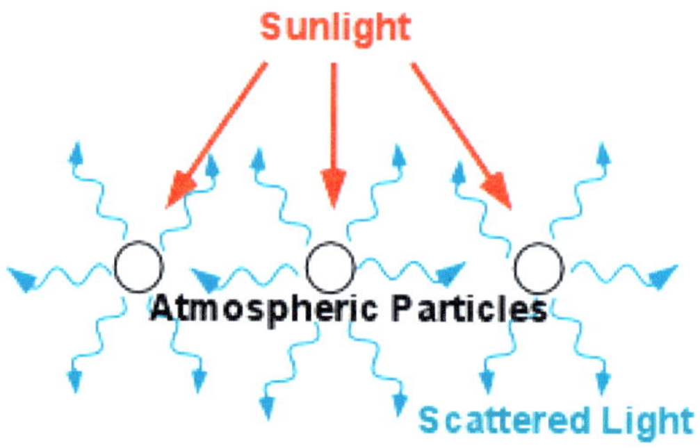

Fig. (3). Interaction of sunlight with atmospheric particles.

Recall that sunlight has various wavelengths that we perceive as different colors. Rayleigh scattering is a term that applies to the explanation Rayleigh provided to account for the interactions of atmospheric components with light. Light from the sun is perceived as essentially white light and is composed of light at all wavelengths in the visible spectrum from wavelengths of about 400 to 700 nanometers [8]. The electromagnetic spectrum is very broad and includes wavelengths as short as gamma rays, and as long as radio waves (Fig. **4**). The visible spectrum is a small region of wavelengths which, as the name implies, allows us to perceive colors. An object that absorbs all visible wavelengths appears black; it reflects none of the wavelengths back to be perceived by our eyes with interpretation by our brain. Thus, we perceive the color of an object by the light wavelengths that are reflected to our eyes. A blue sky means that a preponderance of the blue wavelengths of light is reflected and other wavelengths are absorbed (Fig. **3**). This reflection is caused by relatively small particles, molecules of nitrogen and oxygen, in the atmosphere. The story is a bit more complicated because of Rayleigh scattering.

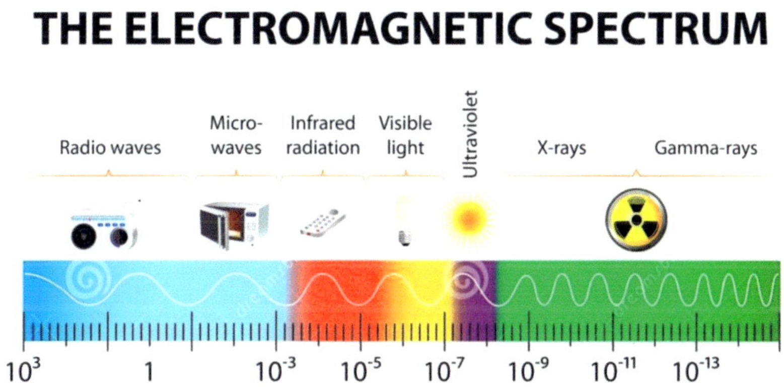

Fig. (4). The electromagnetic spectrum, in relation to the sun; wavelengths and frequencies including the visible spectrum from about 400 to 700 nanometer; and comparative sizes depicted from subatomic to Mt. Everest.

Rayleigh Scattering in More Detail

Red light has a wavelength that is approximately twice as long as blue light. The red light is not very much affected by the small nitrogen and oxygen molecules in the atmosphere. The shorter wavelengths of blue light react more with the small atoms of nitrogen and oxygen by bouncing off (they are scattered). More technically, the molecules of nitrogen and oxygen have vibrational frequencies that resonate with the frequencies of blue light (Fig. **5**). The light tends to be absorbed and then released. This is Rayleigh scattering (Fig. **6**) and the scattered blue light is what the eye sees, and the sky appears blue.

Near sunset when the Sun is very near the horizon, for light to reach one's eyes it must pass through a greater thickness of Earth's atmosphere and more of the red than blue wavelengths are absorbed. Additionally, there is often more dust particles and higher humidity (more water) in this layer of atmosphere which enhances this relative removal of the blue (shorter) wavelengths. There is an old maxim: "Red sky at night, sailors' delight; red sky in morning, sailors take warning". Weather fronts tend to go from east to west. Since higher amounts of water vapor are one reason for the red sky effect, this suggests that stormy weather may be coming your way from a red sky in the morning, and that a red sky at night is not likely to affect weather to the east of the setting Sun.

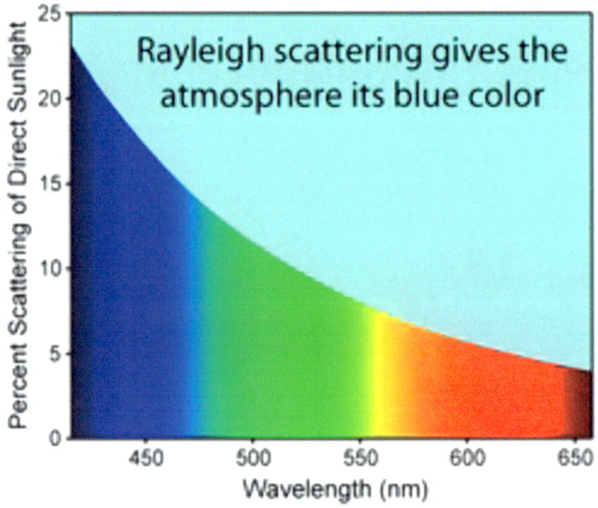

Fig. (5). Rayleigh scattering of light.

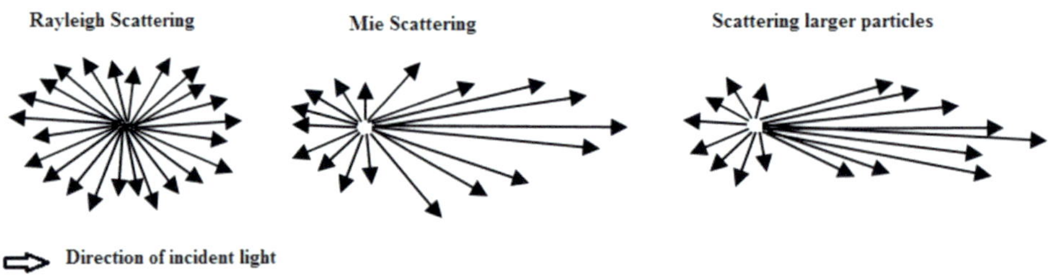

Fig. (6). Larger particles scatter longer wavelengths and make the sky appear red; smaller particles (oxygen and nitrogen molecules), scatter shorter wavelengths making the sky appear blue.

WATER VAPOR AND CLOUD FORMATION

Fluffy clouds spring up on a summer's day because of moisture in the air that condenses as it rises and cools. The clouds appear white because sunlight of all wavelegths is scattered in all directions and the light that reaches our eyes appears colorless (Fig. **7**). The cloud appears because there is a large amount of moisture in the air at the condensation point of water and water vapor turns to small droplets.

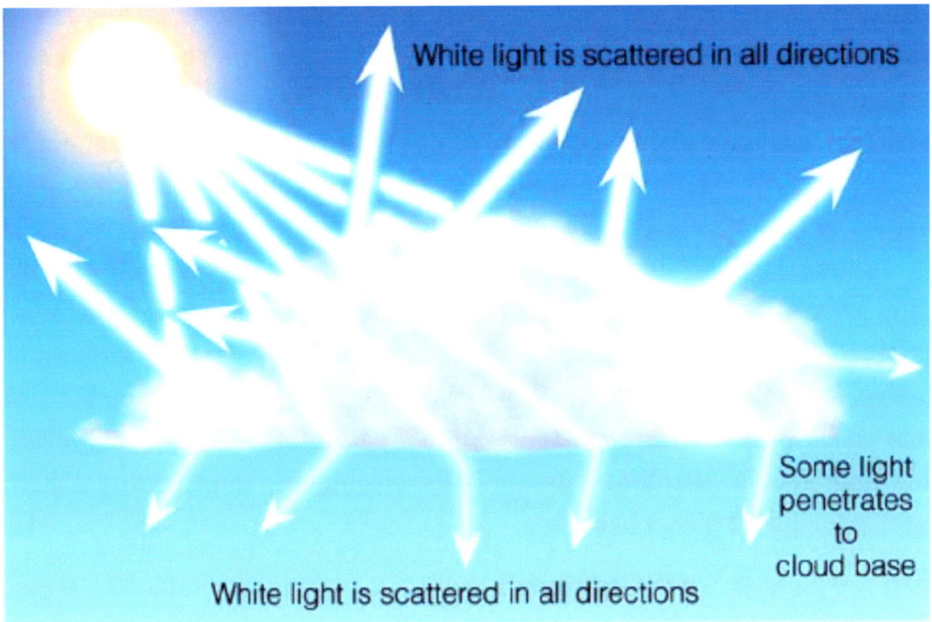

Fig. (7). Clouds appear white because they primarily scatter all wavelengths in all directions.

Rainclouds such as towering nimbocumulus, however, often appear dark in color. This is caused by two factors [9]. These clouds can rise to great hights (50-60 thousand feet) and are very dense with a large amount of water vapor and droplets that have condensed and will include ice crystals at higher altitudes. This dense air and its water content absorbs and reflects more light, some of which is bouncing around in the cloud itself and does not ever reach our eyes. In addition, the cloud appears darker because of a phenomenon called "chromatic induction"– when dark objects (in this case the cloud) are seen against a brighter background of sky they appear even darker.

Something more must be said about the absorption of sunlight by water in the atmosphere. Water absorbs in a broad range of the electromatic spectrum with rotational translations and intermolecular vibrations responsible for absorption in the microwave range (approximately 1 mm to 10 cm) and far-infrared (approximately 10 microns to 1 mm); intramolecular vibrational translations in the infarared (approximately 1 micron to 10 microns); and electronic transitions in the ultraviolet region (less than 200 nanomeers) [10]. Water is the main absorber of sunlight in the atmosphere to provide warmth to the earth; without it Earth would be a frozen planet. There are approximately 13 x 10^{12} tons of water in the total atmosphere which is approximately 0.33% by weight. This compares to approximately 0.04% carbon dixiode, the concentration recently reached. Water vapor is responsible for about 70% of all atmospheric absorption of radiation, and this occurs mainly in the infrared. The average relative humidity of the atmosphere is about 75% at ground level and decreases to about 45% at 5000 meters altitude [10]. Water vapor has a very complex absorption spectrum, including a very small moment of inertia on rotation that produces a large number of combined vibrational and rotational spectra (there are tens of thousands to millions of absorption lines) [10]. Water vapor, liquid water, and ice have very different absorption spectra; the excellent publication by Martin Chaplin can be read for more details [10]. Water has an indispensible role in providing a thermal blanket that keeps the planet habitable.

WATER EXPANDS AS IT FREEZES

Water exhibits unusual characteristics when cooled to near its freezing point. Like most substances water becomes denser as the temperature is lowered. For example, water has a density of 0.997 at 25 degrees C and this density increases as the temperature is lowered until the maximum density of 1.000 gram per ml is reached at approximately 4 degrees C. (Technically, water has its greatest density at 3.98 degrees Celcius (about 39 degrees Fahrenheit)). As water is cooled further on its way to freezing at 0 degrees C, water gradualy expands and becomes less dense. At 0 degrees C (the freezing point) its density is 0.917 grams per ml, and it has lost more than 8% in density. Thus, ice is less dense than water and consequently it floats (Fig. **8**).

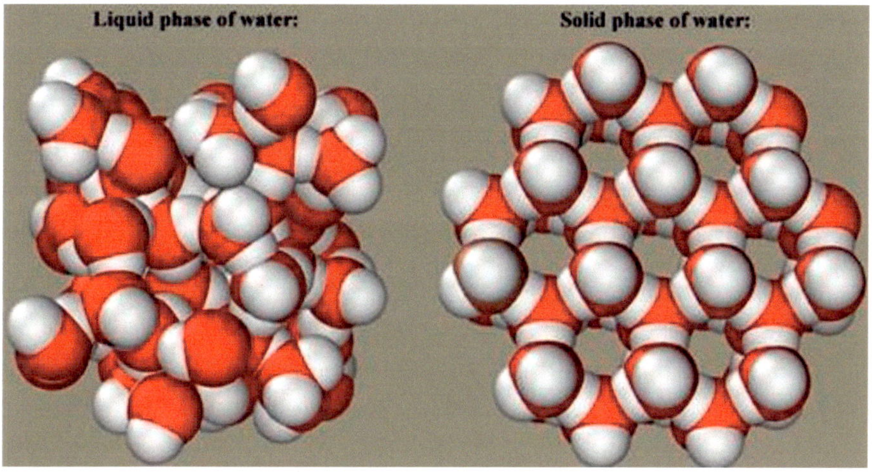

Fig. (8). Space-filling models of a group of water molecules at above freezing temperature (left), and frozen, less dense ice (right).

The mechanisms involved in the behavior of water as it approaches the freezing point are informative. As the system loses energy during cooling, the water molecules move less rapidly. Hydrogen bond formation increases and a quasi-structure increasingly develops that causes the water to be more dense. As more energy is taken out by cooling, the molecules shift into a less dense crystal formation (Fig. **8**) and ice forms. The shift is caused by the hydrogen bonds in ice which align the molecules in a state where they are slightly further apart and the net density is decreased.

This has important biological consequences. Water in streams, ponds, and lakes freezes from the top down and the floating ice is an insulator and unless the water source is very shallow it does not totally freeze as would likely be the case if ice were more dense and sank to the bottom as it froze. It seems reasonable that this feature of water spares the underwater habitat from freezing with benefits for aquatic life in winter. I recall, as a young boy, coming upon a very large turtle under ice near the edge of a farm pond that was frozen over. The turtle was basking in the rays of the Sun that penetrated the ice, which was thin and a buffer to the cold surface temperature while leaving most of the pond's water unfrozen.

Freezing and thawing of water in soil also has further consequences. It can fracture rocks and contribute to soil building; it can lift roots and fenceposts out of the ground; and it can damage roads. Water inside pipes can exert sufficient force to rupture them.

BOILING AND FREEZING POINTS OF WATER

Water has unusually high boiling and freezing points compared to most other compounds of similar size and complexity. For example, di-hydrogen compounds of the metals tellurium, selenium, and sulfur, have descending molecular weights (129, 80, and 34 grams per mole, respectively) and associated lower boiling points than water (-2, -42, and -60 °C, respectively); and associated lower freezing points (-49, -64. and -84 C, respectively) [11]. If you accept that these disulfides form a series that should continue with oxygen as the next in sequence, having a lower molecular weight, would you not predict that water should have a lower boiling point and lower freezing point in a continuation of the trend for the other compounds? Indeed, what would you predict as water's approximate boiling point and freezing point? The boiling point at +100 and freezing point at 0 are very much out-of-character compared to similar compounds. Because water is strongly polar and can form 4 hydrogen bonds with neighbors, it orients and forms a "structure" even as a liquid. This primarily accounts for its "odd", but fortunate for life, behavior.

SURFACE TENSION

Water has the second highest surface tension of any liquid (mercury has the highest surface tension). When molecules of a substance have an attraction for each other, there is a surface effect. Because at the surface of water there are no molecules of water above the air-liquid interface, the attractive force tends to keep the water molecules from flowing apart. The surface is said to be under tension and will support objects that you would expect to sink into it. Hydrogen bonds between water molecules and the polar nature of water molecules are the sources of the surface tension which is 73 dynes per centimeter at 18 degrees centigrade. In comparison, ethanol (ordinary alcohol), a molecule similar in size, does not hydrogen bond and its surface tension is only 24 dynes per centimeter [11]. If there are no other significant forces at work, water will round itself into a ballshapc due to the surface tension (act to minimize the surface area). This happens on a leaf, and is aided by the hydrophilic coating on the leaf (Fig. (**9**), right). Because of the analogy that the surface tension acts as if the water had a "skin" at the surface, objects that are heavier than water can remain (float) on the surface of water. A water strider demonstrates this (Fig. (**9**), left).

Fig. (9). Water strider taking advantage of water's surface tension (left) and water beading on a leaf due to surface tension (right).

THE WATER CYCLE

The physical and chemical properties of water, many of which are unusual and even unique as just described, make water ideal and indespensable for water's function in a cycle in nature as sustainer and restorer of life and all of nature on the Earth (Fig. **10**).

The total global water on Earth is estimated to be approximately 1.423 x 10^{21} kg. It is surprising and non-intuitive that less than 3% of Earth's water is fresh water. Over 68% of the fresh water is found in icecaps and glaciers [12 - 15]. The atmosphere contains an estimated 1.290 x 10^{16} kg of water which is only about 0.001% of the total global water. If all the atmospheric water precipitated at once as rain it would cover the globe to a dcpth of about 1 inch. The total global fresh water is calculated to be approximately 3.5 x 10^{19} kg which is only about 2.5% of the total global water. It is estimated that only about 0.3% of the global fresh water is found in surface waters of lakes, rivers, streams, and swamps.

Have you ever wondered how much a cloud weighs and if it has weight how does it float in air? The United States Geological Society has made estimates of the weights of clouds of various types [12]. Since air has weight, clouds will float if their density is less than air. Clouds form in air because of moisture; thus, the difference in density of the cloud and the surrounding air is due to moisture content. One calculation for a cumulus cloud is approximately 0.5 grams per cubic meter [16]. The calculations for this estimate are quite involved and values were given for various types of clouds with fog at 0.05 to 0.5 grams per cubic meter, and 0.5 to 3.0 grams per cubic meter for various types of clouds. A cumulus cloud 1 cubic km in volume would have a mass of approximately 5 x 10^8 grams of water (about 1.1 million pounds).

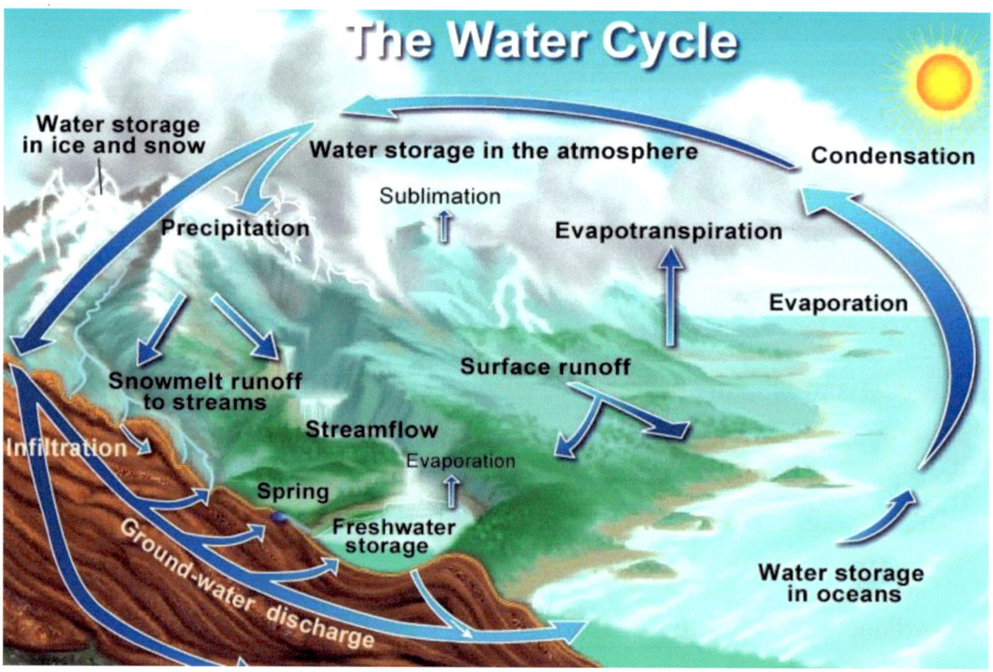

Fig. (10). The water cycle includes the evaporation and condensation of water in nature and it includes water as gas, liquid and solid as it acts to form clouds, is stored in oceans and under ground and also serves to refresh and clean the land while serving the needs of plant and animal life on earth's land and oceans.

The cycling of water (Fig. **10**) results primarily from the unusual boiling and freezing points of water which regulate its condenstion and evaporation at temperatures compatible with life on earth. If water behaved like the dihydride of sulfur (near neighbor in the Periodic Table), it would have boiling and freezing points, respectively, of approximately -60 C and -84 degrees C [11]. Based on these differences alone, a world with H_2S instead of H_2O would be a very different world and not compatible with life as we know it.

During the water cycle, the formation of ice depends on hydrogen bonding to result in the lower denisty of ice as water molecules align as previously discussed. Increased pressure applied to ice will lower its melting point. This causes melting of ice from the pressure of the blade of the ice skate to allow smooth gliding, and the weight of a glacier weakens the crystal lattice of ice and allows glaciers to flow under the influence of gravity [11].

LATENT HEAT OF WATER

The latent heats of fusion and evaportion of water are important for some of water's essential functions in the water cycle (Figs. **10** and **11**). Latent heat is the term for the amount of heat energy absorbed or released by a substance as it changes phases (liquid to solid and the reverse, and liquid to gas and the reverse). Rarely, there is direct conversion of solid to gas: sublimation (Fig. **11**). Heat is released during the process of freezing and it will be absorbed by the surroundings. The condensation of water vapor releases the same amount of heat as was absorbed by water as it evaporated.

The large amount of heat "stored" in steam is released when it condenses and this can be used to advantage in steam radiatiors for home heating. Similarly, a thunderstorm is an example of the rapid release of heat energy in the upper atmosphere from the condensation of hot humid air. Hurricanes also are an example of the huge energy content in the water in the atmosphere with the heat derived from the tropical ocean water [11]. Thus, the driving force for the water cycle is the heat content of the water, derived from the Sun, and made possible by the unusual behavior of water with respect to condensation and evaporation. Surface water, water in lakes, ponds and streams, and derived from soil, as well as very salty sea water, is distilled as it evaporates. The pure water condenses in the uppper atmosphere and falls as precipitation which washes and cleanses the land and also serves to modulate temperatures (Fig. **10**).

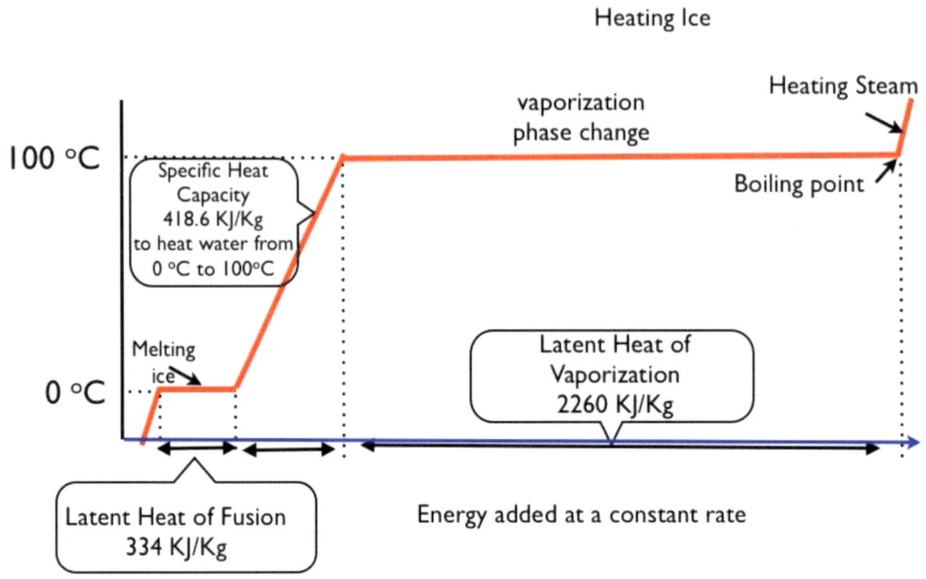

Fig. (11). The heat required to melt ice and boil water involves latent heats of fusion and vaporization.

SNOWFLAKE FORMATION

A unique and one of the most beautiful and unusual forms created by oxygen (in the form of its dihydrogen molecule, water) is the snowflake (Fig. **12**). During the water cycle, snowflakes may form with a large variety of shapes, many with perfect symmetry, including: six-pointed stars, hexagons, needles, prisms, and even branches somewhat like a tree.

Mathematicians have recently (2012) modeled the growth of snow crystals using computers as reported in the *Scientific American* magazine [17]. This source states that the process of snowflake formation and its intricate designs has long been hidden from science, and even Johannes Kepler (1571-1630) unsuccessfully wondered about how they formed [17]. Snowflakes respond very dramatically to even small changes in humidity and temperature in the air during their formation. Although the modeling was successful in generating a variety of snowflake patterns, it was conceded that mysteries remain about how snowflakes form based on fundamental principles of physics. The mystery of their beauty will always remain.

Fig. (12). A snowflake, six-pointed star version.

Indeed, a question I have sometimes been asked is: "Is it really true that no two snowflakes are alike". The most authoritative answer I have found is that given by Kenneth G. Libbrecht of Caltech [18]. Libbrecht says the simple answer is "Yes". Of course, the detailed answer is more complex. Yet, if we only consider snowflakes large enough to be visible to the naked eye, Libbrecht says:

> "And thus it's unlikely that any two complex snow crystals, out of all those made over the entire history of the planet, have ever looked completely alike."

Liebbrecht's reasoning for this begins with consideration of the smallest of snow crystals in the form of hexagonal prisms. He says that:

> "... If you sifted through a reasonable number of Antarctic snow crystals you would find two that were essentially indistinguishable in a microscope."

Note that he is referring to tiny "snow crystals" and not to formed snowflakes. The logic Libbrecht used for calculating the probability that snowflakes are identical starts with the fact that one electron is exactly like every other electron (his statement, not mine) because electrons are elementary particles with no component parts that could differ. However, a water molecule is different. If we consider only water molecules that are made of the most abundant isotope of oxygen (^{16}O) and ordinary hydrogen, such water molecules are exactly alike. But, one of every 5000 naturally-occurring water molecules contains an atom of deuterium in place of one of the hydrogens and about one in 500 contains an atom of the isotope ^{18}O instead of the commoner ^{16}O; thus, they are different. Libbrecht goes on to calculate that based on the fact that a small snow crystal might contain 10^{18} water molecules, about 10^{15} will be different from the others. These unusual molecules will be randomly placed in the crystal giving it a specific design, and he calculates that:

> "Even with 10^{24} crystals per year, the odds of it happening within the lifetime of the Universe is indistinguishable from zero."

REACTIONS INVOLVING WATER

Water participates in a variety of types of reactions; only a brief overview will be provided of some of the most significant types of reactions of water that are most relevant to the functions of water in biological systems. These include: metabolism (anabolism and catabolism), cellular respiration, and photosynthesis, which we have considered in some detail throughout the book. Here, we will briefly address condensation, hydration, and dehydration. Condensation and hydrolysis reaction are opposites and water is removed and added, respectively, to bring two molecules together or to separate them (Fig. **13**). Large molecules can be synthesized and cleaved by such reactions, and are important in biochemistry inside cells. Fig. (**14**) shows a more specific example with an OH functional group attached to a carbon atom on a sugar molecule reacting with H on a carbon of another sugar molecule to form a disaccharide from two molecules of a monosaccharide. Table sugar is a disaccharide made of the monosaccharides glucose and fructose condensed (joined) as shown in Fig. (**14**). Thus, water functions as the milieu inside cells as well as a participant (reactant).

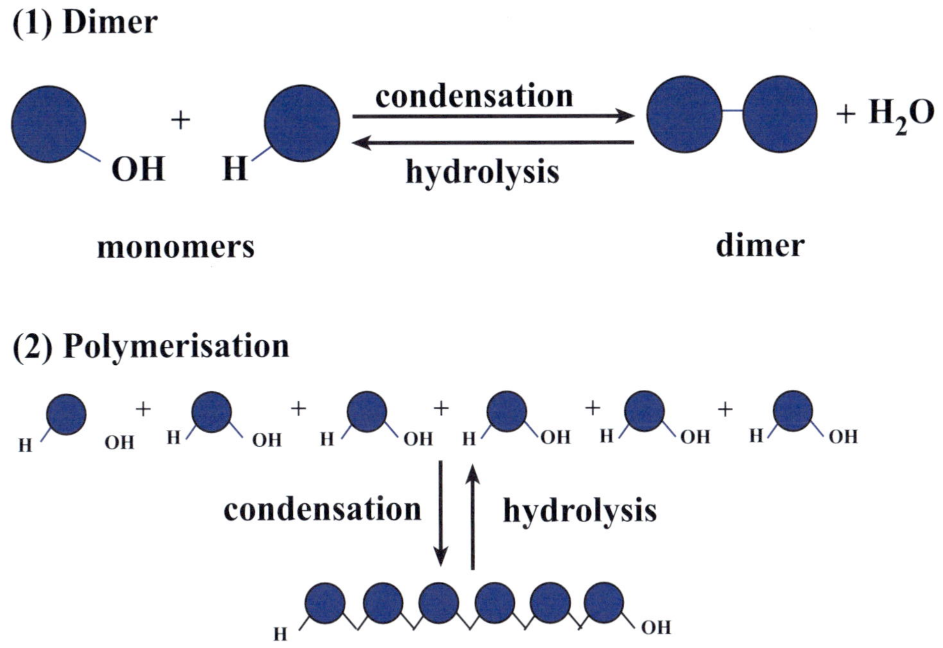

Fig. (13). Generalized depiction of condensation and hydrolysis reactions.

THALES: ALL MATTER BEGINS WITH WATER

I cannot leave the story of water without mentioning Thales of Miletus (circa 635-630 B.C.-543 B.C.) (Fig. **15**). Because he was considered to be extraordinarily intelligent, he earned the distinction of being designated as one of the "Seven Sages of Greece" [19]. He is probably best known for his theorems in geometry which you undoubtedly were supposed to have learned as a child. However, relative to the subject of this chapter, Miletus is said to have believed that all matter originates with water; *i.e.*, "Water is the Primary Principle", and Miletus formed the Milesian school of natural philosophy to teach this and other matters of philosophy [19]. Aristotle[3] credited Thales as the founder of natural philosophy [19], and wrote:

"'Thales says that it [the nature of things] is water."

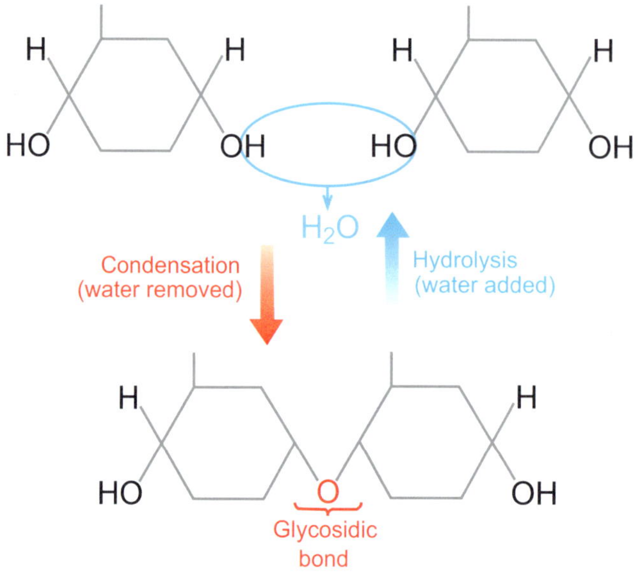

Fig. (14). Condensation of two molecules of a monosaccharide (such as glucose) to create a disaccharide and the reverse (hydrolysis) by the addition of water.

There is difficulty here. For water to be the basis of all matter it is necessary to explain how all things could have come into being from water and that water had the potential to change into all the myriad things of which the universe is made. This would have taken the ancients into theories about the nutritive capacity of water, its ability to change form (solid, liquid, vapor), and their beliefs about spontaneous generation of life. Indeed, they discussed the "moist nature" of all life which they related to water. It is possible that Thales was influenced by observations that the water of the Nile River in flood created new land, as did other river deltas. Atëius is said to have attributed to Thales the idea that:

> "… even the very fire of the sun and the stars, and indeed the cosmos itself is nourished by evaporation of the waters".[4]

It is recorded that Miletus predicted the eclipse of May 28, 585 B.C. and that this stopped the fighting that was going on between the Lydians and the Medes [19].

Fig. (15). Thales of Miletus[5].

Patricia O'Grady [19] relates that Aristotle wrote about an interesting idea attributed to Thales; namely that the earth floats on water:

> "This is the most ancient explanation which has come down to us."

The theory included the analogy of the earth resting on water as wood floats on water but not on air, and he may have equated the earth to ships that he had seen at sea. The importance that ancient philosophers placed on water dates back to water as one of the four basic elements: earth, air, fire, and water.

The pronouncement by Thales that the Nature of all things is water, seems verified, at least poetically, by the view from outer space of the "Big Blue Marble" (Fig. **16**) that is our home, the Earth– dominated by the nearly circular expanse of the Pacific Ocean with a surface area of 63.8 million square miles which is approximately 46% of the Earth's water surface and 33% of the total Earth's surface; larger than all the land areas combined.

Water, made unique by oxygen, is a dominant feature of the planet when viewed from outer space; a dominant feature when viewed from the perspective of a living organism; and yet a dominant feature when viewed from the perspective of molecular chemistry. Water provides beauty and function and is a perfect example of the principle associated with the 20th century architect Louis Sullivan, the "Father of the Skyscraper", credited for saying [20] that: "Form ever follows function".

Fig. (16). The giant, blue marble "Earth" as viewed from space with the grandeur and dominance of water on the planet an obvious feature.

NOTES

[1]*ReligioMedici,* 1.16, 1642, ed. John Addington Symonds, 1886; in "*Quotationary*" Leonard Roy Frank, Ed. Random House, NY, 2001, p 320.

[2]*Codex Leicester* was purchased by Bill Gates who paid at auction $30,802,500, a record price as the most expensive book sold to date.

[3] Metaph. 983 b22.

[4] Atëius, *Placita*, I.3).

[5] He lived from about 635/630 B.C.-543 B.C. and has been called the "Father of Science" and "The Father of Philosophy".

REFERENCES

[1] The World Factbook, Central Intelligence Agency. Available from: https://www.cia.gov/library/publications/the-world-factbook/geos/xx.html#Geo 2016.

[2] Gleick PH. Water in crisis : a guide to the world's fresh water resources. Oxford University Press 1993.

[3] Perlman H. Water Properties, facts and figures about water. Available from: http://water.usgs.gov/edu/water-facts.html 2016.

[4] Capillary action. Available from: http://science.jrank.org/pages/1182/Capillary-Action.htm 2016.

[5] Sadi-Carnot. Blue sky problem. Available from: http://www.eoht.info/page/Blue+sky+problem 2016.

[6] Lilienfeld P. P. Lilienfeld. A Blue Sky History. Available from: http://www.osa-opn.org/home/articles/volume_15/issue_6/features/a_blue_sky_history/ 2016.

[7] da Vinci L. Leonardo da vinci: the codex leicester. Available from: http://www.odranoel.de/index.php?lang=eng&menu=codex&area=wasser&page=17 2016.

[8] Madigan J. What wavelength goes with what color? Available from: http://science-edu.larc.nasa.gov/EDDOCS/Wavelengths_for_Colors.html 2016.

[9] Czarnecki S. Re: Why rain cloud is sometimes black in colour; esp. before thunderstorm? Available from: http://www.madsci.org/posts/archives/1999-10/939766534.Es.r.html 2016.

[10] Chaplin M. Water absorption spectrum. Available from: http://www1.lsbu.ac.uk/water/water_vibrational_spectrum.html 2016.

[11] Riley CD. PDHonline Course C104 (2 PDH) The Unique and Unusual Properties of Water. Available from: www.PDHonline.org 2016.

[12] The water cycle: water storage in the atmosphere. Available from: http://water.usgs.gov/edu/watercycleatmosphere.html 2016.

[13] Earth's Freshwater. Available from: http://nationalgeographic.org/media/earths-fresh-water/ 2016.

[14] Burton D. Common conversion factors for water, ice, sea-level & air. Available from: http://www.sealevel.info/conversion_factors.html 2016.

[15] Gleick PH. Water resources. In: Schneider SH, Ed. Encyclopedia of Climate and Weather. New York: Oxford University Press 1996; Vol. 2: pp. 817-23.

[16] DeCaria. ESCI 340 – Physical Meteorology Lesson 1 – Properties of Clouds.

[17] Cowen R. Snowflake growth successfully modeled from physical laws. Scientific American. Available from: http://www.scientificamerican.com/article/how-do-snowflakes-form/ 2012.

[18] Libbrecht KG. Is it really true that no two snowflakes are alike? Available from: http://www.its.caltech.edu/~atomic/snowcrystals/alike/alike.htm 2016.

[19] Thales of Miletus (c.620 B.C.E.- c. 546 B.C.E). Available from: http://www.iep.utm.edu/thales/ 2016.

[20] Louis Sullivan. Available from: http://www.louissullivanfilm.com/sullivan/ 2016.

CHAPTER 6

Oxygen Therapy, The Early Years

Abstract: Since the discovery of oxygen, there have been swings from emphasis on the essential, beneficial nature of oxygen to focus on its damaging effects. Outright speculations and critically-considered hypotheses have ebbed and waned for the therapeutic use of elevated oxygen and practical applications of gas mixtures for deep-sea diving and pressurized caissons for construction under water. Oxygen as therapy has been proposed as beneficial or even a panacea. In some hands it was harmful and even deadly. The use of elevated oxygen tensions (as compressed air) began even before oxygen was isolated and identified from air. Oxygen was used as therapy and compressed air was extensively breathed by miners and workers in caissons tunneling under water to construct bridges. These practical, non-therapeutic applications of various breathing mixtures containing oxygen also contributed to scientific understanding of the physiology of oxygen in the human body. Elevated oxygen tensions also were used for U.S. Navy diving operations and laboratory experiments with humans and a variety of life forms. In the early years oxygen therapy was used for a wide variety of diseases, many of which were treated without any established connection to a mechanism of action. Thus, the therapeutic use, laboratory experimentation with, and practical applications of compressed air, gas mixtures at and below one atmosphere pressure, and hyperbaric oxygen has had an interesting, long, and colorful development that included charlatans (probably), adventuresome physicians ahead of their times, visionaries, and practical men of industry.

Keywords: Air embolism, Albert Behnke, Atmospheric pressure, Caisson, Carbon monoxide poisoning, CNS toxicity, Compound oxygen, Cunningham, Decompression sickness, Domicilium, Gas gangrene, Henshaw, Hyperbaric pressure, John Bean, John Haldane, Lorrain Smith effect, Paracelsus, Paul Bert effect, Pulmonary toxicity, Spanish flu treatment, USS Squalus.

Olen R. Brown

INTRODUCTION

In Chapter 2, the conclusions of Priestley, who is credited with discovering oxygen, were quoted to emphasize that from that early time forward, the hope that increased oxygen could be therapeutic was constrained by the caution that it could be harmful. After oxygen was available in purified form, and based only on preliminary medical experiments, oxygen unfortunately was elevated by some to the status of a panacea. An extreme example of oxygen therapy was a large iron sphere constructed to house patients in an atmosphere containing an elevated concentration of oxygen as therapy for various conditions. The therapy fell into disrepute and the sphere was razed for its iron content prior to WWII. The necessity to provide breathable atmospheres for: undersea operations (hard-hat diving suits, scuba gear, submarines, and caissons to support construction workers building bridge foundations under water); therapeutic uses; aircraft operating at high altitudes and later for space capsules drove laboratory research. Practical needs often dominated and applications were made before mechanisms were available to explain observations. Basic laboratory studies were generally inadequate to safely place limits on human tolerances to elevated concentrations and pressures of oxygen. Successes and failures in the practice of oxygen therapy gradually increased safety but also disclosed the falsity of hopes that either oxygen itself or antioxidants were cures for all ailments. The period after the late 1960s onward has been a time of increased focus on understanding the mechanisms of oxygen toxicity and biological, oxidant-stress defenses.

HISTORY OF OXYGEN AS THERAPY

Introduction

From the time of its discovery as an element and its isolation from air, the essential nature of oxygen for animal life was apparent. Early experiments by Joseph Priestley (described in Chapter 2) also provided the basis for speculating that oxygen concentrations above the 21% found in air, might be beneficial. Several physicians were quick to treat patients in chambers filled with compressed air which exposed them to elevated oxygen, and some men died. Practical men exposed workers to breathing compressed air (containing elevated oxygen) for projects including tunneling and bridge constructions before it was proven safe, and more men died. Despite this, largely with trial and error and at considerable risk, oxygen was used and misused for therapy. Today, it is safe and acceptable, and indeed, indispensable therapy for various medical conditions. I can think of no other therapy that has had such a wide and colorful history.

The 1600s

The earliest recorded attempts at oxygen therapy came about in the 1600s by increasing atmospheric pressure with air as the gas in a sealed chamber termed a "domicilium" by a British clergyman named Henshaw (I could not find his first name or biography) [1 - 4]. The domicilium was filled by air (oxygen was not discovered until 1773) pressurized using an old organ bellows. The chamber could be operated at above and below normal atmospheric pressure, and it was sealed and unventilated. It must have been frightening for patients. Henshaw treated his patients who suffered from acute conditions with increased pressure, and those suffering from chronic conditions he blessed with lowered pressure. Insight into his approach to therapy is provided by a quotation [4]:

> "In times of good health this domicilium is proposed as a good expedient to help digestion, to promote insensible respiration, to facilitate breathing and expectoration and consequently, of excellent use for prevention of most affectations of the lungs."

Scientists also experimented with the effects of reduced air pressure. In 1659 Robert Boyle (1627-1691) commissioned construction of a device he called the "pneumatic engine" (in today's world, a vacuum pump). He conducted many experiments which he described in a book[2] which has been reproduced [5]. Experiment number 41 (in a series) was designed to demonstrate the requirement of living animals for air to survive. He exposed a variety of life forms including birds, mice, eels, snails and flies. His description of a lark exposed to decreased pressure and lack of oxygen is quaint:

> "...the Bird for a while appear'd lively enough, but upon a greater

> Exsuction of the Air, she began manifestly to droop and appear sick, and very soon after was taken with as violent and irregular Convulsions, as are wont to be observ'd in Poultry when their heads are wrung off."

Joseph Wright (1734-1797) in 1768 painted an oil-on-canvass masterpiece of something like the described event with an audience in a candle-lit scene (Fig. **1**). The on-lookers show a variety of emotions as the bird in the glass vessel is deprived of air. Experiments like these established that air was essential for life.

Fig. (1). An Experiment on a Bird in the Air Pump by Joseph Wright of Derby, 1768, the National Gallery, London.

The 1700s

This period is notable for the discovery of oxygen and early experiments with the pure gas; details were described in Chapter 3.

The 1800s

In France, exposures to air pressurized from two to four atmospheres absolute were used because it was believed that it increased circulation of blood to the internal organs including the brain. Patients were generally described as agreeably responding with a feeling of well-being [2]. A 4-atmosphere chamber was designed by James Watt (1736-1819) who was approached because of his knowledge and reputation as inventor of the steam engine that used a pressurized "boiler". During this time, a large chamber was constructed in France and primarily was used by physicians to treat pulmonary diseases for which there seemed to be a rationale. However, some unfortunate patients who suffered from illnesses including deafness, cholera, rickets, menorrhagia, and conjunctivitis were also subjected to the "chamber" and it is reasonable to assume they suffered without much evidence of success [2]. Paul Bert did important work during this period and this will be described subsequently. The first fatality due to hyperbaric therapy was reported due to a decompression accident in about 1879 [2]. I found no details (gruesome or otherwise) of what illness this patient suffered from or how he died during treatment.

During the mid-to-late 1800s caissons employing hyperbaric pressure were used. The French paleontologist Jacques Triger (1811-1867) first used the caisson for coal mining in water-logged strata in Chalonnes-sur-Loire about 1839 [6]. Caissons were used subsequently in bridge construction. A caisson is a large watertight chamber, open at the bottom, with the water kept out by air pressure, and in which humans can do work beneath the surface of water. Workers complained (apparently with little sympathy) of joint pains, CNS dysfunctions and other ailments [2, 7].

The Eads bridge in St. Louis, now a protected National Landmark, was constructed over the time period 1867 to 1874 at a cost of $6 million for the bridge itself. It had two spans of 502 feet and a center span of 520 feet (Fig. **2**). It required 2,390 tons of steel (used for the first time in a major bridge) and 3156 tons of wroght iron, and was the largest bridge in the world when it was built. For its construction James B. Eads (1820-1889) was the first to use compressed air in an underwater casisson for bridge construction, and it was the largest and deepest cassion ever used [8]. The depth to which workers were required to descend underwater determined the pressure needed to keep the excavating space free from the impending river. The report by Margaret L. McGunegle [8] contains 11 references and provides many interesting details about the bridge and its construction, including the decompression sickness suffered by the workers.

Fig. (2). The Eads bridge, under construction in 1867-1874 that required the largest and deepest caissons ever used.

Eads was born in Lawrenceburg Indiana, had little schooling and at one time peddled apples on the street because his father had suffered "financial reverses" [9]. His future changed when he was hired by a merchant who let Eads use his extensive library [9]. Eads designed a surface craft (he called it The Submarine) which had a diving bell (made from a 40-gallon whisky barrel) and he used it to great profit for salvaging on the Mississipi River and he retired, wealthy, at age 37 [9].

The Civil War interveined and he helped the North by building seven ships that had iron plates. His "ironclads" went into action at Fort Henry in 1862 [9]. There is an interesting history of an "economic war" between Chicago and St. Louis which is beyond our purpose except that it led to the need for a bridge across the Mississippi River at St. Louis to compete for transportation by rail. After the Civil War a group of St. Louis buisnessmen made plans for such a bridge, and Eads presented a bold plan which was approved in 1867 [9]. A most difficult aspect of the bridge construction was the necessity to dig foundations for stone abutments on each shore and for two stone towers on piers in the river. Eads desired these piers to be on bedrock which was covered as deep as 80 feet by mud and 12 feet of water. While on a trip to France, Eads learned about construction of pneumatic caissons and in October of 1869 he constructed a huge caisson that was 82 feet long and 60 feet wide to make the east pier. Seven airlocks were provided for men who worked inside the casisson and a telegraph was installed. The workers received $4 per day, twice the wages for surface bridge workers [9].

According to one account [9], the caissson eventually had to be pressurized to 50 pounds per square inch with air (approximately 3.5 amtospheres overpressure of air). After only one day of work some men began to suffer what was called "the Grecian Bend" so-called because of similarity to a popular 19th Century dance in which dancers bent at the waist [10]. The bends is known to occur when divers ascend too rapidly from deep dives; it can also occur when non-pressurized aircraft rise too quickly to high altitudes. It results because, when breathing air at pressure, excess nitrogen dissolves in the blood. If the diver returns too quickly to lower pressure this gas bubbles out of the blood. If the bubbles form faster than they can be eliminated by the lungs, they can cause various symptoms including skin rashes, severe joint pains, paralysis, emboli, and even death. Eads hired a physician to investigate and a decompression chamber was used but the decompression schedule was inadequate and 15 men died, 2 were permanently disabled, and 77 others were badly injured [9]. The pressuress necessary for operating the caissons were very high: the east tower had to be driven to 93 feet below the river to reach bedrock and the west tower extended 77 feet below the river's surface.

Decompression sickness had been described as early as 1666 by Robert Boyle (to whom we have already referred). Boyle provided what is probably the earliest recorded account of this malady, in a snake, not a human[3]:

> "What I once observed in a viper furriously tortured in our exausted receiver, namely, that it had manifestly a conspicuosus bubble moving to and fro in the waterish humor of one of its eyes."

The "exhausted receiver" refers to the chamber from which the air had been removed.

In 1847 T. J. Watelle and B. Pol, who were French physicians, had noted that recompression produced relief from symptoms produced by compressed air in workers and they recommended a slow decompression, and the French physician, Antoine Foley in 1863 recommended recompression to treat decompression sickness [10].

Use of compressed air for diving opertions was not new; diving bells of primitive sorts had been in use for centuries; and Aristotle wrote about them in 400 BC [10]. Charles and John Deane are credited with having designed the first successful diving helments in the 1820s [10]. They initially intended these to be used as "smoke helments" for firemen but they began to be used by divers around 1828. Initially they were not fixed to water-tight suits and diver had to remain upright. Subsequently, helment were attached to these suits which continued to be developed and various adaptations were made (Fig. **3**). This included provisions for rebreathing air with removal of carbon dioxide by absorption. This came surprisingly early with the first successful re-brather made by Henry Fleuss in 1878. Emile Gagnsan and Jacques Cousteau in 1942-1943 developed the aqua lung [10]. Divers using any of these devices can suffer decompression sickness.

The Brooklyn Bridge was constructed during 1869-1883 and it was reported that several hundred workers suffered decompression sickness and there were 27 reported mortalities. Excavations using caissons for New York's Hudson tunnel during 1882 were reported to have resulted in a horrendous 25% mortality. In 1890, to treat these problems, a decompression chamber was installed and extended time was spent decompressing the workers after each work shift. During eighteen months it was reported that there were "only" 3 fatalities [2, 11]. The tolerance for pain and death during these undertakings on the part of the workers and their employers is difficult to fathom.

Fig. (3). Eight divers of the Royal Navy Admiralth Experimental Diving Unit, Tolworth, Surry, wearing various diving suits and gear used in underwater warfare in WWII.

The first hyperbaric chamber for treating patients with air under pressure on the North American Continent was constructed in 1860 in Oshawa, Ontario, Canada. The first one in the United States was in New York in 1861 and it was said to have been for treating "nervous and related disorders" [12].

In the late 1800s a number of physicians treated patients under conditions that are for the most part poorly described as are the results obtained. Starcky and Palen in Philadelphia described their medical practice, including the use of the ill-defined "Compound Oxygen" in a published book ("Compound Oxygen", Its Mode of Action and Results; first edition, 1881 and second edition, 1888). Quotes taken from the 1988 edition (found in the review by Martin [13]) indicate the nature of the treatment.

> "It is difficult to declare just what our "Compound Oxygen" is. That it contains the elements of the atmosphere we can prove to any one capable of appreciating the value of chemical tests. This is truly a compound of oxygen and other elements. Hence the name of "Compound Oxygen" is a proper epithet to apply to it; somewhat abbreviated for the convenience. But it is more than simply a compound of oxygen and nitrogen. In the atmosphere, there these elements exist in the proportion of one of oxygen to five of nitrogen; and the combination of them is that known as "mechanical mixture", and so, neither element is changed by being mixed with the other. In the Compound Oxygen the proportions of the elements is so altered that the oxygen is greatly in excess. And in the manufacture of it, the oxygen is magnetized; which give it the quality known by scientists as "Ozone"... The change that is thus wrought in the mixture of oxygen and nitrogen is almost too great to be believed".

Surely, this would be called quackery today, but, of course, I cannot know the motivations of the originators of the treatment.

In addition to the above described medical practices, Lawrence Martin [13] also included an extensive list of "oxygen therapy" practices from the late 1800s. These accounts give the language and mindset of physicians of that era, and inform us about the extensive use of oxygen at the time. I highly recommend this reference for its quotations taken from publications during the period from 1774 through 1928. I shall include references and quotations for some of these. I suggest consulting the reference for an enjoyable read for those who want to explore the history of oxygen therapy and especially for physicians. Because I have not read all the original references I will list some sources in footnotes.

In a publication in the 1880s, Dr. Samuel S. Wallian, an ethical practitioner, decried the use of oxygen therapy by certain physicians[4]:

> "Its use has long enough been surrendered to advertising quacks. It is time the regular profession rescued it from the smirch of charlatanry

which has, to some extent, heretofore attached to it."

In a second paper Dr. Wallian disclosed practices he considered unethical[5].

> "Most of these advertisers name their "secret" preparation "compound oxygen"....all are about alike; and it ought to be unnecessary to repeat to intelligent physicians that none of them contain any free oxygen, beyond the inconsiderable quantity always found in common water... (This) will forever set at rest any doubt as to whether possibly there might not be some virtue, some undeveloped "secret," behind all this pretensious (sic) nonsense".

A publication in 1886 by Charles J. Smith tells us much about the way oxygen therapy was being conducted[6].

> "...but chiefly from the difficulty and cost of production of the gas in any quantity, oxygen has never been brought into general use by the profession in this county... while, unfortunately, those persons who have written at all upon the subject have done so in such a manner as to have exposed it to the suspicion of quackery."

Further information about this interesting period is warranted and according to Lawrence Martin [13], in 1887 the great William Osler (1849-1919), one of the four founding professors of Johns Hopkins University, wrote about the use of oxygen therapy[7]:

> "The inhalation of oxygen recommended by Dr. Loomis in the oedema, has an important use whenever, as shown by the lividity, the blood is imperfectly oxidized."

In 1887, Dr. C. E. Ehinger wrote a book[8] that included a series of case histories from "*Medical Current*" (a journal published in Chicago at the time). He concluded that brief, intermittent oxygen therapy was beneficial for a variety of conditions including asthma, indigestion, pneumonia, bronchitis, cough, and menstrual irregularities, and other conditions. Ehinger discussed "Compound Oxygen," and stated that it was:

> "... not oxygen at all, but very dilute laughing gas (N_2O) generated by heating ammonium nitrate and mixed with ferric carbonate or potassium chlorate to give it color and impress the patient."

According to Lawrence Martin [13], Ehinger stated that there is no such thing as compound oxygen and that it had been advertised in an unprofessional manner.

In 1888, the famous Dr. J. H. Kellogg (Superintendant of the Medical and Surgical Sanitarium at Battle Creek, MI) published a strange account of the use of "oxygen" enemas[9]. Kellogg wrote:

> "The importance of... oxygen in the human system has no parallel among the other substances required for the healthy maintenance of the body... A man may live a month without food, a week without water, but dies in a few minutes when deprived of oxygen... an increased amount of oxygen means an increased amount of vital work... The process of digestion...is hence very clearly dependent upon the supply of oxygen. That this is not merely a theoretical conclusion, or a deduction of physiological chemistry, is evidenced by the prodigious digestive powers of the woodsman and the mountaineer, and the correspondingly feeble digestive ability of the persons of sluggish or sedentary habits... It seems to me to be entirely probable that oxygen enemata may be advantageously employed in quite a variety of cases."

In 1890, Albert N. Blodgett published accounts of the use of oxygen as therapy[10] and Lawrence Martin [13] concluded that this article probably was the first record documenting continuous oxygen therapy. The patient was a 37-year-old woman with severe pneumonia, attended to at home by Dr. Blodgett:

> "For a time there was no appreciable benefit from the (oxygen) gas, but gradually the color of the surface improved, the respiration was less embarrassed, the patient could swallow, and the immediate symptoms were again relieved... When I directed the continuous administration of the gas, I did so under the positive conviction that the patient was irrevocably doomed, and the best result that I looked for, was simply relief to the sensation of suffocation, and not any curative action... At this time I had only employed the gas in the manner ordinarily directed, that is, two or three gallons at a time, several times daily. I now directed its use without cessation, and to my great surprise, the patient not only obtained the relief desired, but was enabled to carry on the function of respiration... The amount of gas employed was not far from two hundred gallons in twenty four hours. The dealer who supplied the gas was astonished at the amount required, and... sent me a cautionary message, implying that no human being could possibly stand so great an amount of oxygen."

W. L. Conklin, MD. in 1899 described that oxygen therapy was gaining wider acceptance but with still considerable doubt among physicians about its efficacy[11].

This account documents the state of affairs regarding oxygen therapy as the 18th Century drew to a close.

> "I have used it (oxygen) in a variety of pathological conditions accompanied by dyspnoea, with uniform relief... in nearly every case... I have always used the mixture prepared by the Walton Oxygen Works... made according to the formula of the London Oxygen Hospital and consisting of one part of oxygen and two of nitrogen non-oxide. The latter gas has anaesthetic properties which make it of value, and the mixture is not irritating to the lung tissue, as

> pure oxygen is said to be. In looking over current medical literature I find many cases reported in which the results obtained from oxygen inhalations are so evidently favorable as to furnish, it seems to me, strong clinical proof of its therapeutic value. Believing, as I do, that there are no facts in physiological chemistry which really discredit this clinical evidence, or prove that the use of oxygen is unscientific, I feel confident that it will steadily grow in favor and come in time to be considered of definite and positive therapeutic value by all who will give it a fair trial."

Thus, as medicine entered the Twentieth Century, oxygen's medical use was on firmer ground for the most part.

The 1900s

Two tunnels under the East River in NY, during 1906-1908, required workers to be exposed to pressures up to 3.9 atmospheres, sometimes twice daily, and there were several thousands of cases of decompression sickness with 20 fatalities from a reported total of approximately 500,000 exposures [2]. The psychological impact on both the employers and the workmen enduring this ordeal can only be imagined.

Dr. Orville James Cunningham (1882-1937) is both famous and infamous for dominating both the publicity and application of hyperbaric therapy in the early 1900s. He was a graduate of Rush Medical College, Chicago, IL, and was professor of anesthesia at the University of Kansas for 15 years. He died of cerebral thrombosis on February 23, 1937. (I am grateful to Lynn Duchez Bycko of the Michael Schwartz Library, Special Collections, of Cleveland State University for providing me with information from his obituary). Cunningham noted that patients with circulatory and heart diseases had difficulty when at higher altitudes. He felt that such patients and those with respiratory disorders might benefit from treatment with air at increased pressure. He took the opportunity to test this in 1918 using a resident physician suffering from the "Spanish" flu which was pandemic at the time. He placed the patient in a chamber, used for animal studies, at 2 atmospheres and successfully treated a breathing crisis [14]. Cunningham continued to use hyperbaric oxygen to treat patients who had the flu that was epidemic after WWI, and also for cyanotic and comatose patients [12]. The costs were reported to be $3.00 for an a. m. or p. m. session and an entire night $6.00 was charged [2]. A mechanical failure one night caused a rapid, total loss of pressure and all the patients under treatment died. He continued to treat patients and is reported to have believed (falsely) that anaerobic infections had a role in syphilis, hypertension, diabetes mellitus, and cancer.

Fig. (4). Cunningham spherical sanitarium; image courtesy of Cleveland State University, Michael Schwartz Library, Special Collections.

Later, in Cleveland, OH (with financial support of $1 million from Timken of the Timken Ball Bearing Co. of Canton OH) he built the largest hyperbaric chamber ever known to have been constructed (Fig. **4**). The steel ball was five stories high and 64 or 65 feet in diameter [15 - 17]. It weighed 900 tons and was made of half-inch steel plates. Timken gave his support after a friend had been treated successfully with hyperbaric therapy by Cunningham [15].

Inside the ball (Fig. **4**), on five floors there were 38 rooms (some accounts say 36) with 350 ten-inch portholes. Dining and recreational facilities were provided [2, 12, 15] in a three-story sanitarium/hotel (Fig. **5**, left). Inside the sanitarium were lavish facilities including a dining room with a vaulted ceiling (Fig **5**, right). The sphere itself could accommodate forty patients and it was pressurized by several steel tanks [15, 17]. It was built by the Melbourne Construction Co [15], and it was stated that it was the only functional hyperbaric chamber in the world at that time [12]. Patients were treated at 30 pounds per square inch of air pressure (about two atmospheres) [15, 16].

Fig. (5). The 40-bed sanitarium (left) and dining room (right); images courtesy of Cleveland State University, Michael Schwartz Library, Special Collections.

Cunningham failed to document his medical claims for hyperbaric treatments as requested by the American Medical Association (AMA). He apparently made little effort (other than one paper published 1927) to establish or document his claims and was censured by the AMA in 1928 which wrote [12]:

> "Under the circumstances, it is not to be wondered that the Medical Profession looks askance at the 'tank treatment' and intimates that it seems tinctured much more strongly with economics than with

scientific medicine. It is the mark of the scientist that he is ready to make available the evidence on which his claims are based."

Failing to present evidence of efficacy, and because of the economic depression, the facility was dismantled in 1930; the iron "bells" were scrapped for only $25,000; and the metal was ultimately said to have been used to build tanks for WWII [2, 12].

The first use of hyperbaric oxygen (previously, hyperbaric air was used to provide elevated oxygen tensions) was said to have been by Alvaro Ozorio de Almeida who was Professor of Physiology at the University of Rio de Janeiro who published in 1934 relative to treatment for cancer, leprosy and gas gangrene [2]. The U.S. Navy also used hyperbaric oxygen in the late 1930s to enhance inert gas elimination during decompression after dives. Breathing pure oxygen during staged decompression reduced the time required to ascend to the surface [2]. Research revealed that supplementary oxygen was helpful in treating decompression sickness. Breathing oxygen under pressure washes out excess nitrogen dissolved in tissues [1].

Churchill-Davidson in London used hyperbaric oxygen to sensitize tumors to radiation therapy in 1955. The long-term outcomes were not promising and this application decreased in the 1970s [2]. Dr. Ite Boerema was the first to report the use of hypothermia as an aid to cardiac surgery; and in 1956 his research showed that dogs could tolerate cardiac arrest for much longer when breathing 3 atmospheres of oxygen [18]. In 1960 he also proposed the concept of "life without blood" relative to artificially supplying oxygen through fluids that had greatly elevated oxygen concentrations. He had early and significant results with cardiac surgery on infants [2]. During this time, it became popular for gurus to "treat" people with little or no scientific basis and main-stream medicine frowned on it. However, in about 1960, W. H. Brummelkamp *et al.* successfully treated anaerobic infections (gas gangrene) with hyperbaric oxygen [19]. Also, in 1962 G. Smith and G. R. Sharp described great benefits using hyperbaric oxygen at two atmospheres of pressure for carbon monoxide poisoning [20].

No account of the medical history of the use of oxygen would be complete without chronicling the work of Captain Albert R. Behnke (1903-1992). I first met him at a scientific hyperbaric oxygen conference in the late 1960s with many military people in attendance. I found him to be delightful and very approachable. He was respected as the grand-old-man of Navy diving. He was a physician and responsible for establishing the U.S. Naval Medical Research Institute. He is celebrated for having identified the symptoms of arterial gas embolism as separate from the symptoms of decompression sickness and he was an advocate of giving oxygen as recompression therapy (Fig. **6**). I recall that, with a grin, he said he did not like me emphasizing the "toxicity" of oxygen which was the focus of my work because the benefits of oxygen to his divers were paramount to him.

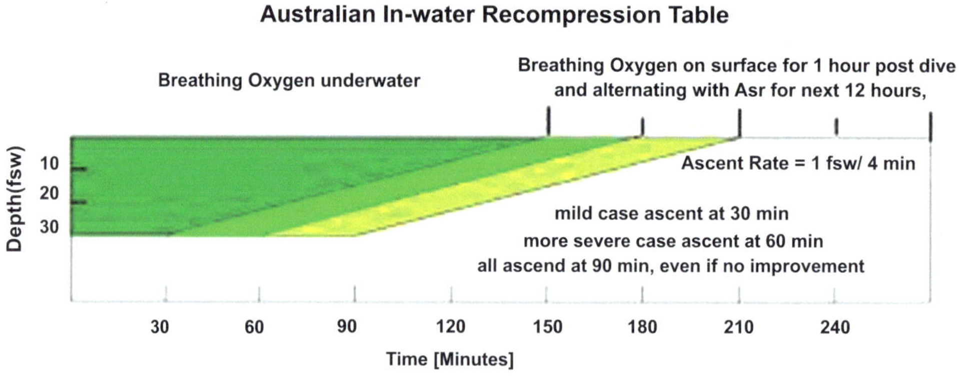

Fig. (6). Example of a recompression schedule for a diver with symptoms of decompression sickness. Depth in feet of sea water (Y-axis) *vs.* time in minutes (X-axis).

Behnke joined the Navy to fulfill a requirement by Stanford Medical School for obtaining a medical doctorate degree [21]. In 1932, the Navy sent him to the Harvard School of Public Health. He had a life-long interest in deep-sea diving and served on the USS Holland and Submarine Division 20 in San Diego under the command of Chester Nimitz (1885-1966). Nimitz, later, in 1942, was in command of the U. S. forces that stopped the Japanese invasion of Midway with the legendary Battle of Midway [22]. Behnke also served on the USS Ortolan, a submarine rescue ship where he made his first "hard hat" dive [21, 23]. Behnke was sent to Pearl Harbor in 1935 and assigned for submarine escape training. With others he researched the use of oxygen for recompression therapy following deep-sea dives [24, 25]. This use of oxygen is now standard. In 1937 Behnke introduced the no-stop decompression tables [24, 25]. When I met him in the late 1960s, he was probably most widely-known for these tables which were in wide use.

In 1939 with Yarborough, Behnke published research documenting that gases other than nitrogen could induce narcosis [26]. Also in 1939 [21] he was involved with the rescue of 33 surviving crew members of the sunken submarine USS Squalus which was world-wide news for days [27]. The divers involved in the rescue used the then recently developed diving schedules based on a mixture of helium and oxygen which had been developed based largely on Behnke's theories about nitrogen's involvement in narcosis for deep dives. It is said that his involvement with the Squalus rescue was instrumental in the approval he obtained from President Franklin Roosevelt in establishing the U.S. Naval Medical Research Institute (NMRI) [28] in 1942 shortly after WWII began. Behnke remained at NMRI until 1950 and directed research in "pioneering studies" [21] of hydrostatic weighing of humans to assess percentage fat which related to his interests in the roles of fitness and fat content in effective gas elimination after deep dives [29]. In 1950 he earned the Navy and Marine Corps Medal for saving the life of a civilian diver who surfaced too quickly from an ocean dive. Behnke spent two days in a decompression chamber with the recovering diver [21]. Behnke had a long career after retiring from the Navy in 1959. This included designing safety programs for pressurized gases in the caissons used to build the Bay Area Rapid Transit system in California. Duke University Medical Center has archived a collection of research-related items from Behnke's career [30].

Modern Uses of Hyperoxic and Hyperbaric Medicine

In 1966 the National Academy of Sciences published a report that favorably "endorsed" hyperbaric medicine [31] signifying that the basic principles were well established. In 1967 the Undersea Medical Association was formed (I recall joining soon thereafter) and by 1970 it began successful efforts to restore credibility for hyperbaric medicine and in 1977 a list was published of medical diagnoses for which hyperbaric oxygen therapy was recommended and major medical insurance companies began to accept this therapy [2]. In 2000, the American Board of Medical Specialties approved undersea and hyperbaric medicine as a subspecialty of emergency medicine and of preventative medicine [1]. Currently, it has entered the realm of veterinary medicine and the Hagyard Equine Medical Institute even treats horses with hyperbaric oxygen using a large facility for delivering oxygen [32].

In today's world of human medicine, hyperbaric therapy has successful, approved applications for several conditions. The Food and Drug Administration has oversight and its current site states: "Thirteen uses of a hyperbaric chamber for HBOT have been cleared by FDA". (HBOT is hyperbaric oxygen therapy). However, the site does not actually list 13. They include treatments of air or gas embolism (dangerous "bubbles" in the bloodstream that obstruct circulation); carbon monoxide poisoning; decompression sickness (often known by divers as "the bends"), and thermal burns (caused by heat or fire)" [33]. Consequentially, I found that other sources provide lists that are in general agreement that there are 13 conditions that the FDA has approved for treatment of by HBOT. One website lists FDA approved uses [34] as follows with how 13 is accounted for, not clear:

- Air or gas embolism.
- Carbon monoxide poisoning.
- Clostridial myositis and myonecrosis (gas gangrene).
- Crash injury, compartment syndrome, and other acute traumatic ischemias.
- Decompression sickness.
- Enhancement of healing in selected problem wounds (ref. cited) such as: (Diabetically derived illness, such as diabetic foot (ref. cited), diabetic retinopathy, diabetic nephropathy).
- Exceptional blood loss (anemia); Intracranial abscess.
- Necrotizing soft tissue infections (necrotizing fasciitis).
- Osteomyelitis (refractory).
- Delayed radiation injury (soft tissue and bony necrosis).
- Skin grafts and flaps (compromised).
- Thermal burns (ref. cited).

MRI (magnetic resonance imaging) and SPECT (single-photon emission computed tomography) are being used pre- and post-hyperbaric oxygen therapy to research mechanisms and assess benefits. Current, cutting-edge investigations include applications of hyperbaric therapy for brain injuries, stroke and other neurologic conditions [35].

OXYGEN TOXICITY

Introduction

The work of Paul Bert and Lorrain Smith established that elevated oxygen concentration could be toxic for the brain and lungs, respectively. Subsequently, the work of many researchers with studies continuing today at ever deeper biochemical levels have shown that oxygen is essential but can be harmful. Thus, mechanisms of oxygen toxicity were first documented at the whole animal level (including in humans); shown to be manifested first in certain organs; and found, at elevated concentrations, to be toxic for all cells. Understanding how oxygen interacts chemically, biochemically, and biologically is a complex story that involves many disciplines from pure chemistry and physics to biochemistry, molecular biology, and toxicology.

Indeed, the basic founding principle of toxicology, expressed by Paracelsus[12], that the "dose makes the poison" could not be more true than for oxygen. Therefore, a sidebar to explore Paracelsus (Fig. **7**) is warranted.

Paracelsus, born Philippus Aureolus Theophrastus Bombastus von Hohenheim (1493-1541), was of Swiss and German nationality, and he was a philosopher, botanist, astrologer, and general occultist who is credited as the founder of toxicology based partly on his stated conclusion "Solely the dose determines that a thing is not a poison" [36]. This concept underpins not only toxicology but also pharmacology and relates to the therapeutic use of synthesized chemicals and natural substances as well, and even to foods. It is fundamental to our story of oxygen as therapy that it is unsafe to breathe this gas at only two- to three-times the concentration found in air for more than a few hours. Failure to appreciate this "dose effect" and a lack of understanding of the relationship between oxygen concentration and pressure of the breathing gas mixture (Henry's Law *vs.* Boyle's Law), caused much of the difficulties that have plagued therapy with hyperbaric air and with hyperbaric oxygen from their beginnings.

Fig. (7). Theophrastus von Hohenheim (public domain).

Paracelsus' given name "Bombastus" characterizes his personality. It is said [36] that beginning at an early age he learned much from his physician father who also was an alchemist. Paracelsus studied medicine at several European universities and obtained the baccalaureate in 1511 and assumed the name "Paracelsus" (meaning- para: beside or beyond, and Celsus: a well known Roman physician). He began the practice of medicine by challenging much of what was common practice. He felt his university experience was lacking and travelled widely and became an itinerant physician and surgeon. His interests and contemplations extended beyond medicine to considerations of the meaning of life, death, health, and disease, and the relationship between man and God. He is generally described by those who have written about him as a: "free thinker, iconoclast, and theosophist"; and as the "Luther of Medicine". A sense of his interest in and contributions to medicine, toxicology, philosophy and theology is present in the following quotation in the paper by J. F. Borzelleca [36]:

> "The physician must pass through the examination of nature, which is the world, and all its causation. And what nature teaches him he must commend to his wisdom, not seeking anything in his wisdom, but only in the light of nature."

He is said to have believed that medicine was a divine calling and that character was more important than skill. His understanding of the use of toxic chemicals to treat illnesses presaged our understanding of the idea of the therapeutic index which allows toxic agents to be used at appropriate doses [36]. Oxygen is a chief example of this principle, which leads to great harm when not obeyed.

Early Understanding of Oxygen Toxicity

Paul Bert (1833-1886) (Fig. **8**) was perhaps the first to document the toxic effects of pure oxygen and he experimented on himself and on various life forms. His *"La Pression Barometrique: Reserches de Physiologie Experimental"*, published in 1878, is his most important writing [37].

Fig. (8). Paul Bert.

He reported the effects of elevated and lowered pressures and reported that oxygen became toxic for many life forms including insects, mollusks, earth worms, fungi, germinating seeds, and birds [37]. Thus, one hundred years ago, the exceptional mind and writings of Paul Bert contained essentially all that was known about the toxic effects of barometric pressure and of oxygen.

The *gran mal* seizures (effect on the CNS) which are induced by exposure to elevated oxygen (hyperoxia) became known as the "Paul Bert effect". We have already noted the work in the 1930s of Behnke and his associates which helped establish the general understanding that 3 hours of exposure to oxygen at 3 atmospheres absolute or 30 minutes at 4 atmospheres were approximate limits for safe tolerances by humans, based on CNS signs and symptoms. This work established the effects of hyperoxia at the intact animal and whole organ level, but did not provide information about cellular or sub-cellular mechanisms of oxygen toxicity. Bert largely missed observing the effects of oxygen on the lungs, which we shall describe separately.

Paul Bert's interests were wide. He was a French zoologist, physiologist and politician and is sometimes known as the "Father of Aviation Medicine". [38]. He entered the *Ecole Polytechnique* in Paris with the intent to become an engineer; changed to law; and then studied physiology. He was professor at two French universities, and in 1870 he became openly involved in politics [39]. His views were extreme leftist and he was an advocate of: "liberating national education from religious sects, while rendering it accessible to every citizen" [39]. It is reported that he "created a great sensation" with a lecture on modern Catholicism, by ridiculing the chief religious tracts and handbooks being circulated in the south of France at that time. He died of dysentery in Hanoi in 1886.

James Lorrain Smith (1862-1931) did research in the laboratory of John Scott Haldane and Smith is credited with establishing that oxygen was toxic to the lung; this became known as the "Lorrain Smith effect" [40]. The lung toxicity occurred at lower pressures than the more obvious convulsion that dramatically manifested the CNS effects.

As a young man, he studied the arts, turned to medicine at the University of Edinburg, and was successively appointed at several universities. Prior to the First World War, he was involved with the antiseptic "Eusol" and he suggested the idea of using charcoal in gas masks [40]. Eusol was an antiseptic solution prepared from chlorinated lime and boric acid that was used in treating war wounds. The name was derived from e(dinburg) u(niversity) sol(ution).

John Scott Haldane (1860-1936) (Fig. **9**) also contributed to the story of oxygen by his bold laboratory work. Haldane was a Scottish physiologist noted for experimenting on himself (he also experimented on his son) in relation to the effects of gases and their uses in underwater diving. Haldane is said to have "... locked himself in sealed chambers breathing potentially lethal cocktails of gases while recording their effect on his mind and body" [41]. He also personally investigated the causes of various mining disasters, and when Germany began using poison gases in WWI, he went to the front lines and personally attempted to identify the gases being used in the field of battle. He is credited with inventing the first "box respirator". He was forced to stop assisting with the efforts to thwart poison gases because of alleged German sympathies. He began helping victims of gas warfare and produced the first pure oxygen which began to be used to treat patients. He investigated mine disasters and identified carbon monoxide as the toxic component in combustion gases, and noted the skin of victims was cherry-red from carboxyhemoglobin. He designed respirators for use by rescue workmen. He even tested the effects of carbon monoxide on himself in a closed chamber. He introduced the use of small animals (white mice and canaries) for miners to detect dangerous levels of carbon monoxide. Mice and canaries have a rapid metabolism and show effects of toxic exposure before it is critical for humans which serves as an "early warning" system. It is interesting to see the degree to which Haldane was personally involved in the field as well as in the laboratory, often using himself as the test subject [41, 42].

Haldane attended Edinburg Academy, Edinburg University and the Frederick Schiller University of Jena and graduated in medicine (Edinburg University) in 1884. He was a lecturer, honorary Professor (he received numerous honorary degrees), and President of the English Institution of Mining Engineers, and a Fellow of The Royal Society and other Royal Societies. He was a contributor to the development of the mechanism involved in regulation of breathing and discovered the "Haldane Effect" (increased oxygenation of hemoglobin promotes dissociation of carbon dioxide). He designed the first decompression chamber to make diving safer and he drafted the first decompression tables based on experiments with animals. He was considered to be an expert on silicosis, a pulmonary condition affecting miners [41, 43, 44].

Fig. (9). John Scott Haldane.

John W. Bean (1901-1987) in 1945 wrote a long article (167 pages) that detail most of what was known at the time about the biological effects of oxygen at increased concentrations and pressures [45]. I recall reading this as foundational when I first became motivated toward research on the biology and chemistry of oxygen in the mid-to-late1960s.

John Bean was associated with the Department of Physiology of the University of Michigan for forty-seven years and his research focused on respiration and circulation, and especially the physiological effects of high pressure oxygen [45].

The comprehensive review by Bean [45] provides us with a good understanding of the state of knowledge and research in 1945 regarding the biology of oxygen. It was organized in three parts:

- Part I. The introduction, included two sections: "The effects of increased oxygen tensions not in excess of one atmosphere, as induced by pure oxygen; and hyperoxygenated air, or normal air at increased O_2."

- Part II. "Possible complicating variables introduced by compressed air and artificial gas mixtures".
- Part III. "Concerning the effects of oxygen tensions in excess of one atmosphere, as induced by pure oxygen, hyperoxygenated air or normal air at high pressure (OHP)". More than nine pages of references were cited in the 147-page publication.

The treatise was historical, descriptive for many of the effects of oxygen, comprehensive for the time, and lacking in mechanisms which documents that not much was known at the cellular and biochemical level when this was published in January of 1945.

Historically, it is interesting to note that Bean, in his review [45], quoted Priestley as follows:

> "From the greater strength and vivacity of the flame of a candle, in this pure air, it may be conjectured that it might be peculiarly salutary to the lungs in certain morbid cases... But, perhaps, we can also infer from these experiments, that though pure dephlogisticated air might be very useful as a *medicine,* it might not be so proper for us in the usual healthy state of the body: for, as a candle burns out much faster in dephlogisticated than in common air, so we might, as may be said, *live out too fast*, and the animal powers be too soon exhausted in this pure kind of air. A moralist, at least, may say, that the air which nature has provided for us is as good as we deserve... The feeling of it to my lungs was not sensibly different from that of common air; but I fancied that my breast felt peculiarly light and easy for some time afterwards. Who can tell but that, in time, this pure air, may become a fashionable article in luxury. Hitherto only two mice and myself have had the privilege of breathing it."

These thoughts and words of Priestley have motivated and guided the directions of research into the biology of oxygen at the intracellular level that is the frontier of knowledge today.

Transition to a Focus on Biochemical Mechanisms of Oxygen Toxicity

During the period between the mid 1940s and the late 1960s, there was a transition by laboratory investigators toward an increasing focus on elucidating mechanisms by which elevated concentrations of oxygen poisoned biological systems at the cellular and intracellular levels. There was a fundamental understanding that lipids could be oxidized (become rancid) and that this could lead to functional damage to biological membranes which would be significant for the known effects on brain and lung. There was a general belief that enzymes, especially those containing sulfhydryl groups, were very sensitive to oxidative damage. Protection by "reducing" agents and by enzymes like catalase and peroxidases was an acceptable theory to many. A seminal event, for which the field of laboratory investigations was fully ready, was the discovery of the function of a known copper-zinc protein, superoxide dismutase, by McCord and Fridovich with publication in 1969. My own work at this time challenged the idea that enzymes in general were easily poisoned by hyperbaric oxygen.

The review by Niels Haugaard provides an excellent perspective of the status of knowledge about the mechanisms of oxygen toxicity when it was published in 1968 [46]. Laboratory experiments with small animals and plants had been followed by studies using tissue slices and homogenates and subsequently enzyme preparations. Some of these experiments gave conflicting results because many factors other than the concentration or pressure of oxygen have ultimately been found to affect results. Haugaard and coworkers in 1944 [46], and J. W. Bean in 1945 [45] published what is probably the first significant evidence that oxygen exposure could inactivate certain enzymes. In the years 1944 to 1946 ten additional publications were coauthored by Haugaard and two were authored by F. Dickens reporting that some enzymes appeared to be resistant to oxygen, but many others were poisoned by it, and others were inactivated by the oxygen in air. (The review by Haugaard can be consulted for these references.) Thus, in a short period of time in the mid-1940s enzymes began to be focused on as significant sites of oxygen toxicity within cells.

Despite the conclusive evidence that some enzymes were inactivated by hyperoxia, not all researchers were convinced. This early work sometimes was with crude extracts of cells or other preparations and results were subject to some differences in interpretation, variability in exposure times and oxygen concentrations, sometimes with multiple enzymes involved, and the effects of metals and reducing agents were neither understood nor controlled. In addition, the time required to significantly inactivate enzymes was generally much longer or required higher oxygen pressures than those which created symptoms in exposed intact animals. In addition, Haugaard pointed out [46] that "No conclusive demonstration had been made that any enzyme, in any organ or the intact animal, was inhibited during exposure of the animal to oxygen at an elevated pressure". However, Haugaard [46] also stated that: "These earlier views may have been too cautious... Evidence is now also beginning to appear that some enzyme reactions are indeed inhibited *in vivo* during hyperbaric oxygenation."

Research with tissue slices reported in the era 1942-1946 showed that overall tissue metabolism was impaired by exposure to hyperoxia and Haugaard [46] cited five publications. This pioneering work was not designed to identify specific enzymes in most cases. In 1957, Haugaard and coworkers reported that 1 atmosphere of oxygen was toxic for metabolic oxidation of glucose and pyruvate in heart homogenates. This impairment was traced to the enzyme glyceraldehyde phosphate dehydrogenase and resulted in decreased ATP which was reported by Haugaard and associates in two papers published in 1965 and 1966, and described in the review by Haugaard [46]. This is an example of pinpointing a specific enzyme affected by hyperoxia.

Bacteria and fungi were reported in the review by Haugaard [46] to be inhibited by hyperoxia by several groups of investigators including J. Callwell in 1963 and 1965, S. F. Gottlieb and colleagues in 1964, W. L. Hopkinson and A. G Towers in 1963, T. A. McAlester and colleagues in 1963, and S. M. Rob in 1966. These reports appeared before I began studies at the University of Missouri on the toxic effects of oxygen using *Escherichia coli.*

Haugaard concluded: "The possibility that autoxidation of SH groups occurs during hyperbaric oxygenation deserves consideration" [46]. This was logical since mono-thiols could form di-thiols and di-thiols such as lipoic acid (a biological molecule with two nearby thiols) could form an internal S-S bridge by oxidation reactions. Ferrous ions would be a likely catalyst for such oxidations. C. M. Lyman and E. S. Barron, as far back as 1937 (as reported by Haugaard [46]) found that oxidation of reduced glutathione by air was strongly catalyzed by cupric ions, which also accelerated the oxidation of ascorbic acid. E. S. Barron also discovered that high pressure oxygen oxidized various SH-containing compounds of biological interest, including dimercaptopropanol, mercapto-ethylamine, reduced glutathione, coenzyme A and cysteine, as reported by Haugaard [46]. However, Haugaard, after reviewing the literature extensively, concluded: "Much of the information about the autoxidation of sulfhydryl compounds is contradictory and there is a great need for systematic studies of the reaction mechanism of metal-catalyzed oxidation of sulfhydryl compounds by molecular oxygen".

Oxygen Free Radicals

Research emphasizing the role of oxygen free radicals did not really begin until the discovery of superoxide dismutase near the end of the 1960s by McCord and Fridovich which will be the starting focus of Chapter 7. However, peroxidation of lipids was described by Haugaard [46] as a mechanism of oxygen toxicity; and malonaldehyde formation was already being used to detect peroxidation, but actual detection of free radicals was not being done. However, D. L. Gilbert, R. Gershman, J. Coehn, and W. Sherwood, as early as 1957 reported formation of hydrogen peroxide during oxygen exposure of DNA solutions (as reported by Haugaard [46]). The perhydroxyl radical was presumably involved. However, the state of knowledge was such in 1968 that Haugaard [46] stated: "It is not known to what extent free radicals or reactive molecules such as H_2O_2 are involved in producing the toxic effects of elevated pressures of oxygen *in vivo*"; and he also wrote: "Free radicals may be an important part in oxygen toxicity but it is probably the end result, *i.e.* the loss of an enzyme or coenzyme, that is the cause of the damage to cell function, no matter what the nature of the intermediary reactions". In the same review, Haugaard cited B. Chance and coworkers in a 1965 paper as demonstrating unequivocally in subcellular preparations: "... that one or more sites in or associated with the electron-transport chain are sensitive to inactivation by oxygen at elevated pressures".

NOTES

[1] Baron Gottfried Wilhelm von Leibnitz (1646-1716). Quoted in *Bulletin of the New York Academy of Medicine.* Vol. 5 [1929], p.752.

[2] New Experiments Physico-Mechanical, Touching the Spring of the Air, and its Effects (Made, for the Most Part, in a New Pneumatical Engine).

[3]*New Pneumatical Experiments about Respiration.* Robert Boyle, 1666.

[4] Further Report: On oxygen as a therapeutic agent.1885. Samuel S. Wallian, M.D. The Medical Record, October 31, pages 483-488, 1885.

[5] A further word on oxygen treatment and oxygen charlatans. The Medical World. Samuel S. Wallian, M.D., page 97, 1886.

[6] Clinical notes: Oxygen in therapeutics. Charles J. Smith. Published in The Lancet, December 18, Page 1172, 1886.

[7] The treatment of pneumonia, in *Medical News*, page 324, 1887.

[8] Oxygen in therapeutics. A treatise explaining the apparatus, the material, and the processes used in the preparation of oxygen and other gases with which it may be combined, also its administration and effects, illustrated by clinical experience of the author and others. C. E. Ehinger, Chicago, W. A. Chatterton & Co., 1887 [46].

[9] Original article: Oxygen enemata as a remedy in certain diseases of the liver and intestinal tract. Journal of the American Medical Association, 11: 258-262, 1888.

[10] Original article: The continuous inhalation of oxygen in cases of pneumonia otherwise fatal, and in other diseases. Albert N. Blodgett, M.D., Boston Medical and Surgical Journal, Vol. CXXIII, No. 21, pages 481-484, 1890. (The Boston Medical and Surgical Journal was the forerunner of The New England Journal of Medicine.)

[11] The therapeutic value of oxygen. W.L. Conklin, M.D., New York State Medical Journal, September 2, pages 338-341, 1899.

[12] Paracelsus is famous for the quote, foundational to the field of toxicology: "The dose makes the poison" (*Sola dosis facit venenum*) appearing in Paracelsus, *dritte defensio*, 1538.

REFERENCES

[1] Morrison DS. History of Hyperbaric Medicine and Oxygen Therapy. Available from: http://www.quackwatch.com/01QuackeryRelatedTopics/HBOT/hm01.html 2016.

[2] Richmon J. History of hyperbaric medicine and applications in head and neck surgery. Johns Hopkins Medicine. Available from: http://www.gbmc.org/workfiles/HeadNeck/HBOGrandRounds1_10_14.pdf 2014.

[3] B T, Arjunan K. Hyperbaric oxygen therapy concepts and myths. WebnetCentral: ENT Scholar. Available from: http://www.webmedcentral.com 2016.

[4] Henshaw & the Domicilium. Available from: http://www.londondivingchamber.co.uk/index.php?id=history&page=5 2016.

[5] West JB. Robert Boyles landmark book of 1660 with the first experiments on rarified air. J Appl Physiol 2005; 98(1): 31-9. [http://dx.doi.org/10.1152/japplphysiol.00759.2004] [PMID: 15591301]

[6] Acott C. A brief history of diving and decompression illness. South Pacific Underw Med Soc J 1999; 29(2): 98-109.

[7] Triger M. Letter to Monsieur Arago. Compte Rendus L'Academie des Sci Paris 1845; 20(334): 789-801.

[8] Mcgunegle ML. Construction of the Eads Bridge. Available from: http://web.mst.edu/~rogersda/umrc-ourses/ge342/EadsBridge.pdf 2004.

[9] James B. Eads and His Amazing Bridge at St. Louis. Available from: https://www.nps.gov/jeff/learn/historyculture/upload/eads.pdf 2016.

[10] Bends, Caissons, Hyperbaric Oxygen, and the Adelaide Hyperbaric Unit. Website for the Virtual Museum. Available from: http://www.samhs.org.au/Virtual Museum/Medicine/caissons/caissons.html 2016.

[11] Moir E. Tunnelling by comeed air. J Soc ARTS 1896; 44(567): 249-301.

[12] American History, Hyperbaric Chambers - A history of hyperbaric chambers and dive medicine, London Recompression & Hyperbaric facilities - The London Diving Chamber. Available from: http://londondivingchamber.co.uk/index.php?id=history&page=7 2016.

[13] Martin L. Oxygen therapy: the first 150 years (continued). Available from: http://www.lakesidepress.com/pulmonary/papers/ox-hist/ox-hist2.html 2016.

[14] CFF, Jr. DT (Tim) C, NREMT-I DDDFD. Hyperbaric oxygen therapy: the history. Available from:

dvm360.com 20(3): 284-8.

[15] Cunningham Sanitarium- The Encyclopedia of Cleveland History. Available from: http://ech.cwru.edu /ech-cgi/article.pl?id=CS6 2016.

[16] Martin L. Monster steel ball forms air cure hospital. Available from: http://www.lakesidepress.com/ pulmonary/hyperbaric/steelball.html

[17] Choffin M. The Cunningham Sanitarium. Available from: http://clevelandhistorical.org/items/show/3-78#.V1V4gvkrIdV 2016.

[18] Boerema I, Huiskes JW, Kroll JA, Kroon B, Lokin E, Meyne NG. High atmospheric pressure as an aid to cardiac surgery. Arch Chir Neerl 1956; 8(3): 193-211.
[PMID: 13373586]

[19] Brummelkamp WH, Hogendijk J, Borema I. Treatment of anaerobic infections (Clostridial myositis) by drenching tissues with oxygen under high pressure. Surgery 1961; 49(3): 299-302.

[20] Smith G, Sharp G. Treatment of coal gas poisoning with oxygen at two atmospheres pressure. Lancet 1962; 21(1): 816-9.

[21] Albert R. Behnke - Alchetron, The Free Social Encyclopedia. Available from: http://alchetron.com/ Albert-R-Behnke-1331588-W 2016.

[22] Buchanan PJ. The sad suicide of Admiral Nimitz. Available from: http://www.wnd.com/2002 /01/12433/

[23] Bornmann R. Dr. Behnke, founder of UHMS, dies. Press Newsl Undersea Hyperb Med Soc 1992; 21(2): 1-4.

[24] Yarbrough O, Behnke A. The treatment of compressed air illness using oxygen. J Ind Hyg Toxicol 1939; 21: 213-8.

[25] Berghage T, Vorosmarti J, Bernard E. Recompression treatment tables used throughout the world by government and industry. US Nav Med Res Cent Tech REport NMRI 1978; 78(16): 1-80.

[26] A B, Yarborough O. Respiratory resistance, oil-water solubility and mental effects of argon compared with helium and nitrogen. Am Jorunal Physiol 1039; 126: 409-15.

[27] Behnke A. Log of diving during rescue and salvage operations of the USS Squalus: diving log of USS Falcon, 24 May 1939- 12 September 1939. Reprinted by Undersea and Hyperbaric Medical Society in 2001. Undersea Hyperb Med 2001.

[28] Teven L. NMRI's new hyperbaric research complex dedicated to Dr. Behnke. Newsl Undersea Hyperb Med Soc 1981; 10(4): 1-2.

[29] Behnke AR. Physiologic studies pertaining to deep sea diving and aviation, especially in relation to the fat content and composition of the body. Bull N Y Acad Med 1942; 18(9): 561-85.
[PMID: 19312284]

[30] Albert R. Behnke Collection | Duke Medical Center Archives. Available from: https://archives.mc. duke.edu/blog/albert-r-behnke-collection 2016.

[31] Fundamentals of hyperbaric medicine, Publication #1298. 1966.

[32] Hyperbaric Medicine, Hagyard Equine Medical Institute. Available from: http://hagyard.com/

Hyperbaric-Medicine.html 2016.

[33] Comissioner O of the. Consumer Updates - Hyperbaric Oxygen Therapy: Don't Be Misled. Office of the Commissioner. Available from: http://www.fda.gov/forconsumers/consumerupdates/ucm364687.htm 2016.

[34] FDA approved uses for Hyperbaric Oxygen Therapy. Available from: http://www.woundsource.com/print/blog/fda-approved-uses-hyperbaric-oxygen-therapy 2016.

[35] Modern World Hyperbaric Chambers- A History of Hyperbaric Chambers. Available from: from: http://londondivingchamber.co.uk/index.php?id=history&page=9 2016.

[36] Borzelleca JF. Paracelsus: herald of modern toxicology. Toxicol Sci 2000; 53(1): 2-4. [http://dx.doi.org/10.1093/toxsci/53.1.2] [PMID: 10653514]

[37] Dejours P, Dejours S. The effects of barometric pressure according to Paul Bert: the question today. Int J Sports Med 1992; 13 (Suppl. 1): S1-5. [http://dx.doi.org/10.1055/s-2007-1024577] [PMID: 1483742]

[38] Hitchcock FA. Paul Bert and the beginnings of aviation medicine. Aerosp Med 1971; 42(10): 1101-7. [PMID: 4937813]

[39] Chisolm H. Bert, Paul Encyclopedia Britannica 3. 11th ed., Cambridge University Press 1911.

[40] Smith JL. The pathological effects due to increase of oxygen tension in the air breathed. J Physiol 1899; 24(1): 19-35. [http://dx.doi.org/10.1113/jphysiol.1899.sp000746] [PMID: 16992479]

[41] Acott C. JS Haldane, JBS Haldane, L. Hill, and A. Siebe: a brief resume of their lives. South Pacific Underw Med Soc J 1999; 29(3): 160-5.

[42] Obituary. Professor J.S. Haldane. Available from: http://www.dmm.org.uk/archives/a_obit20.htndmm.org.uk 1936.

[43] Boycott AE, Damant GC, Haldane JS. Prevention of compressed air illness. J Hyg (Lond) 1908; 8(3): 342-443. [http://dx.doi.org/10.1017/S0022172400003399] [PMID: 20474365]

[44] Hellemans A, Brunch B. The timetables of science. Simon &; Schuster 1988.

[45] Bean J. Effects of oxygen at increased pressure. Physiol Rev 1945; 25(1): 1-147. Available from: http://physrev.physiology.org/content/25/1/1.abstract

[46] Haugaard N. Cellular mechanisms of oxygen toxicity. Physiol Rev 1968; 48(2): 311-73. [PMID: 4870377]

CHAPTER 7

Oxygen Biology, Boon and Bane

Abstract: Oxygen paradoxically exhibits both strongly positive and violently negative effects in biological systems. Irwin Fridovich succinctly characterized this as the "boon and bane" of oxygen. Joe M. McCord in Fridovich's laboratory, in 1969, had discovered the enzymatic activity of copper-zinc superoxide dismutase that fueled a revolution in research into the mechanism used by the body to defend itself against the toxicity of oxygen. Superoxide dismutase converts the biologically-active oxygen radical, superoxide, into molecular oxygen and hydrogen peroxide and the latter is a substrate for catalase which is abundant in cells and converts hydrogen peroxide into oxygen and water. Oxygen has been shown to be a vital participant in extremely diverse biological functions including phagocytosis, athletic performance, ageing, mitochondrial energetics, and a host of oxidant stress-related and degenerative diseases which are areas of continuing research today. The mechanisms of oxygen toxicity at the cellular level were primarily speculative until near the end of the 1960s when the significance of oxygen "free radicals" became the dominate research theme. The means to examine the mechanisms of oxidative damage to enzymes, membranes and DNA became available and scientists began to probe into intracellular sites and mechanisms of oxygen toxicity. The realization that various electronically "excited" (free radical forms of oxygen) were the culprits that damaged the vital and sometimes delicate machinery of life was brought into sharp focus by the discovery of superoxide dismutase and its exploration by both pure and applied researchers in medicine, chemistry, biochemistry, and molecular biology. There were controversies, primarily about which oxygen radicals were most significant, but this led to realization that a host of factors including the chemistry and biology of the many oxidized species, iron and other transition metals as electron donors, and complex cellular defense mechanisms were essential parts of the new story that began in the late 1960s and is continuing today.

Keywords: Ageing, Diffusion limited, Fenton reaction, Hydrogen peroxide, Hydroxyl radical, Iron, Irwin Fridovich, Joe McCord, Lipid peroxidation, Mitochondrial respiration, Oxidation-reduction, Oxygen free radical, Peroxidase,

Olen R. Brown

Redox cycle, Singlet oxygen, Superoxide, Superoxide dismutase, Transition metals, Vitamin A, Vitamin C, Vitamin E.

INTRODUCTION

The end of the decade of the 1960s was the beginning of a new era in laboratory studies focused toward understanding the mechanisms of oxygen toxicity at the intracellular level. This research was driven by a desire to understand how oxygen was toxic at the most basic level. Research into toxic mechanisms had progressed that far. It was also necessary to pursue oxygen's toxic mechanisms to better understand what was already discovered about oxygen's role in energy metabolism and, of course, for its safe therapeutic application. The studies being undertaken were also relevant to probe how various chemicals, including therapeutic drugs such as some used to treat cancer, produced harmful side-effects. Oxygen radicals appeared to be involved in how both ionizing and non-ionizing radiation damage cells. Oxidant stress and oxygen free radicals were surely involved in lipid peroxidation, mutations in DNA and damage to protein structure. Understanding the nature and formation of reactive oxygen species— "free radicals"— became an obvious requirement for a deeper understanding of the chemistry of oxygen within cells.

After a brief excursion as prelude to explore catalase, hydrogen peroxide, hydroxyl radical, and superoxide dismutase (their stories are intertwined) I will focus this chapter on cellular mechanisms of the toxicity of oxygen and the biological defense strategies, beginning with the discovery of the function of the enzyme superoxide dismutase.

Hydrogen peroxide is thermodynamically unstable and decomposes to water and molecular oxygen. Biologically, however, it is decomposed with great efficiency by the enzyme catalase which Louis Jacques Thenard (1777-1857) is credited with discovering in 1818. Thenard speculated that an "unknown substance" was responsible for the observed rapid breakdown of hydrogen peroxide. Thenard was the son of poor French peasants but managed to become educated at the academy of Sens and went to Paris at age sixteen to study pharmacy. It is said that he was unable to pay the small regular fee of twenty francs a month and he received a position in the laboratory of the prominent chemist, Vanquelin, only at the intercession of Vanquelin's sisters. He did so well that he became a teacher of chemistry and secured a position at the Ecole Polytechnique and the Faculte des Sciences. After his death a statue was erected in his memory at Sens and subsequently the name of his native village was changed to Louptiere-Thenard. He is famous for Thenard's blue, a coloring matter, based on cobalt, which he discovered that "was capable of withstanding the heat of a porcelain furnace" [1].

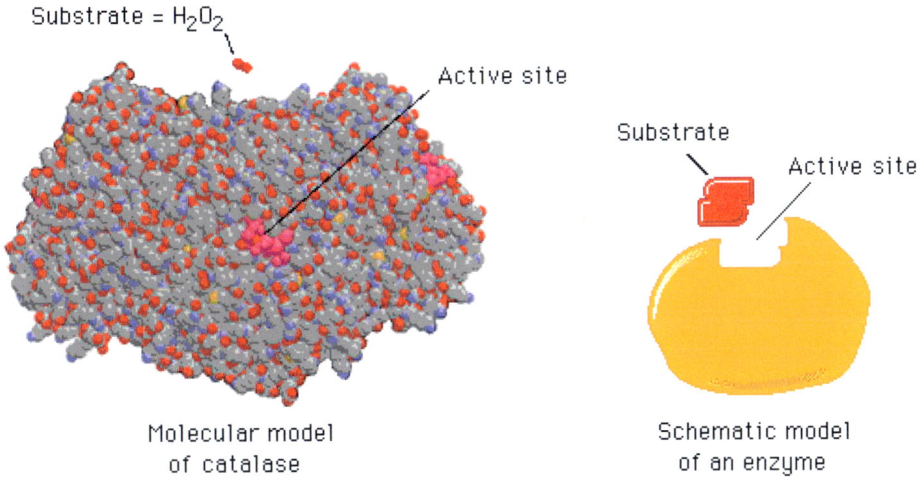

Fig. (1). Catalase (left) is a large protein molecule with a three-dimensional structure containing an active site where the substrate H_2O_2 fits, and is catalytically cleaved to water (diagrammatically shown on the right).

Thenard's biological substance, because of its strong effervescent, was known to be present abundantly in tissues, but its function was not understood. It was named catalase by Oscar Loew in 1900 [2] and it is central to understanding the oxidation-reduction reactions that are fundamental to cell metabolism. It is a catalyst for the breakdown of hydrogen peroxide with the formation of water and oxygen at a greatly accelerated rate (Fig. **1**). This catalytic rate is reported to be up to five million molecules of hydrogen peroxide converted per minute. However, its magic can be better understood by simply saying that this enzyme functions at the maximum theoretically possible rate; the rate dictated by the rate of arrival of hydrogen peroxide molecules by diffusion to the active site of the enzyme and not on the rate of cleavage of hydrogen peroxide by the enzyme. There are more than

300 known catalases with slightly different protein sequences, and some are bi-functional (catalase and peroxidases activity). Catalases also show great divergence when rates are measured. They do not follow Michaelis-Menton kinetics except at very low substrate concentrations. Values range from 54,000 to 833,000 molecules of hydrogen peroxide broken down per second in the paper by P. Chelikani *et al.* [3].

Catalase made the "Molecule of the Month" designation on the Protein Data Base site for September 2004 [4]. A quote from the site, popular to the non-scientist, is instructive:

> "Living with oxygen is dangerous... it is easily converted into other reactive compounds. Inside our cells, electrons are continually shuttled from site to site by carrier molecules... if oxygen runs into one of these carrier molecules, the electrons may be accidentally transferred to it. This converts oxygen into dangerous compounds such as superoxide radicals and hydrogen peroxide, which can attack the delicate sulfur atoms and metal ions in proteins. To make things even worse, free iron ions in the cell occasionally convert hydrogen peroxide into hydroxyl radicals. These deadly molecules attack and mutate DNA, One theory, still controversial, is that this type of oxidative damage accumulates over the years of our life, causing us to age."

The slight oversimplifications here and the difficulties in hydroxyl radical reaching DNA sites will be discussed in due course.

Another tissue-derived substance, orgatine, is part of the story. Orgatine was used by some at the margins of human and animal medicine to designate an unknown substance that was abundantly present in extracts prepared from liver. Liver extracts were extolled as having unusual curative powers. I recall being approached by an earnest entrepreneur who was captivated by the potential of orgatine for human and veterinary medicine and told some wondrous stories of remarkable cures. The surviving vestiges of orgatine are traceable to a patent application [5] and a mention as PEG-SOD (PEG-orgatine; Sterling-Winthrop) with the mechanism of action designated as free radical scavenger in a table titled: "Compounds that have failed to produce an overall improvement in Phase III (or large Phase II) TBI (traumatic brain injury) trials" [6]. I have not seen literature that conclusively traces the future of orgatine; however, I believe the active principle was what came to be known as superoxide dismutase. Officially, the trail leading to superoxide dismutase was research interest in the blue copper-containing protein isolated by T. Mann and D. Keilin in 1938 (from liver and other sources) [7].

MECHANISMS OF OXYGEN TOXICITY AND OXIDANT DEFENSES

Introduction

Because of their catalytic activities, enzymes are natural candidates of interest as sites of oxygen damage capable of explaining the toxicity of oxygen. This results, in part, because inactivation of each molecule of enzyme is amplified by virtue of the fact that enzymes are catalysts. Damage to one molecule of protein that is an enzyme obviously has an extended consequence, compared, for example to damage to the product of that enzyme. In addition, damage to enzyme molecules is likely to be significant because most enzymes have an essential function (produce an essential product). There was another reason to suspect enzymes. All enzymes are proteins (at least this was believed to be true in the 1960s, a time before the discovery of the catalytic action of certain RNA molecules). Proteins are basic to all life, and all life forms are subject to oxygen poisoning. Some enzymes catalyze oxidation-reduction reactions. Some enzymes require metals as cofactors and transition metals participate in oxidation-reduction reactions that can be intimately associated with enzyme catalytic sites. Individual enzymes may be vital for functions that are, themselves, essential to life. Indeed, enzymes often are indispensible elements in complex chains of reactions or cycles that cease to function when a single enzyme link in the chain is lost. Enzymes have large, complex, three-dimensional structures whose specific topology must be intact for proper functioning. Minor changes in atoms of amino acids, the primary unit of protein structure, can have large consequences and even cause total loss of function. Some enzymes are essential; some enzymes are required but not essential; some enzymes are present in large amounts, and some in only small amounts. Thus, effects of oxygen at the enzyme level might explain a variety of symptoms expressed at the whole animal level.

To apply basic research findings, there is the need to relate finding from the most detailed chemical level back to the whole animal level; and for medical uses, it must be shown relevant to the human. Experiments with various species, various organs, cell-free homogenates and extracts, purified enzymes, coenzymes, and cofactors provided a systematic basis for discovering the effects of oxygen toxicity and documenting its relevance. Basic research, even when the goal is to understand a human disease or toxicity, is sometimes made easier by using simpler systems, including bacteria, when the focus of interest is at the enzyme level. Basic principles often can be more easily investigated in simpler systems but interpretations require comparative knowledge of various life forms including functions from the molecular to the organ to the intact species level. Because of my background and interests, I chose to use *Escherichia coli* for my research that began in the late 1960s into oxygen toxicity mechanisms.

The sulfhydryl (SH) group was a known functional group on proteins, including some enzymes. This SH group was known to be essential in the reaction site of some enzymes and the SH group is quite subject to oxidation. Thus, the SH group was a presumptive candidate for oxygen toxicity. Iron and other metals were known to possess accessible electrons and to participate in oxidation-reduction reactions. Metals also, therefore, were suspected participants in oxygen toxicity.

Transition metals are defined by the International Union of Pure and Applied Chemists (IUPAC) as an element whose atom has a partially filled d sub-shell, or which can give rise to cations with an incomplete d sub-shell [8]. This fundamentally describes elements in groups 3 to 12 of the periodic table (sometimes referred to as d-block elements). A full description is beyond the scope needed here; however, transition elements are typical metals and transmit electrons well and form compounds with several oxidation states due to the relatively low energy differences between the different possible oxidation states. Iron, copper, and manganese are the most important biologically because they have electrons that are accessible to exchange reactions and they are (comparatively) small in size and diffuse into and fit into reaction sites. As is apparent from this brief description, transition metals can participate in oxidation-reduction chemistry and can create radical oxygen species.

SUPEROXIDE DISMUTASE, THE START OF SOMETHING NEW

Joe McCord, working as a graduate student in the laboratory of Irwin Fridovich in 1969, discovered the catalytic properties of erythrocuprein (hemocuprein) [9] which now is known as the Cu-Zn form of the enzyme superoxide dismutase (Fig. **2**). The publication of this discovery, from my perception, is the demarcation point for the unfolding of new directions, some controversy, and for the focus of this chapter describing oxygen radical biology and biochemistry.

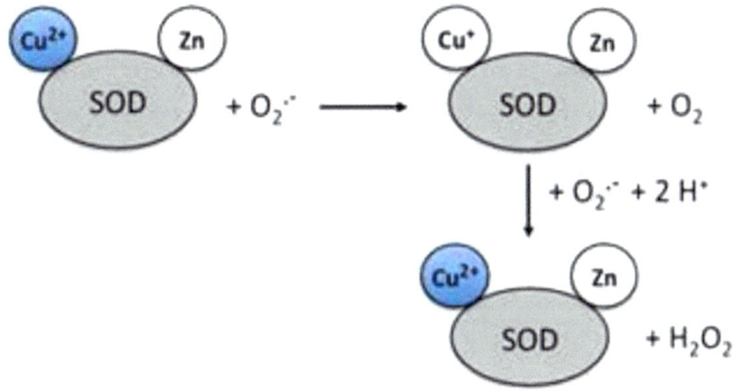

Fig. (2). Simplified function of the enzyme superoxide dismutase.

Superoxide dismutase, *via* reduction of copper present in the enzyme, transfers the electron to reduce superoxide to oxygen, while oxidizing another molecule of superoxide (with addition of two protons to re-oxidizes the copper), and produces hydrogen peroxide (reduced oxygen).

Joe McCord today is recognized as a Redox Pioneer [10] because (in the words used, in part, to describe this honor): "... he has published at least three articles on antioxidant/redox biology as first/last author that have been cited over 1000 times and has published at least 37 articles each cited over 100 times". This article further describes that his research, while being a graduate student in the laboratory of Irwin Fridovich, demonstrated that superoxide is "… essential to aerobic life, yet, if left unchecked by SOD, toxic to a multitude of systems". A longer quotation by McCord reveals his spirit and soul [10]:

> "My advice to students is this: Science requires the same creativity, inventiveness, and passion that we expect from artists, composers, and writers. When you feel it, you know it. If you don't feel it, then science probably isn't the best career choice for you. This has little to do with whether science seems easy or difficult. It always seemed difficult to me as a student, as it probably should."

McCord's research related to discovery of SOD will be noted here and his following contributions will be placed in context with other discoveries in following sections of this chapter. SOD was purified from red blood cells from the cow. Copper (2 equivalents per mole of enzyme) was present and required for activity but could be reversibly removed. The cow SOD was shown to be identical with previously described human erythrocuprein and cow hemocuprein and it was shown to be widely distributed within mammals. The oxidation of epinephrine to adrenochrome by the enzyme xanthine oxidase was mediated by the superoxide radical. Quotations from the paper are in order:

> "The recognition of superoxide dismutase as an enzymic activity and its final identification as erythrocuprein (hemocuprein) have provided explanations for many observations… The nature of the superoxide radical is such that it may act either as a reductant or as an oxidant. In the dismutation reaction, half is oxidized to O_2 by the other half, which is in turn reduced to H_2O_2… The abundance of superoxide dismutase activity in the variety of animal tissues assayed suggests that the enzyme might play a significant, even vital, role in protecting the organism against the damaging effects of the superoxide radical."

This last statement expressed a sentiment that was to prove prophetic. The scientific paper, and more factually the work it represented, started the revolution of research into mechanisms of oxygen toxicity and defenses against this toxicity. My own contributions in this field began with the publication with students in my laboratory of three papers in the late 1960 describing the effects of hyperoxia on the action of antibiotics and the role of carbon dioxide deficiency for lethality of pure oxygen for bacteria [11 - 13].

A search of the literature using Medline (Ovid), a worldwide database of biomedical literature covering 4000 journals from 70 countries, with the quite specific search term "oxygen toxicity" yielded 1478 papers in the database from 1940 through 2016 (as of 07-23). Analysis by decades showed that research defined as related to oxygen toxicity increased significantly in the 1960s, more than tripled in the 1970s, almost doubled again in the 1980, and again in the 1990s, and decreased somewhat since then.

OXYGEN TOXICITY DISCOVERIES IN THE 1970S

As a means of managing the great volume of literature on this subject, in this chapter I have chosen to discuss my own research by decades in context with what I consider to be the most significant progress relative to understanding intracellular: (1) sites of oxygen toxicity, and (2) defenses against oxygen toxicity. The many published reviews on these subjects are sources of more complete and detailed information. My goal includes the historical perspective, as has been the case throughout this book.

The decade of the 1970 was a time of great interest and progress in research relating to the biology of oxygen, and from 1970-1979 there were 207 papers published dealing specifically with oxygen toxicity, as recovered by the Ovid database (as previously described). This research often was focused on the superoxide radical and on the enzyme superoxide dismutase. The increased research was not without controversy. Some biochemists remained unconvinced about the protective benefits to the organism for an enzyme that resulted in converting the toxic superoxide species into the toxic hydrogen peroxide species. Some even questioned the extent of the toxicity of superoxide and its stability even in the absence of enzymatic dismutation. The almost universal presence of the enzyme, subsequently discovered to be in multiple forms with different metal ions, and studies with aerobic, aerotolerant and anaerobic microbes helped in understanding the role of this enzyme. Important publications in the 1970s from many laboratories reported roles for: superoxide dismutase in phagocytosis [14 - 19]; other superoxide dismutases [20 - 22]; and the relationship of superoxide dismutase to obligate anaerobiosis [23], as examples.

A very significant review was published at the end of the decade (1979) by Britton Chance, Helmut Sies, and Alberto Boveris [24], with 561 cited references.

After introductory sections for: history, enzymes that metabolize oxygen-reduction products, and methods for measuring the oxygen species there were the following sections relevant to this chapter: cellular sources of hydrogen peroxide, intracellular regulation of oxygen-reduction products, oxygen toxicity and defense against oxygen-reduction products, and other sections. Some quotations (Section IV) are historically interesting and reveal the knowledge state of that time:

> "Gerschman *et al.* proposed that an increase of oxidizing free radicals to toxic levels in cells and tissues was the cause of the toxicity of both X irradiation and oxygen. Recent data on the rates of generation of decomposition of the intermediates of oxygen reduction and on their specific effects on cellular components fill some of the gaps in the earlier work... In spite of microvascular control of the tissue oxygen levels, oxygen pressurization increases the generation of O_2^- (implicated by Fridovich) and H_2O_2, leading to an increased steady-state concentration of these intermediates."

Lipid peroxidation and the role of reduced glutathione were discussed. A section entitled: 1. Toxicity of hydrogen peroxide and superoxide anion is most relevant and informative, especially since Britton Chance is an author. Only a direct quotation will suffice:

> "The toxic action of H_2O_2 and O_2^- apparently is due to the capacity of these intermediates of oxygen reduction to generate other reactive

> oxygen species such as the hydroxyl radical and singlet oxygen, which then initiate a radical chain reaction leading to extensive lipid and organic peroxide formation... Hydrogen peroxide itself seems to be quite unreactive, it does not induce lipid peroxidation, high-amplitude swelling, or inhibition of respiratory control in isolated mitochondrial membranes. On the other hand, H_2O_2 reportedly inactivates transforming DNA...; it induces chromosomal aberrations...and produces base liberations and backbone breakage of DNA. These points may be relevant to theories of aging."

In a section entitled: "Defense against toxic products of oxygen reduction", the state of knowledge is revealed by the following ("oxygen reduction" was used to describe the process by which oxygen was reduced by a sequential gain of electrons, which could be one at a time):

> "The principal defense against toxic products of oxygen reduction lies at the level of the tissue oxygen tension, which is maintained at rather low levels... the arterioles are sensitive to high oxygen pressure and cause a decrease in the blood circulation ... so that the effects of hyperbaric oxygen are greatly decreased in organs other than the lung.... the high affinity of cytochrome oxidase for oxygen, together with the relatively large intercapillary distances for active organs such as brain and heart, leads to oxygen tension gradients from capillary to the "lethal corner" ... that may diminish the oxygen concentration 100-fold or even 1,000-fold from that of the blood vessel."

This 78-page review [24] further stated that "the evidence suggests" that cytochrome oxidase retains the intermediates of oxygen reduction until water is formed and that the enzyme has "pockets" where a number of oxygen molecules are retained; therefore, in mitochondria, the ubiquinone-cytochrome b region of the respiratory chain (in a particular state they defined as "state 4") generated O_2^- (superoxide). The review stated that the enzymes that protected cells from reduced oxygen species that were released from mitochondria and other intracellular sources were: superoxide dismutase, catalase, and glutathione peroxidases. A key feature of these defenses was that they "seemed" to be located at sites where the appropriate intermediates accumulated. A thoughtful appraisal of the literature of that time indicates that the focus was on understanding the nature of, site(s) of formation, reactivity, likely concentrations, pathology implicated, and defenses against the identified reactive oxygen species.

The research from my laboratory was more narrowly focused and was on areas amenable to investigation using bacteria as a tool. Bacteria are more like mitochondria than intact animal cells, and although the major goal of my laboratory was to uncover information that ultimately would apply to humans, bacteria offered the means to conveniently study enzymes and even entire pathways provided due consideration was given to comparative physiology and biochemistry. By the end of the decade using *Escherichia coli*, Fred Yein from my laboratory and I published a short abstract [25] which summarized that exposure to hyperoxia reversibly inhibited growth by reversibly impairing biosynthesis of: branched-chain amino acids, aromatic amino acids, NAD-niacin, thiamine, RNA, and by inhibiting reverse glycolysis. Significantly, genetic "stringency" was induced with production of a regulatory compound (pp-guanin--pp) with shut-down of significant pathways of biosynthesis. Two enzymes, phosphoribosyl pyrophosphate synthetase and quinolinate transferase were identified as potential, specific sites of oxygen poisoning. It was pointed out that hyperoxic-inhibition of these enzymes could have relevance to the effects of hyperoxia on the brain, including convulsions.

In 1971, R. L. Brunker in my laboratory reported [26] that exposure to 1 atmosphere of oxygen: (a) increased the generation time of *Escherichia coli* by 33%; (b) NAD was decreased by 30%; (c) oxygen at 6 atmospheres was immediately bacteriostatic; (d) two atmospheres of exposure to oxygen produced a 62% decrease in NAD while NADH, NADP and NADPH were unchanged, and (e) these coenzymes were as stable in buffer solution at 15 atmospheres of oxygen as were solutions exposed only to air. It was noted that the reduction in NAD concentration could have caused the bacteriostasis; however, NADPH, as required for reductive steps in biosynthesis, was not limiting. In 1979, research jointly by four workers in my laboratory identified quinolinate phosphoribosyl transferase as the oxidant sensitive step in NAD biosynthesis [27]. The enzyme is essential for *de novo* biosynthesis of NAD in *Escherichia coli* and in humans. It was noted that niacin enters the biosynthesis pathway below the oxygen-poisoned enzyme step and therefore could be protective.

By 1978, with F. Yein in my laboratory, we were able to report that exposure to hyperbaric oxygen rapidly inactivated dihydroxyacid dehydratase (an enzyme required for biosynthesis of branched-chain amino acids) but not the other enzymes in the pathway [28].

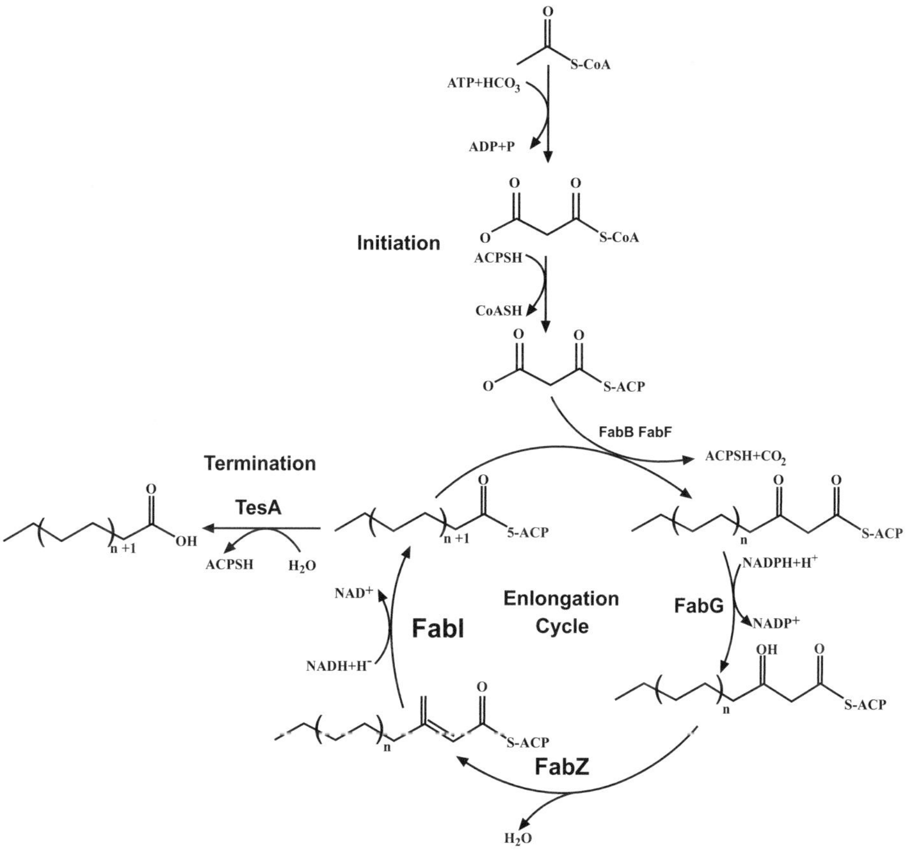

Fig. (3). Fatty acid synthesis in *Escherichia coli.*

As early as 1971, I published evidence that exposure of *Escherichia coli* to 6 atmospheres of oxygen stopped growth and reduced respiration rate but did not significantly impair oxidative phosphorylation efficiency in cell extracts where the rate was 34,000 ± 9,700 of ATP produced per bacterium per second. However, bacteria in exponential growth had approximately 59-times the oxygen uptake rate shown by extracts [29]. Subsequently, R. Mathis in my lab showed that although exposure to 4.2 atmospheres of hyperbaric oxygen stopped growth of *Escherichia coli* within 5 minutes, significant changes in ATP did not occur until after growth inhibition; therefore, lack of ATP was not the cause of inhibition of growth [30].

Jack Stees and Harvey Howitt from my laboratory investigated the effects of hyperoxia on fatty acid synthesis (Fig. **3**).

This metabolic site was selected for in-depth study because of several factors. Fatty acid synthesis is essential for growth and reproduction. Fatty acids are integral parts of lipids and lipids had been proposed (almost universally) as suspect sites for oxygen toxicity. The lipid biosynthesis pathway had a dedicated sequence of enzymes and cofactors some of which possessed the SH functional group that was believed to be oxidant-sensitive, and several enzymes did not have this SH relationship. In addition, the overall functioning of the pathway in hyperbaric oxygen, compared to individual steps, could be assessed, and the pathway of lipid biosynthesis in various living species bears remarkable similarities. We chose first to examine fatty acid biosynthesis using the microbe *Escherichia coli* with funding from the Office of Naval Research and the Dalton Cardiovascular Research Center of the Graduate School of the University of Missouri. Growth and fatty acid synthesis were inhibited by exposures to oxygen at 1 atmosphere and stopped almost instantly at 4.2 atmospheres of oxygen. The effects were reversible, and fatty acid synthesis returned to normal when cells were re-incubated with air as the gas phase. The spectrum of fatty acids synthesized was unchanged by oxygen pressures that inhibited growth rates; and re-initiation of synthesis did not result in production of aberrant fatty acids. The activity of the specific pathway enzyme fatty acid synthase was unchanged in cells exposed to 3.2 and 15.2 atmospheres of oxygen. The spectrum of fatty acids synthesized by cell free extracts during incubation at 4.2 atmospheres of oxygen was not changed compared to controls incubated with air as the gas phase. *In vitro* fatty acid synthase assays in hyperbaric oxygen included NADPH, acyl carrier protein, mercaptoethanol [31], and malonyl coenzyme A; therefore, damage other than reversible sulfhydryl oxidation, to the apoenzymes of fatty acid synthase were ruled out. In was a significant advance to learn that all these enzymes in this important pathway were not irreversibly inactivated by hyperbaric oxygen exposure. Note that, significantly, SH groups are required in the reduced form and there was abundant speculation in the scientific community, based on chemistry, that these sensitive, chemical entities would be attacked by oxygen radicals, perhaps irreversibly.

In continuing this research, Jack Stees and I chose to study fatty acid synthesis using bakers' yeast. This microscopic organism (the same as used in baking bread) has an intricate enzyme complex, as large as some viruses, that coordinates the individual steps in fatty acid synthesis in a unifying sequence that is very efficient because each synthesized intermediate is simply "handed-off" to the next enzyme in the step and eventually a very long-chained fatty acid can be made with addition of two carbons at a time (Fig. **4**).

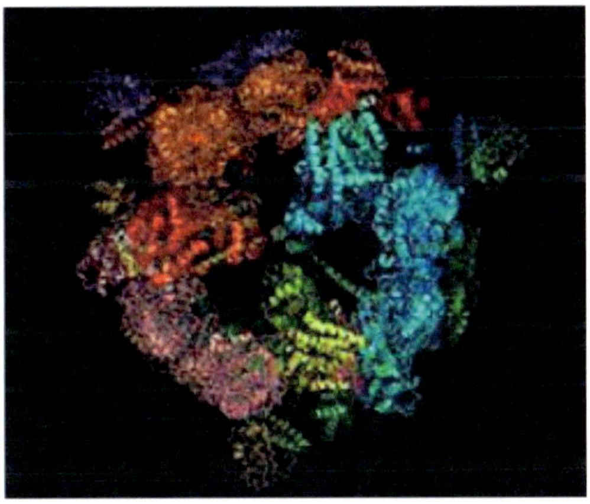

Fig. (4). Fatty acid synthetase complex of bakers' yeast. It is as large as some viruses and has all the enzymes necessary for synthesizing fatty acids by adding two carbons at a time.

The fatty acid synthetase complex was isolated from bakers' yeast and purified 120-fold. We observed the complexes by microscopy and designed equipment to allow exposure to oxygen at 40.2 atmospheres with provisions for sampling. After 1 hour at 20 degrees C and 40.2 atmospheres absolute oxygen, none of the component enzymes were detectably impaired in specific activity. Note that the fatty acid synthetase complex has thiol (SH) groups that are required for enzyme activity (Fig. **5**). Several relevant things should be noted. There are 6 different enzymes in the complex: KS (ketoacyl synthase), MAT (malonyl/acetyl transferase), DH (hydroxyacyl dehydrase, ER (enoyl reductase), KR (ketoacyl reductase), TE (thioesterase), in order from head to foot counterclockwise. The substrates for all these enzymes are bound to acyl carrier protein *via* sulfur atoms. Two-carbon fragments beginning as CH_3-(C=O)~S- CoA, are sequentially added to make the growing fatty acid straight chain of molecules. The fragments are transferred from the~S group of Coenzyme A to the~S group of the acyl carrier protein which then transfers them to the growing chain with the aid of the enzyme activity of the complex. These "sulfur" functional groups must be in a reducing environment but they are readily oxidized to a non-functioning state (for the purpose of serving as the attaching site). As further explanation, these fragments are carried attached to sulfur of a particular coenzyme called coenzyme A. This molecule allows each enzyme in the continuing sequence to be in position to catalyze the next step in lengthening the chain of carbon. The enzymes in the complex "recognize" and also "hand-off and receive" the growing chain. Thus, the growing chain, that is the fatty acid being made, does not diffuse away and have to be "re-collected" by the next enzyme in the complex.

Our research showed that the anticipated inactivation by hyperoxia did not inactive this "oxidatively-sensitive" complex as was reasonably expected; at least, no inactivation that was not readily reversed, occurred [32].

OXYGEN TOXICITY DISCOVERIES IN THE 1980S

For the decade of the 1980s, a search of Medline (Ovid) using the term "oxygen toxicity" yielded 376 publications. A review entitled "Biochemistry of reoxygenation injury" in 1988 by Lars Ernster [33] cited 60 references and this author stated that it: "... summarizes current knowledge on the biochemistry of oxygen toxicity in general and the ischemia-reoxygenation tissue injury in particular". Primarily, toxicity of oxygen radicals was said to result in "lipid peroxidation... believed to be initiated by perferryl radical... with secondary effects... due to disturbance in cellular calcium homeostasis". Possible therapy approaches were said to include: superoxide dismutase, catalase, calcium entry blockers, iron chelators, xanthine oxidase inhibitors, and agents to prevent leukocyte adhesion. There were sections devoted to "Pathways of Generation and Removal of Oxygen Radicals". Formation of superoxide *via* xanthine oxidase was reviewed and it was said that this enzyme was absent from the human heart. It was known that NADPH oxidase, present in polymorphonuclear leukocytes and able to convert oxygen to superoxide, could be activated in endothelium by attachment of bacteria as part of the bactericidal mechanism; however the adherence of bacteria to neutrophils is not sufficient to stimulate superoxide formation in healthy endothelium. Neutrophils were found to be involved in tissue damage… "at foci that are already damaged". The author pointed out that: (1) nonenzymatic formation of superoxide could occur *via* autoxidation of semiquinones and other redox-labile compounds including some Fe^{2+}chelates; (2) quinones could undergo 1-electron reduction *via* NADH or NADPH-linked dehydrogenases in microsomes or mitochondria (*via* cytochrome linkage) with formation of superoxide; and that this source of superoxide in mitochondria was less than 2% of the total oxygen consumed *via* respiration but that this could be increased with some agents including the cancer chemotherapeutic Adriamycin; (3) "it is now evident" that superoxide can serve as an electron donor for the mobilization of iron from ferritin; (4) hydrogen peroxide formed through SOD is removed either by catalase and/or various peroxidases (primarily glutathione peroxidases); (5) most of the catalase in cells in peroxisomes; and (6) formation of hydroxyl radical may occur non-enzymatically through Fenton chemistry.

In a section titled: "Effects of Oxygen-Radical Generation on Cellular Components and Functions", two categories were described: effects caused by oxygen radicals directly on cellular components and functions and disturbances in cellular Ca^{2+} homeostasis. The latter leads to activation of cytosolic proteases and phospholipases further resulting in breakdown of cytoskeleton and activation of enzymes including xanthine oxidase and NADPH oxidase, giving rise to further oxygen radicals. It was stated that it was generally believed that lipid peroxidation was initiated by a perferryl radical with the oxidized iron bound as a chelate. It was stated that the current view of the mechanism of NADPH-linked microsomal peroxidation involved the extraction of a hydrogen atom from a lipid to create a free radical that subsequently led to formation of a lipid peroxide with a cyclic continuation of oxidation. Other mechanisms were also described whereby lipid peroxidation could be initiated by superoxide generated by hypoxanthine and xanthine oxidase plus chelated oxidized iron. The relevance of lipid peroxidation to oxidation of membrane phospholipids leading to "severe cell injury" was discussed. "Primary effects" by hydroxyl radical were also discussed, with oxidation of "... certain amino acids residues and purine and pyrimidine bases in proteins and nucleic acids, respectively, thereby catalyzing enzyme inactivation and DNA strand-break". It was noted that the former are implicated in ischemia-reoxygenation damage to brain. Considerable description was devoted to oxygen radical disturbance of Ca^{2+} homeostasis.

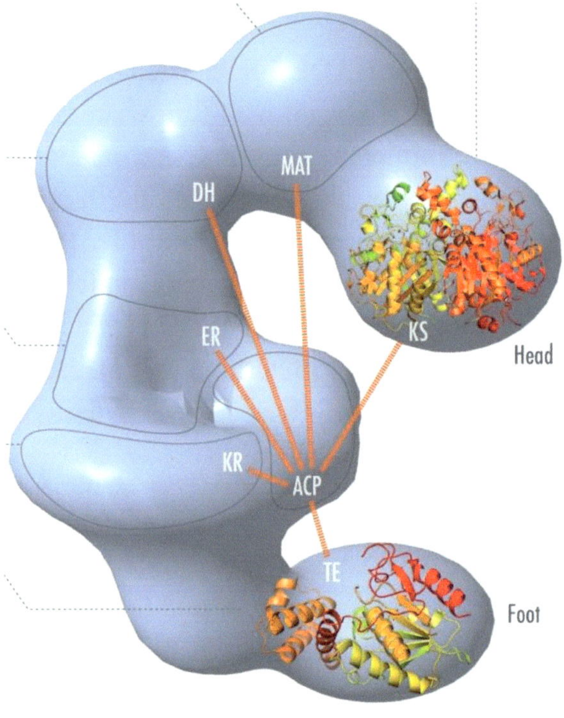

Fig. (5). Yeast fatty acid synthase complex.

This review [33] further states that oxidative stress may upset the intracellular calcium distribution by depleting GHS (reduced glutathione) with the effect of release of Ca^{2+} from the endoplasmic reticulum *via* inactivation of an ATPase plus release of calcium from mitochondria. The latter is proposed to result from oxidation of NAD transhydrogenase and activation of mitochondrial calcium efflux to increase cytosolic calcium concentration. This may be further exacerbated by inactivation of the plasma membrane calcium-ATPase (because of depletion of reduced glutathione). The increased Ca^{2+} was proposed to have significant consequences: activation of cytosolic calcium-dependent proteases and phospholipase A_2. The latter can result in damage to the cell cytoskeleton. Also, activation of phospholipase A causes release of arachidonic acid from phospholipids (primarily from membranes) creating a "cascade" that stimulates biosynthesis of prostaglandins, leukotrienes, and thromboxanes that can result in increased leukocyte adhesion to the endothelium; and when the endothelium is damaged, activation of NADH oxidase can generate additional superoxide.

In a final section on "Protective Mechanisms Against Reoxygenation Injury" the author described that cytosolic and mitochondrial SODs are located in cellular compartments where superoxide is likely to be formed, and, therefore, works in concert with catalase and peroxidases to destroy the resultant hydrogen peroxide [33]. It is pointed out, also, that phospholipase A_2 has "recently" been shown to preferentially attack peroxidized membrane phospholipids and that this may start the process necessary for repairing oxidatively damaged membranes, it is stated that "An enzyme of growing interest during the last three decade in connection with oxygen toxicity is DT diaphorase, a dimeric flavoprotein present in the cytosol of various animal tissues". This enzyme is said to be unique among NADPH-oxidizing flavoproteins because it transfers two electrons (as a quinone reductase) and, therefore, is important in preventing one-electron transfers with autoxidation of semiquinones and superoxide generation. It is stated that induction of DT diaphorase: protects against the toxicity of various xenobiotics including polycyclic hydrocarbons, polychlorinated dioxins and anthracycline antibiotics; and that in the brain DT diaphorase may protect against one-electron transfer to produce free radicals from oxygen.

A useful publication for understanding the status of research into oxidant stress defense mechanisms in the decade of the 1980s, is the paper by Joe McCord and Irwin Fridovich published in 1988: "Superoxide dismutase in the first twenty years (1968-1988)" [34]. This "Historical Perspectives" paper is recommended for insight into the research stimulated by the discovery of superoxide dismutase.

Research from my laboratory in the decade of the 1980s focused on five continuing areas of investigation relating to oxygen toxicity: (1) induction of genetic stringency, (2) niacin metabolism, (3) pyridine nucleotide biosynthesis, (4) oxidant-sensitivity of dihydroxyacid dehydratase, and (5) a minor interest in an approach to malaria chemotherapy.

Because enzymes required for the biosynthesis of certain amino acids by *Escherichia coli* were poisoned by exposure to hyperoxia, I proposed, based on work including others in my laboratory, in 1978 that the stringent genetic response would be induced [28, 35]. In 1981, Richard Seither and I published work that stringency was induced by exposure to hyperoxia and that guanosine tetraphospate (the mediator of stringency) began to accumulate within 1 minute of exposure to oxygen at 4.2 atmospheres [36]. Within 10 minutes guanosine tetraphosphate was 6-times the concentration prior to hyperoxic exposure. The stringent response occurred in the stringent strain, but not the relaxed strain of *Escherichia coli.* Decay of guanosine tetraphosphate occurred when cells recovered growth upon removal from hyperoxia and re-incubation with air as the gas phase. Stringency was induced because of inhibition of amino acid biosynthesis; thus, stringency is causal for the growth inhibition in minimal medium and also provides significant protection.

In 1981, in work done in my laboratory by Cheng-Shu Song, niacin, but not quinolinate, reduced the impairment in growth induced by exposure of *Escherichia coli* to hyperbaric oxygen, and prevented the decline in pyridine nucleotide coenzymes [37]. Both compounds are in the biosynthetic pathway for pyridine nucleotides. These findings support previous evidence that hyperbaric oxygen poisons phosphoribosyl transferase an enzyme required in both prokaryotes and eukaryotes (including humans) for *de novo* synthesis of nicotinic acid dinucleotide. Therefore, niacin could be beneficial for humans exposed therapeutically to potentially toxic oxygen concentrations.

In my laboratory, Michael Heitkamp found that when niacin was included in the culture medium it significantly reduced the growth-inhibitory effects of 1 mM paraquat for *Escherichia coli* [38]. Niacin is an intermediate in the salvage pathway for pyridine nucleotide biosynthesis, and the nicotinic acid dinucleotide

(NAD) concentration was rapidly and significantly decreased during paraquat poisoning. While niacin prevented the decline in NAD during paraquat poisoning, quinolinate (which is an NAD biosynthetic intermediate prior to niacin in the pathway) did not. Therefore, paraquat poisons the *de novo* biosynthesis of pyridine nucleotides. It was noted that similar results were seen for poisoning by hyperbaric oxygen. Additional work, published in *Science*, by Heitkamp and Cheng-Shu Song extended this finding to rats [39] and niacin was found to reduce paraquat toxicity. Deaths began approximately 30 hours later; the time required to reach LD-50 increased from 60 to 120 hours in niacin-treated rats; and the death rate was reduced from 75% to 55%.

In 1989, work in my laboratory by Richard Seither determined that paraquat inhibits NAD biosynthesis at a specific enzyme, quinolinate synthetase [40]. Thus, the oxidant-poisoned site was determined to be the initial and rate-limiting step catalyzed by the quinolinic acid synthetase complex.

Earlier, in 1980, with James Fee, A. Lees, P. L. Block, P. L Gililand, we had reported that the reversible, growth-inhibitory effects for *Escherichia coli,* of oxygen and aerobic paraquat occurred by the same mechanism: inhibition of the enzyme dihydroxyacid dehydratase which is required for the biosynthesis of branched-chain amino acids [41]. This dehydratase enzyme has an iron-sulfur structure in the active site (Fig. **6**). This site is required for activity and the iron atoms are attached (bonded) to sulfur atoms of amino acids (part of the enzyme protein structure) in the manner shown, and iron can receive electrons to change oxidation states as shown, with loss of enzyme activity and even with loss of iron that disrupts the iron-sulfur structure.

By 1983, published work in my laboratory with Richard Seither had extended findings with oxygen and paraquat to the redox-active drug nitrofurantoin [42]. Using data obtained primarily from experiments with *Escherichia coli,* sites of toxicity in common were discovered for hyperoxia, paraquat and nitrofurantoin. Sites in common (for hyperoxia and paraquat) were found to be: inhibition of biosynthesis of specific amino acids, induction of genetic stringency *via* unloaded transfer RNAs caused by amino acid deficiencies due to blocked biosynthesis, decreased thiamine content, and impaired biosynthesis of pyridine nucleotide coenzyme biosynthesis. Inhibition of specific amino acid biosynthesis and induction of stringency also was observed for nitrofurantoin. RNA and DNA biosynthesis were also impaired by hyperoxia; this was not yet tested for paraquat or nitrofurantoin; however, it was noted that a biochemical basis and preliminary data supported inhibition of NAD biosynthesis as an element of toxicity of these agents for mammals, and possibly for humans.

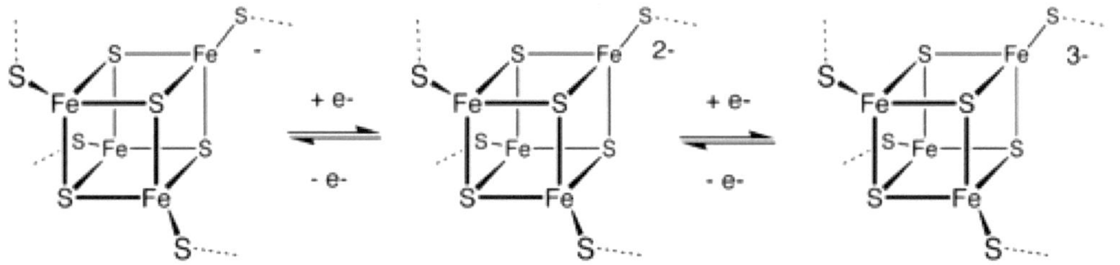

Fig. (6). The iron-sulfur cluster, found in the enzyme dihydroxyacid dehydratase which is oxidant-sensitive.

In 1987 Ted Green and I reported results of testing five nitrofurantoin derivatives (obtained from Norwich Eaton Pharmaceuticals), menadione, juglone, alloxan, paraquat, mitomycin C, and nalidixic acid for inhibition of malaria parasites *in vitro* [43]. The novel nitrofurantoin derivatives were chosen by researchers at Norwich Eaton as representative of a large number of nitrofurans they had prepared during investigations of antimalarial properties of drugs. The nitro group on these nitrofurans can accept electrons singly and can transfer electrons to form reactive oxygen species. Nitrofurantoin itself is capable of redox-cycling. Of the compounds tested, mitomycin-C was the most active; mitomycin-C is an anti-cancer agent with a quinone structure and can redox-cycle with formation of a one-electron reduction product, a semiquinone. Some of the compounds can be reduced by both one- and two-electron paths, and futile cycles can occur biologically; thus, greater than stoichiometric amounts of NADPH can be oxidized. Sources of oxidant stress in this system also included spontaneous oxidation of hemoglobin which reasonably could be accelerated in parasitized cells; conversion of oxyhemoglobin to methemoglobin (which generates superoxide) and activity of red blood cell methemoglobin reductase during regeneration of oxyhemoglobin. These results established that drugs that work *via* oxidant stress could have applications for therapy of malaria.

OXYGEN TOXICITY DISCOVERIES IN THE 1990S

A paper by Irwin Fridovich in 1998 reflected on the effect of the discovery of the enzymes known as superoxide dismutase, on the scientific understanding of oxygen toxicity [44], including the following as a quotation:

> "During its reduction to water, O_2 readily gives rise to dangerously reactive intermediates. This threat is diminished by families of defensive enzymes which include the superoxide dimutases, catalase and peroxidases. Free radical chain reactions are controlled by antioxidants, such as ascorbate and the tocopherols, and oxidative damage, which occurs in spite of these defenses, is largely repaired or is nullified by *de novo* biosynthesis. Yet some damage is sustained and it contributes to mutagenesis, to senescence and to numerous pathological processes."

A section of the paper is devoted to paramagnetism and its relationship to the univalent pathway of reduction of oxygen and Fridovich provides an exceptionally clear and instructive description [44]. The pairing of electrons with opposite spins, as in orbitals, prevents the penetration of a magnetic field and this is the "diamagnetic" state of most substances. Oxygen is unusual (is paramagnetic) because of unpaired electrons with spin sense not cancelled. Dioxygen has two unpaired electrons with the same spin state which is a barrier to acceptance of a pair of electrons because the incoming electron pair would need to join the parallel-spinning, unpaired electrons of the oxygen and one electron would have the same spin state as the electron it would be joining in an orbital (a forbidden state; forbidden by the Pauli exclusion principle). The way around this barrier is to add (have oxygen accept) electrons singly which "works' because spin states of electrons can be inverted by interaction with nuclear spins. This is a slow process at the atomic level, compared to other relevant processes including the lifetime of molecular collisions; however, when electrons are added one at a time there is time between collisions for the inversions of electron spin states, and this provides a route for reduction of dioxygen *via* a series of univalent steps.

There is a section describing the intermediates of the univalent pathway, including the hydroxyl radical. Fridovich stated:

> "It is sobering to contemplate the ease with which (hydroxy radical) can be produced... (Hydroxy radical) is an extraordinarily powerful oxidant, which attacks most organic compounds at diffusion-limited rates."

Hydroxy radical can be produced by the reduction of hydrogen peroxide by metal cations such as iron 2+ or copper 1+. Fridovich points out that the superoxide radical is less (more selective) in reactivity than hydroxy radical. He states that, paradoxically, this makes it biologically more dangerous/damaging, and explains as follows:

> "Hydroxy radical will react with something, perhaps something expendable, within a very small radius of its site of generation, whereas (superoxide) can diffuse a considerable distance before it encounters a suitable, and possibly critical, target."

I have expressed this elsewhere in this book and agree; however, it must be kept in mind that there is a large difference in reactivity between hydroxy radical and superoxide radical.

There are very instructive sections describing defenses against the superoxide radical (CuZnSODs, MnSODs, FeSODs, and NiSOD– which we have not previously mentioned; an unusual SOD in the microbe *Streptomyces griseus*); and defenses against hydrogen peroxide (catalases and peroxidases). I recall that my major professor for my graduate studies at OU, J. B. Clark, was investigating the mechanism by which this microbe was, somewhat miraculously, able to resurrect itself after what should have been a lethal dose of ionizing radiation by exposure to visible light. It is speculation, but I wonder if superoxide had something to do with this. It is also well-known that there is "purposeful" production of superoxide, including in phagocytic white blood cells during what is termed "the respiratory burst" following engulfment and destruction of bacteria (Fig. 7). In this process a membrane-bound NADPH oxidase (enzyme) is activated and dioxygen is reduced to superoxide and oxidatively attacks engulfed bacteria.

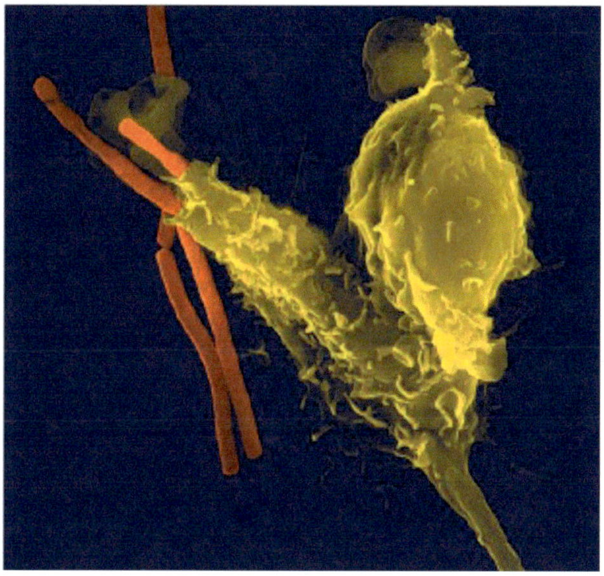

Fig. (7). Phagocytic white blood cell engulfing anthrax bacilli (red rods).

There is a section of "Defense in depth- the soxRS regulon" in the review [44]. I have not previously addressed this subject. The soxRS regulon in *Escherichia coli,* when the microbe is under oxidant stress, is activated and approximately a dozen genes are "turned on" that produce an in-depth defense. The oxidant sensor is the SoxR protein. It has a functional group, 2F-2S, that is readily oxidized and reduced and in its oxidized state it is the signal that turns on the soxS gene which activates all the other genes in the regulon (which includes genes for MnSOD, glucose phosphate dehydrogenase- to supply NADPH for glutathione reductase and alkyl hydroperoxide reductase; fumarase C that serves as a stable replacement for fumarases A and B- that are readily inactivated by superoxide; aconitase A- also oxygen sensitive; endonuclease IV– which repairs oxidatively damaged DNA; ferredoxin-flavodoxin reductase that functions in reductive reactivation of fumarases A and B, aconitases A and B, and the other (4-Fe-4S)-containing dehydratases that are oxygen-sensitive. I shall address the latter in relation to my own work during this period.

Discoveries in my laboratory in the 1990s involved research with *Escherichia coli* and with rats exposed to hyperbaric oxygen and to aerobic paraquat, with a focus on intracellular sites and mechanisms of toxicity and cellular defenses. Research showed that plasmid-containing strains that overproduced Mn-SOD by 4- and 5-fold, and iron-SOD 7-fold, were not more resistant than parent strains to 1 mM paraquat (a known generator of superoxide) when tested by the consequences measured as survival, growth inhibition, or induction of stringency. This showed that SOD did not mitigate the effects of paraquat for this microbe under the test conditions; thus, other factors dominated [45].

In a study involving various selected strains of *Escherichia coli*, various nutritional conditions, with and without chloramphenicol (a known inducer of genetic stringency), and other manipulations to induce or permit stringency, the effects of 4.2 atmospheres of oxygen were quantified for: net synthesis of protein, lipid, RNA, and DNA [46]. Due to stringency induction as the toxic mechanism, net lipid synthesis failed indirectly. Protein biosynthesis declined due to stringency and due to lack of ability to synthesize specific amino acids. RNA biosynthesis was impaired from a direct toxic effect and indirectly due to stringency, and DNA biosynthesis was indirectly impaired.

Significant, collaborative research with Dennis Flint, and Joseph Tuminello, and Elizabeth Smyk-Randall and Bozena Draczyska-Lusiak in my laboratory, provided new information about the oxygen sensitivity mechanism for dihydroxyacid dehydratase, an enzyme containing the iron-sulfur (Fe-S) cluster [47]. It had been known for about 20-years that enzymes containing the Fe-S cluster participated in redox-reactions, and more recently that several enzymes denoted as hydro-lyases (including mammalian aconitase and spinach dihydroxy-acid dehydratase), *Escherichia coli* dihydroxy-acid dehydratase and *Escherichia coli* fumarase A and fumarase B contain the Fe-S cluster (with the Fe-S cluster essential for enzyme activity). *Escherichia coli* dihydroxyacid dehydratase functions in branched-chain amino acid biosynthesis by catalyzing the conversion of 2,3-dihydroxyisovaleric acid and 2,3-diydroxy-3-methylvaleric acid to the corresponding 2-keto acids. We showed that oxygen inactivation was due to the destruction of the (4Fe-4S) cluster (Fig. **8**). The inactivated protein remained intact and activity recovered after cells whose growth was stopped by hyperoxia, were reincubated with air as the phase; thus, the enzyme was reactivated. The enzyme can cycle between forms in which the (4Fe-4S) cluster is either present or absent which is compatible with its regulatory role in addition to the catalytic role.

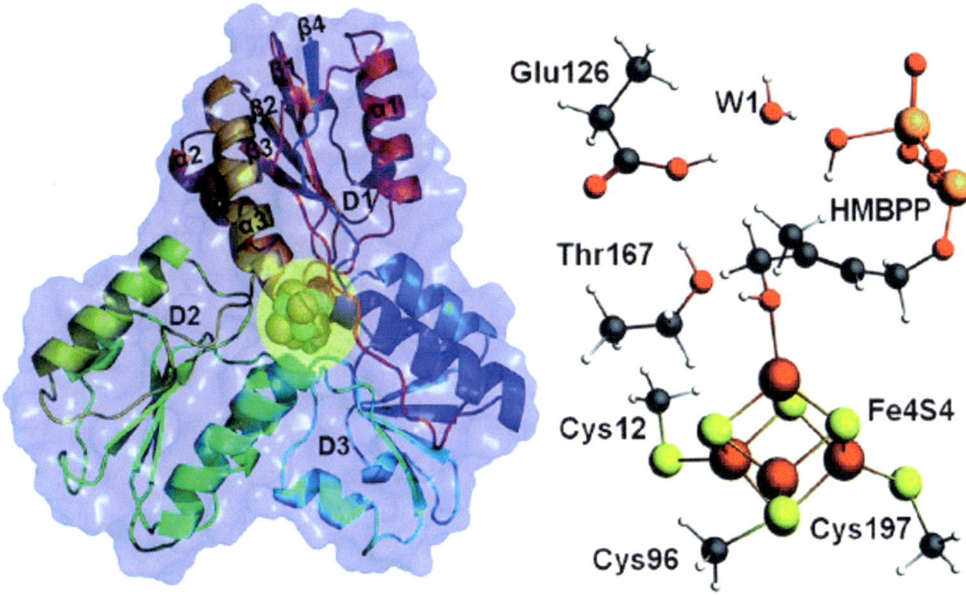

Fig. (8). Ribbon diagram of 4Fe-4S4 cluster-containing dihydroxyacid dehydratase enzyme (left) with the 4Fe-4S cluster in the center in light green, left panel, and relationships of the protein sulfur containing amino acids to the 4Fe-4S cluster (right).

OXYGEN TOXICITY DISCOVERIES IN THE 2000S

E. P. Manning reviewed the topic of central nervous system oxygen toxicity and hyperbaric seizures in 2016 [48]. The authors pointed out that currently (in 2016) the use of hyperbaric oxygen as therapy was accompanied by risk of CNS toxicity; that increased partial pressures of oxygen "appears" to saturate protective enzymes and "unfavorably shift protective reactions in the direction of neural network overstimulation"; "central regions of the CNS appear more susceptible"; and "failure to decrease the elevated pO_2 can result in a tonic-clonic seizure and death". The author called for further study. This dramatically indicates that one of the most toxic, therapeutic consequences of hyperbaric oxygen (seizures and possible death) remains unresolved.

In 2013, Frederick E. Domann [49] wrote the published editorial: "Aberrant free radical biology is a unifying theme in the etiology and pathogenesis of major human diseases". He connected what he termed "the seemingly disparate areas of oxygen toxicity, radiation exposure, and aging..." with the "... aberrant production and/or removal of biologically derived free radicals and other reactive oxygen and nitrogen species". The author cited 24 articles, including 10 review articles, from authors in 11 different countries, and identified connections with reactive oxygen and nitrogen species for the following pathologies: vascular diseases, diabetes, neurodegenerative disorders, lymphoma, esophageal squamous cell cancer, pulmonary and cerebral vascular endothelial dysfunctions, lupus, epilepsy, and chronic regional pain syndrome. The topics most studied were: mitochondrial function, oxidative stress, nitric oxide, and cellular signaling. This report brought attention to more continuing problems than answers, and called for more research.

The article: "Two faces of nitric oxide: implications for cellular mechanisms of oxygen toxicity", published in 2009 [50] (but first appearing in print on-line in 2008) summarized as follows: "It is now emerging... that the anatomical localization of NO release, which depends, in part, on whether the oxygen exposure is normobaric or hyperbaric, strongly influences whether toxicity emerges and what form it takes". Topics well-covered included: Reactive Oxygen Species and NO; NO and Cerebral Blood Flow; Cerebral NO Levels in HBO_2, SOD Regulates Cerebral NO, Sources of the NO that Modulate Cerebral Vascular Tone, Role of nNOS and Peroxynitrite in HBO_2, Normobaric *vs.* HBO_2, and Summary and Future Directions. It is significant that these authors concluded:

> "A new paradigm is emerging in place of the 19th century dichotomy in which oxygen toxicity is manifested as either a pulmonary insult in normobaric exposures, the Lorraine Smith effect or CNS injury in hyperbaric exposures, the Paul Bert effect. It is now apparent that the effects of oxygen on lung and brain are part of a pathophysiological continuum in which a common factor is NO and its interactions with other reactive species. Thus, just as acute lung injury is now known to occur in HBO_2 exposures along with CNS effects, it seems likely that further work will demonstrate that subtle CNS effects occur in

> normobaric hyperoxia, along with pulmonary injury… The fact that NO production may either exacerbate or mitigate the toxic effects of oxygen… affords new opportunities for pharmacological interventions... precisely targeted interventions… in the NO signaling systems could increase the safety and efficacy of hyperbaric therapy and improve the operational capabilities of military, commercial, and recreational divers."

An earlier review in 2004: "Hyperbaric oxygen: its uses, mechanisms of action and outcomes" focused on: "History of hyperbaric medicine, physiological basis of hyperbaric oxygen therapy", and included an extensive section on "Indications and uses for hyperbaric oxygen therapy". These applications included: arterial gas embolism, carbon monoxide poisoning, Clostridial myositis and myonecrosis (gas gangrene), crush injuries, compartment syndromes and other acute traumatic peripheral ischemias, decompression sickness, enhancement of healing in selected problem wounds, exceptional blood-loss anemia, intracranial abscess, necrotizing soft-tissue infections, refractory osteomyelitis, delayed radiation injury (soft tissue and bony necrosis), skin flap and grafts (compromised), and thermal burns. The length and diversity of this list indicates the progress that is being made but when the problems and successes are compared, it indicates the extent to which future research is needed. The authors concluded (in part):

> "HBO is a relatively safe treatment, but does carry some risks, due to the increased pressure and hyperoxia. The commonest effect of oxygen toxicity is a progressive, reversible myopia, thought to be due to physical lens deformation … CNS toxicity may occur… but the Undersea and Hyperbaric Medical Society (UHMS) feel this is not justified within well-defined oxygen tolerance limits… Middle ear and sinus barotraumas are preventable by equalization techniques… There have been some concerns that HBO could stimulate malignant growth by increasing tumour oxygenation. This was not supported… The only absolute contraindication to HBO is an untreated tension pneumothorax… relative contraindications include impaired pressure

equalization, and cardiac disease... Accidents are a risk due to the enriched oxygen and inaccessibility, with over 50 reported deaths due to fire in the last 20 years."

A 2015 review addressed mitochondrial-targeted antioxidants [51]. The authors stated that: "Mitochondria-localized antioxidants are widely studied because the mitochondria, the major producers of intracellular ROS, have been linked to the cause of aging and other chronic diseases". I have addressed this at some length. They discuss two agents: MitoQ (a synthesized derivative of ubiquinone) chemically conjugated to a lipophilic agent, and tiron (a non-toxic iron chelator that penetrates the mitochondrial membrane and has anti-oxidant capabilities). Conclusions included: "Although a growing body of evidence has demonstrated the beneficial effects of exogenous antioxidants, in particular mitochondria-targeted antioxidants in human and animal skin models (refs. provided), controversy still surrounds the use of these agents to prevent or delay age-associated chronic diseases and UVR-mediated oxidative damage".

A 2014 paper [52] assessed: "Interplay between oxygen and Fe-S cluster biogenesis: insights from the Suf pathway". I have previously described research from my laboratory and other work that brought to focus the significance of the iron-sulfur cluster in various enzymes as sites of oxidant toxicity. This paper, a recent assessment, makes the significant points that: iron-sulfur cluster metalloproteins have essential cellular functions in essentially all forms of life; the iron-sulfur cluster is sensitive to oxidation by oxygen; and this can result in enzyme inactivation when the cluster is part of the active site of an enzyme, or oxidation can be an intracellular signal (sensor) that triggers a chain of responses.

Fig. (**9**) shows the interrelationships of many of the intracellular sites and mechanism that I have discussed in this chapter. Note the involvement of metals, mitochondrial cytochrome "spillage" of oxygen radicals, impaired amino acid biosynthesis, the Fenton reaction and DNA damage, lipid peroxidation and a central role for free radical oxidative damage to the iron-sulfur cluster of specific enzymes. Oxygen radical inactivation of specific enzymes is a significant oxidant stress mechanism of damage and the (4Fe-4S) cluster is both a site of damage to important biosynthetic enzymes and also serves as a sensitive link to genetic induction of protective mechanisms.

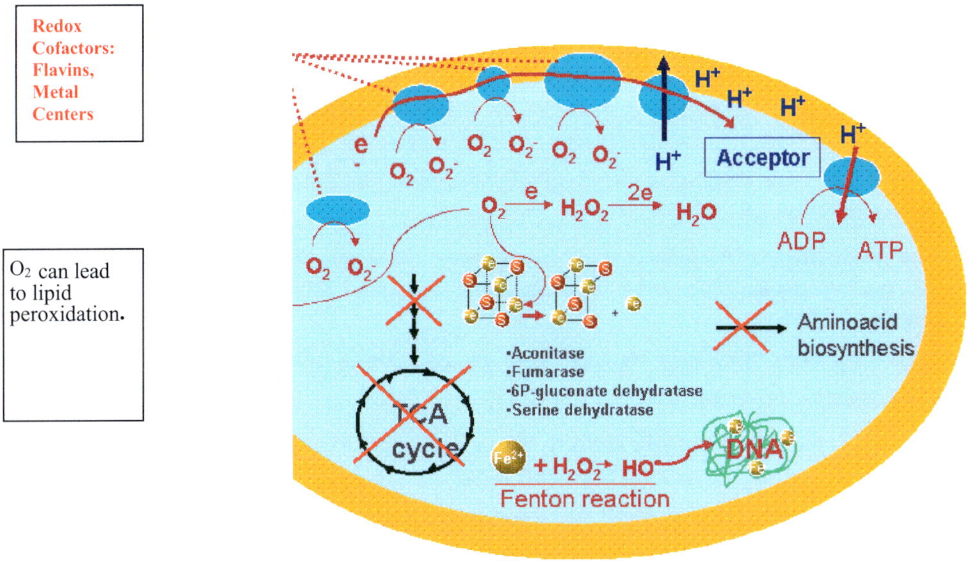

Fig. (9). Oxygen toxicity, a chain of responses with a central role for the oxidant-sensitive (4Fe-4S) cluster.

NOTES

[1] Alexander Pope (1688-1744) *Essay on Criticism.*

REFERENCES

[1] Louis-Jacques Thenard. NNDB Tracking the Entire World. Available from: http://www.nndb.com/people/898/000100598/ 2014.

[2] Loew O. A new enzyme of general occurrence in organismis. Science 1900; 11(279): 701-2. Available from: http://www.ncbi.nlm.nih.gov/pubmed/17751716.

[3] Chelikani P, Fita I, Loewen PC. Diversity of structures and properties among catalases. Cell Mol Life Sci 2004; 61(2): 192-208. [http://dx.doi.org/10.1007/s00018-003-3206-5]

[4] Goodsell DS. Catalase. RCSB Protein Data Bank. Available from: http://pdb101.rcsb.org/motm/57 2004.

[5] Yabuki A, Iwashita Y. Superoxide dismutase combined with a poly(alkylene oxide). US Patent EP0200467 A3, 1986.

[6] Hall ED. Overview of Therapeutics for TBI Available from: https://www.nationalacademies.org/hmd/~/media/ Files/ Activity%20Files/

Nutrition/TBINutrition/Hall%20Ed.pdf.

[7] Keilin D. Hemocuprein and hepatocuprein - Cu protein compounds of blood and liver in mammals. Proc Roy Soc 1938; 126: 303-15. [http://dx.doi.org/10.1098/rspb.1938.0058]

[8] McNaught A, Wilkinson A. Compendium of Chemical Terminology. 2nd ed., International Union of Pure and Applied Chemists 1997.

[9] McCord JM, Fridovich I. Superoxide dismutase an enzymic function for erythrocuprein Hemocuprein 244(22): 6049-65.

[10] Schnell DM, Clair DS. Redox pioneer: Professor Joe M McCord antioxid redox signal Mary Ann Liebert, Inc 2014; 20(1): 183-8.

[11] Brown OR, Huggett DO. Effects in hyperoxia upon microorganisms. I. Membrane culture technique for exposing cells directly to test atmospheres. Appl Microbiol 1968; 16(3): 476-9. [PMID: 4967757]

[12] Brown OR, Silverberg RG, Huggett DO. Synergism between hyperoxia and antibiotics for *Pseudomonas aeruginosa*. Appl Microbiol 1968; 16(2): 260-2.

[13] Brown OR, Stees JL, Mills DF, Davis R, Major S. Killing kinetics of *Escherichia coli* in a carbon dioxide deficient, pure oxygen environment. Microbios 1969; 3: 267-72.

[14] Babior BM, Kipnes RS, Curnutte JT. Biological defense mechanisms. The production by leukocytes of superoxide, a potential bactericidal agent. J Clin Invest 1973; 52(3): 741-4. [http://dx.doi.org/10.1172/JCI107236] [PMID: 4346473]

[15] Babior BM. Oxygen-dependent microbial killing by phagocytes (first of two parts). N Engl J Med 1978; 298(12): 659-68.

[16] Curnutte JT, Whitten DM, Babior BM. Defective superoxide production by granulocytes from patients with chronic granulomatous disease. Massachusetts Medical Society. Available from: http://dx.doi.org/101056/NEJM197403142901104 2010.

[17] McCord J. Superoxide radical as a phagocyte-produced chemical mediator of inflammation. Advances in Inflammation Research. New York: Raven Press 1979; pp. 273-80.

[18] Salin ML, McCord JM. Superoxide dismutases in polymorphonuclear leukocytes. J Clin Invest 1974; 54(4): 1005-9.

[19] Salin ML, McCord JM. Free radicals and inflammation. Protection of phagocytosine leukocytes by superoxide dismutase. J Clin Invest 1975; 56(5): 1319-23.

[20] Gao B, Flores SC, Leff JA, Bose SK, McCord JM. Synthesis and anti-inflammatory activity of a chimeric recombinant superoxide dismutase: SOD2/3. Am J Physiol Lung Cell Mol Physiol 2003; 284(6): 917-25.

[21] Hernandez-Saavedra D, Zhou H, McCord JM. Anti-inflammatory properties of a chimeric recombinant superoxide dismutase: SOD2/3. Biomed Pharmacother 2005; 59(4): 204-8. [http://dx.doi.org/10.1016/j.biopha.2005.03.001] [PMID: 15862716]

[22] Keele BB, McCord JM, Fridovich I. Superoxide dismutase from *Escherichia coli* B. A new manganese-containing enzyme. J Biol Chem 1970; 245(22): 6176-81.

[23] McCord JM, Keele BB, Fridovich I. An enzyme-based theory of obligate anaerobiosis: the physiological function of superoxide dismutase. Proc Natl Acad Sci USA 1971; 68(5): 1024-7. [http://dx.doi.org/10.1073/pnas.68.5.1024]

[24] Chance B, Sies H, Boveris A. Hydroperoxide metabolism in mammalian organs. Physiol Rev 1979; 59(3): 527-605. [PMID: 37532]

[25] Brown OR, Yein F. Sensitivity to and site of oxygen poisoning in *Escherichia coli*. Am J Clin Nutr. American Society for Nutrition 1979; 32(1): 267-7.

[26] Brunker RL, Brown OR. Effects of hyperoxia on oxidized and reduced NAD and NADP concentrations in *Escherichia coli*. Microbios 1971; 4(15): 193-203.

[27] Brown OR, Yein F, Boehme D, Foudin L, Cheng Shu Song CS. Oxygen poisoning of NAD biosynthesis: A proposed site of cellular oxygen toxicity. Biochem Biophys Res Commun 1979; 91(3): 982-90. [http://dx.doi.org/10.1016/0006-291X(79)91976-4]

[28] Brown OR, Yein F. Dihydroxyacid dehydratase: The site of hyperbaric oxygen poisoning in branch-chain amino acid biosynthesis. Biochem Biophys Res Commun 1978; 85(3): 1219-24.

[29] Brown OR. Resistance of oxidative phosphorylation in *Escherichia coli* to hyperoxia. J Bioenerg 1971; 2(3): 217-20.

[30] Mathis RR, Brown OR. ATP concentration in *Escherichia coli* during oxygen toxicity. Biochim Biophys Acta - Bioenerg 1976; 440(3): 723-32. [http://dx.doi.org/10.1016/0005-2728(76)90054-2]

[31] Brown OR, Howitt HF, Stees JL, Platner WS. Effects of hyperoxia on composition and rate of synthesis of fatty acids in *Escherichia coli*. J Lipid Res 1971; 12(6): 692-8.

[32] Stees JL, Brown OR. Stability of yeast fatty acid synthetase component enzymes to irreversible inactivation by hyperbaric oxygen. Microbios 2016; 8(31): 247-56.

[33] Ernster L. Biochemistry of reoxygenation injury. Crit Care Med 1988; 16(10): 947-53. [http://dx.doi.org/10.1097/00003246-198810000-00005] [PMID: 2844473]

[34] McCord JM, Fridovich I. Superoxide dismutase: the first twenty years (1968-1988). Free Radic Biol Med 1988; 5(5-6): 363-9.

[35] Brown OR, Yein R, Boehme D. The Red Cell. New York: Alan R. Liss, Inc 1978; 21: pp. 701-14.

[36] Seither RL, Brown OR. Induction of stringency by hyperoxia in *Escherichia coli*. Cell Mol Biol 1982; 28(3): 285-91.

[37] Brown OR, Song C-S. Pyridine nucleotide coenzyme biosynthesis: A cellular site of oxygen toxicity. Biochem Biophys Res Commun 1980; 93(1): 172-8. [http://dx.doi.org/10.1016/S0006-291X(80)80262-2]

[38] Heitkamp M, Brown O. Inhibition of NAD biosynthesis by paraquat in *Escherichia coli*. Biochim Biophys Acta - Gen Subj 1981; 676(3): 345-9. [http://dx.doi.org/10.1016/0304-4165(81)90169-0]

[39] Brown OR, Heitkamp M, Song CS. Niacin reduces paraquat toxicity in rats. Science 1981; 212(4502): 1510-2.
[http://dx.doi.org/10.1126/science.7233236]

[40] Brown OR, Seither RL. Paraquat inhibits NAD biosynthesis at the quinolinic acid synthetase site. Mdical Sci Res 1989; 17: 819-20.

[41] Fee JA, Lees AC, Block PL, Gilliland PL, Brown OR. Oxygen stasis of bacterial growth: Analogy between the stasis of *E. coli* by hyperbaric oxygen and by aerobic paraquat. Biochem Int 1980; 1(4): 304-11.

[42] Brown OR, Seither RL. Oxygen and redox-active drugs: shared toxicity sites. Fundam Appl Toxicol 2016; 3(4): 209-14.

[43] Brown OR, Green TJ. Oxidant-stress: a possible approach to malaria chemotherapy. Med Sci Res 1987; 15: 563-5.

[44] Fridovich I. Oxygen toxicity: a radical explanation. J Exp Biol 1998; 201(Pt 8): 1203-9.

[45] Siwecki G, Brown OR. Overproduction of superoxide dismutase does not protect *Escherichia coli* from stringency-induced growth inhibition by 1mM paraquat. Biochem Int 1990; 20(1): 191-9.

[46] Brown OR. Mechanisms of hyperbaric-oxygen inhibition of growth and net biosynthesis of RNA, DNA, protein and lipids in *Escherichia coli*. Microbios 1990; 64(260-261): 135-51.

[47] Flint DH, Smyk-Randall E, Tuminello JF, Draczynska-Lusiak B, Brown OR. The inactivation of dihydroxy-acid dehydratase in *Escherichia coli* treated with hyperbaric oxygen occurs because of the destruction of its Fe-S cluster, but the enzyme remains in the cell in a form that can be reactivated. J Biol Chem 1993; 268(34): 25547-52.

[48] Manning EP. Central nervous system oxygen toxicity and hyperbaric oxygen seizures. Aerosp Med Hum Perform 2016; 87(5): 477-86.
[http://dx.doi.org/10.3357/AMHP.4463.2016]

[49] Domann FE. Aberrant free radical biology is a unifying theme in the etiology and pathogenesis of major human diseases. Int J Mol Sci 2013; 14(4): 8491-5.
[http://dx.doi.org/10.3390/ijms14048491]

[50] Allen BW, Demchenko IT, Piantadosi CA. Two faces of nitric oxide: implications for cellular mechanisms of oxygen toxicity. J Appl Physiol 2009; 106(2): 662-7.

[51] Oyewole AO, Birch-Machin MA. Mitochondria-targeted antioxidants. FASEB J 2015; 29(12): 4766-71.
[http://dx.doi.org/10.1096/fj.15-275404]

[52] Boyd ES, Thomas KM, Dai Y, Boyd JM, Outten FW. Interplay between oxygen and Fe-S cluster biogenesis: insights from the Suf pathway. Biochemistry 2014; 53(37): 5834-47.

CHAPTER 8

Oxygen Transport to Lung to Blood to Spark of Intelligence in the Brain

Abstract: The human body contains coordinated, complex components that work together to efficiently extract oxygen from the air and deliver it to the intracellular sites where it is essential for life. Oxygen's journey for life begins in the lung where it is transported by a system of branching tubes of decreasing size, the bronchus, bronchioles and bronchi. The lungs are made of a wonderful collection of cells that are organized into a unique, spongy tissue that contains the bronchial tubes with air sacs, the alveoli, at their termini. The interior surface area of the alveoli, collectively, is equal to the area of a tennis court. An intricate system of blood capillaries, arteries, and veins envelops each alveolus and circulates a large supply of blood that arrives deoxygenated and leaves oxygenated. The lung works efficiently by using air pressure and diaphragm muscles to expand and decrease lung volume during breathing. Oxygen is not sufficiently soluble in blood to supply the body's needs. However, a complex oxygen-carrying molecule, hemoglobin, is packaged into special cells, the red corpuscles where truly magnificent biochemistry reigns. The process works, by analogy, like a subway system to load and unload oxygen and carbon dioxide from alveoli to cells throughout the body. Hemoglobin does double duty and reverses the sequence by loading carbon dioxide from tissues and unloading it in alveoli. The heart supplies the pumping force that moves the red cells which arrive at ever smaller arteries and then to capillaries which intercalate with tissues that must have a continuous supply of oxygen to function. All cells require oxygen to supply the bulk of their energy needs, and some for other purposes, but the brain is especially oxygen hungry and uses approximately 25% of the total resting needs by the body for oxygen. Working muscle is provided with hemoglobin's cousin, myoglobin, which can store oxygen and release it to contracting muscle cells. Oxygen is nearly perfect for bioenergetics and it also has the right stuff to be efficiently transported.

Olen R. Brown

Keywords: Alveolus, Anoxia, Aorta, Asthma, Atelectasis, Bronchial tubes, Glyceraldehyde phosphate, Hemoglobin, Hypoxic, Myoglobin, Oxygen, Oxygen radicals, Plasma oxygen concentration, Red corpuscles, Regulation, Respiration, Surfactant, Vena cava.

INTRODUCTION

Oxygen is ideally suited for cellular respiration. No other molecule has the unique set of characteristics that provide the balance that allows oxygen to be transported to cells, to safely (for the most part) participate in oxidation-reduction reactions, and to capture and remove waste carbon as CO_2. Chlorine and fluorine are small, similar, gaseous molecules but they poison biological systems. Oxygen, as a diatomic molecule is generated by photosynthesis and maintains a steady concentration in the atmosphere. The lungs of multicellular land animals remove it efficiently from air and the gills of aqueous creatures extract it effectively from water. The energy required for this is a manageable loss. Oxygen is insufficiently soluble in water, but hemoglobin and myoglobin transport it efficiently, with provisions to attract oxygen in the lungs, and release it in tissues. The circulatory system can move oxygen, when it is safely bound to hemoglobin packaged in red cells, about the body and deliver it to cells. At the intracellular level, oxygen has the properties of "biological dynamite"; safe enough to use, but capable of a controlled, "cold" burning (oxidation) of fuel to provide (as ATP) the energy needs of cells. At the smallest subatomic scale, the electron arrangement of oxygen permits efficient flow of electrons to sustain energetics with production of water and CO_2 as safe, biologically non-polluting, waste products. The above is an oversimplification but establishes that oxygen, and no other atom or molecule, is capable of sustaining life as we know it on earth.

THE LUNG

Oxygen begins its journey to sustain biological respiration in land animals including humans as an inspired gas from air. The circumstances are initially different for creatures that live in water, but we will focus on air-breathing; the subsequent biochemical processes are similar for all multicellular aerobic creatures. The human lung is efficient; respiration is driven by a simple movement of diaphragm muscles that expand the chest cavity and air pressure does the rest to fill the lungs. Relaxation of the diaphragm muscles expels the air efficiently (Fig. **1**). Oxygen, a low molecular weight gas, is respired efficiently. Its air concentration (approximately 21%) allows efficient metabolism; slightly lower oxygen concentrations significantly impair metabolism and at approximately 2- to 3-times the air concentration, oxygen is toxic. Also, many substances would spontaneously combust or burn at room temperature if air were pure oxygen.

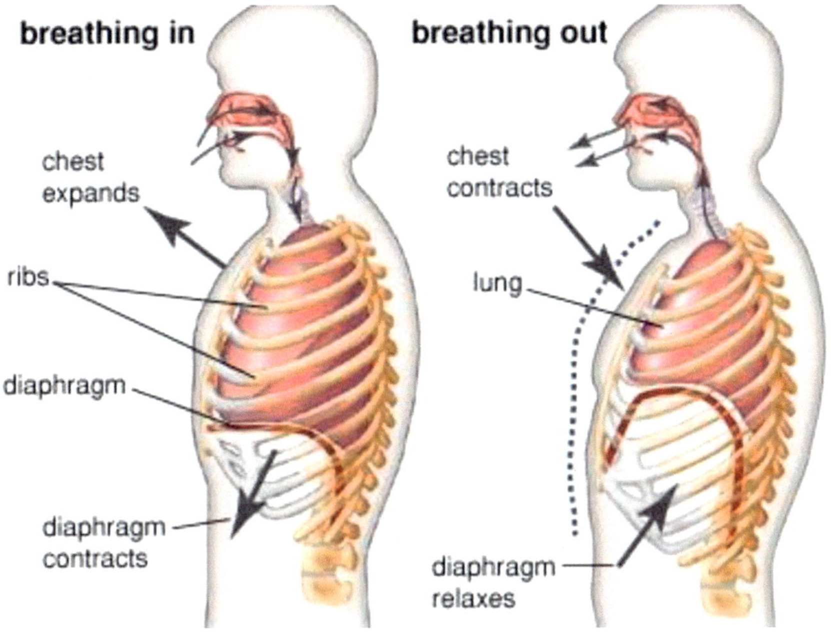

Fig. (1). The mechanism of human respiration.

During human respiration, air is drawn in down a series of branching tubes that decrease in size to the level of the alveolus in the lung (Fig. (**2**), left panel). A debilitating disease, asthma, occurs when bronchioles become inflamed and swollen with narrowing of the passageway for oxygen transport (Fig. (**2**), right panel). The inflammation may involve oxidant stress mechanisms.

Relationships among anatomical features of the respiratory system are detailed in Fig. (**3**) on the following page. Inspired air arrives at the termini of the smallest, branching bronchioles and enters small sacs, the alveoli, which are shown in greater detail in Fig. (**4**). Oxygen and carbon dioxide are exchanged in the alveoli. The epithelium of the alveolus is coated with "surfactant", a substance that regulates surface tension to keep the very small alveoli from collapsing [1]. Pulmonary surfactant is a complex mixture of lipids and associated proteins that provides a stable film within the alveolus [1]. Alveolar capillaries are situated near the alveolar epithelium, beneath the surfactant film. Machiko Ikegami and co-workers hypothesized that the shape of alveolar capillaries, and thus, their oxygenation, was affected by surfactant surface tension in the alveolus [1]. They tested the thesis by altering the surface tension by regulating conditional expression of a surfactant protein in mice which inhibited the surface-tension lowering properties of the surfactant. They documented that increased surfactant tension resulted in narrowing of alveolar capillaries and oxygen desaturation. Loss of the surfactant or damage to it also affects phagocytic activity of alveolar macrophages *in vivo* [2]. Loss of effective surfactant functioning is relevant to malfunctioning of the lungs of neonates and of adults with respiratory distress syndrome.

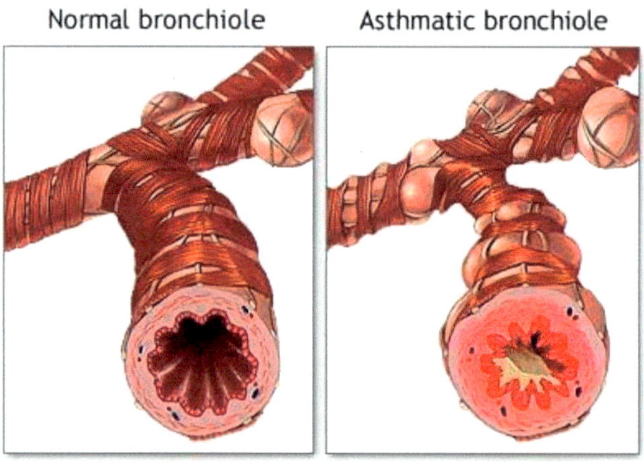

Fig. (2). Healthy *vs.* diseased bronchiole.

This surfactant is subject to damage from oxygen stress and is one sign of oxygen toxicity when hyperbaric therapy is administered. Damage to the surfactant can result in "atelectasis" from oxygen therapy at even relatively low partial pressures, especially in the newborn. Extensive capillary networks connect the arterial and venous blood supply to alveoli.

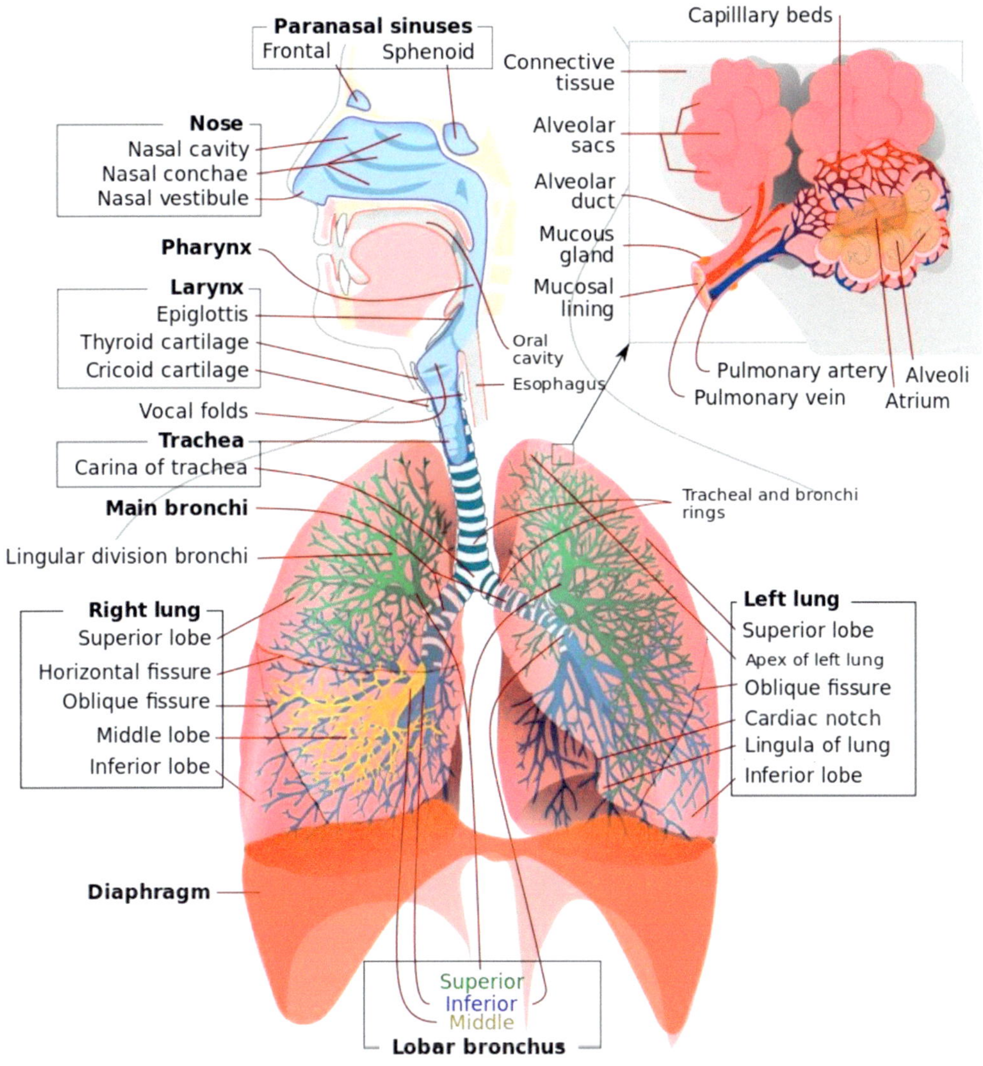

Fig. (3). The respiratory tree with details of alveoli (inset).

Alveoli, in addition to supplying oxygen to oxygenate red blood cells, extract carbon dioxide that diffuses into the exhaled air. Oxygen enters at approximately 21% and the exhaled gas has approximately 15%. Carbon dioxide is inhaled at approximately 0.04 (today's estimated, average global concentration, see Chapter 4) and is exhaled at approximately 4.5% (variable). The amount of oxygen required by a human depends greatly on the activity but 0.8 kg of oxygen (1.5 x 10^{25} molecules) daily is a reasonable calculated amount based on inspired air volume, the amount of oxygen extracted from the air and the molecular weight of oxygen.

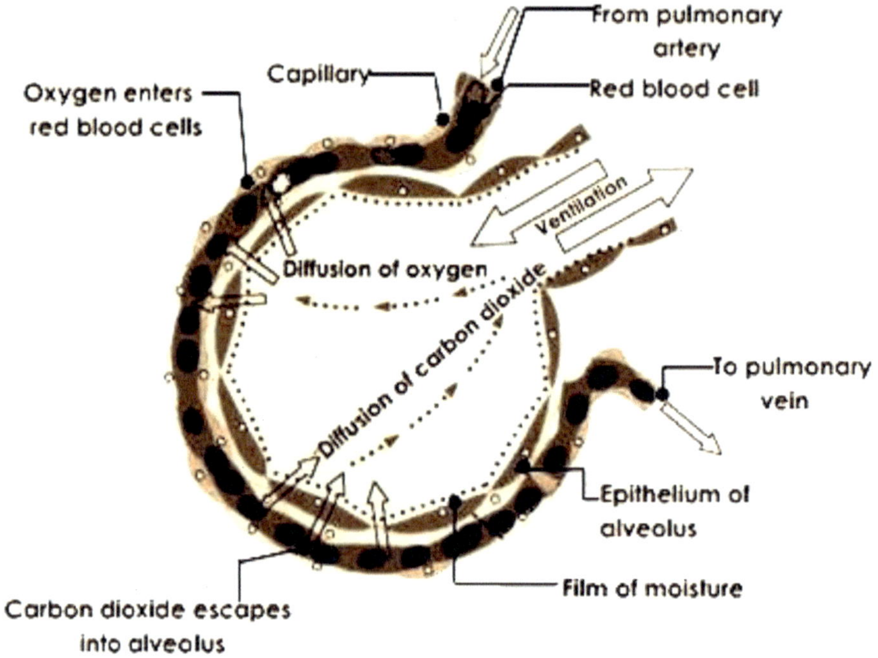

Fig. (4). Details of alveolar gas exchange in the lung.

The human possesses two lungs with a total capacity of about 4 to 6 liters. The left lung is divided into two lobes (superior and inferior), and the right lung into three lobes (superior, middle and inferior) (Fig. **3**). In total, the lungs contain approximately 1500 miles of airways and 300 to 500 million alveoli with a surface area commonly stated to be the size of one side of a tennis court. It is commonly said that if all the capillaries surrounding the alveoli were stretched out they would have a length of about 620 miles [3].

The U. S. Occupational Health and Safety Administration (OSHA) states: "Paragraph (d)(2)(iii) of the Respiratory Protection Standard considers any atmosphere with an oxygen level below 19.5% to be oxygen-deficient and immediately dangerous to life or health" [4]. OSHA also states the following: "At concentrations of 16 to 19.5 percent, workers engaged in any form of exertion can rapidly become symptomatic as their tissues fail to obtain the oxygen necessary to function properly… Concentrations of 12 to 16 percent oxygen cause tachypenea (increased breathing rate), tachycardia (accelerated heartbeat), and impaired attention, thinking, and coordination… even in people who are resting… At oxygen levels 10 to 14 percent, faulty judgment, intermittent respiration, and exhaustion can be expected even with minimal exertion… Breathing air containing 6 to 10 percent oxygen results in nausea, vomiting, lethargic movements, and perhaps unconsciousness. Breathing air containing less than 6 percent oxygen produces convulsions, then apnea (cessation of breathing), followed by cardiac standstill". [4].

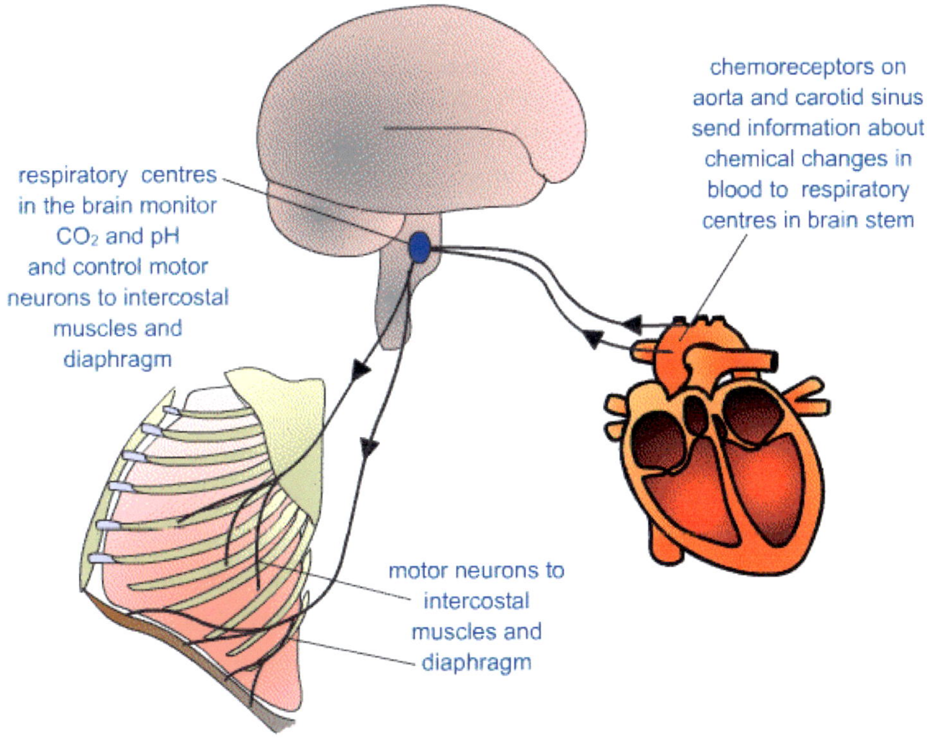

Fig. (5). The brain, heart and lung are intricately connected to regulate gas exchange in the lung.

Respiratory centers in the brain connect to neurons in the intercostal muscles and the diaphragm, and to chemoreceptors in the aorta and carotid sinus and chemical signals about blood changes are monitored and used to control and regulate respiration and match it with blood flow for optimum efficiency over a wide demand range (Fig. **5**).

THE CIRCULATION

Arterial (oxygenated) blood on its way to tissues, and venous (deoxygenated) blood returning from tissues is connected by a system of capillaries where off-loading to tissues and uptake of oxygen from lung occurs (Fig. **6**).

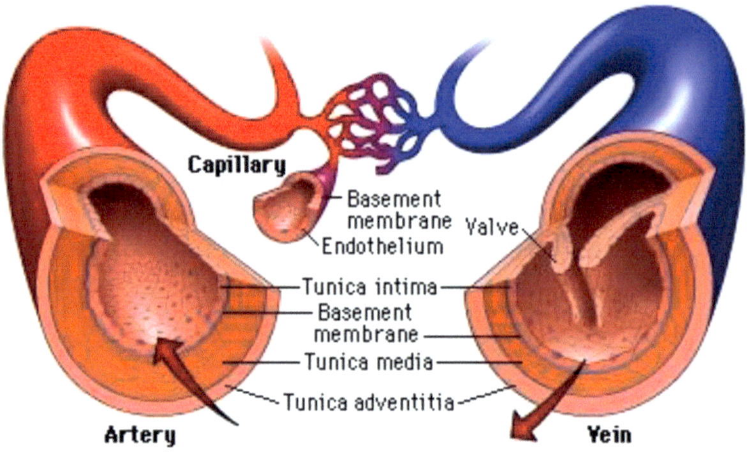

Fig. (6). Arterial system (red) is connected to venous system (blue) by smaller vessel, the capillaries.

THE RED BLOOD CELL

At the level of the red blood cell, oxygen is absorbed and bound to hemoglobin (Fig. **7**). The intracies of the bonding to hemoglobin and the attachment and release mechanisms are not apparent but are required for respiration to function. The limited solubility of oxygen in water (and therefore in blood plasma) means that only a small amount of oxygen can be carried dissolved in the blood at body temperature, pressure, and oxygen tension.

The body has approximaely 5 liters of blood volume; however, even if it were saturated, but without hemoglobin, it could not nearly carry the supply of oxygen that the body demands.

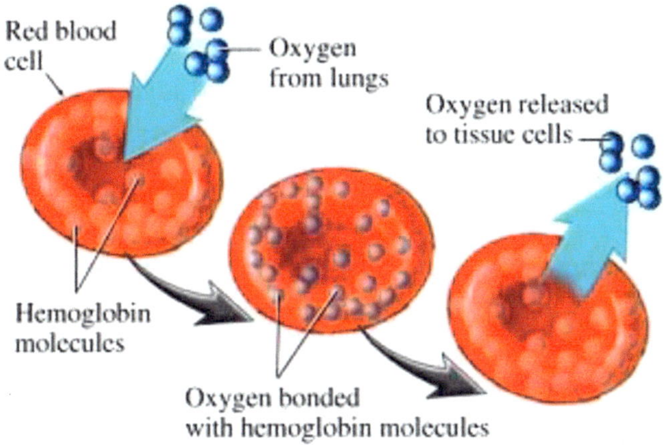

Fig. (7). Red blood cells are small, flexible, biconcave disc loaded with hemoglobin that binds oxygen.

The partial pressure of oxygen decreases steadily (in mm of Hg) from 149 in moist tracheal air to 104 in alveolar gas, to 100 in arterial blood, to 40 in mixed venous blood [5]. The total pressure is, of course, 760 mm Hg. Carbon dioxide gas is very different: 0.3 in moist tracheal air, 40 in alveolar gas and in arterial blood, and 46 in mixed venous blood. Water vapor is steady at 47 mm Hg. The solubility, relative to oxygen assigned as 1.0, is 24 for carbon dioxide. Diffusion of gases (in other gases) obeys Graham's Law and the rate is inversely proportional to the square roots of the molecular weights of the gases. Thus, the relative rates of diffusion of carbon dioxide and oxygen are equal to the square root of 32/44 which is 0.85. In the liquid phase, diffusion rates of gases generally are approximately 10,000 times smaller than in the gas phase because the mean free path of molecules between collisions (primarily solvent molecules) is low [5, 6]. However, this is partially compensated for because the distances over which gas transfer occurs in the liquid phase generally is up to 100 times shorter in the lung [5]. The lung manages exchange of these two very different gases that are present in very different concentrations, with magnificent efficiency.

Fick's Law governs the amount of gas transferred per unit of time across membranes and the math indicates that the rate of carbon dioxide diffusion is 20-times as rapid as for oxygen when the partial pressure difference is the same [5]. When all gas exchange factors are considered, lung gas exchange of oxygen is limited by blood flow [5]. To assess the transport of oxygen into the blood, the plasma and the red blood cells must be considered. The hematocrit (the fractional volume of blood that is occupied by red blood cells) is about 45% for males and 40% for females. When breathing air, about 2% of the total oxygen is carried as gas dissolved in the blood plasma and about 98% is carried bound to hemoglobin in red blood cells [5, 7]. When breathing 100% oxygen at 1 atmosphere compared to breathing air, the amount of oxygen in the plasma increases according to Henry's Law. The oxygen dissolved in the plasma will be directly proportional to the partial pressure of oxygen in the breathing gas mixture. Oxygen's solubility is 0.03 ml/L of plasma per mm Hg; the partial pressure of arterial blood is 100 mg Hg; thus, 1 liter of plasma contains 3 ml of oxygen [5].

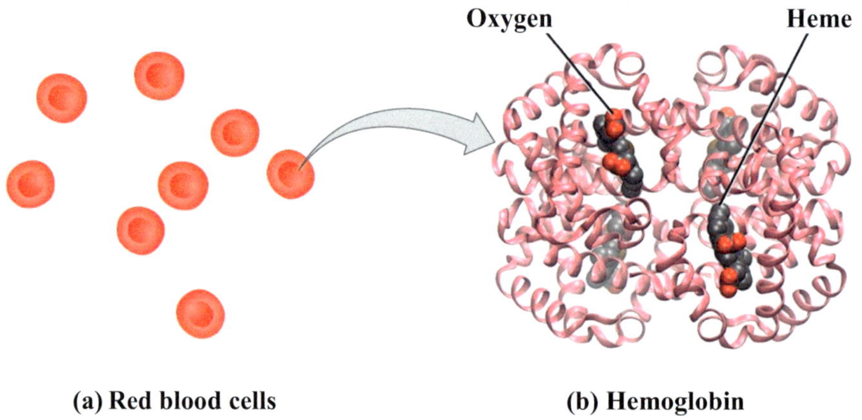

Fig. (8). Hemoglobin of red cells (**a**), binds 4 molecules of oxygen (double red balls), at heme iron sites associated with globular proteins shown as ribbon diagram (**b**).

Hemoglobin

Hemoglobin is a complex, molecule. It carries most of the oxygen in the blood. Hemoglobin has four subunits, each with a heme group plus a globin chain (Fig. **8**). The heme groups have a porphyrin ring containing an atom of iron. This iron can change redox states; it is the +2 (ferrous iron) state which binds oxygen.

An adult human has at least 6 different species of hemoglobin molecules (related to 6 different genes), all with the same function and nearly identical structures. There are, however, approximately 200 structurally different hemoglobin varieties known [5]. Hemoglobin has two α-chains and two β-chains and a molecular weight of 64,400 Daltons. The Dalton is a unit of mass equal to 1 proton or 1 neutron (approximately 1.66×10^{-27} kg). Each α-chain has 141 amino acids and each β-chain has 146 amino acids. The red cell has a hemoglobin content of approximately 150 grams per liter (approximately 7×10^{21} molecules) [5, 8, 9].

The complete structure of hemoglobin was elucidated by x-ray crystallography by Max Perutz (1914-2002) (Fig. **9**) and associates [9, 10]. Perutz won the Nobel Prize in Chemistry in 1962, shared with John C. Kendrew (1917-1997) [11]. Perutz was the right man at the right time with the right technology. He devoted most of his career to deciphering the molecular structure and biological function of hemoglobin. He was the founder and first chairman of the Medical Research Council Laboratory of Molecular Biology at Cambridge which became home for several Nobel Prize winners [12]. Perutz was born into an affluent family of textile industrialists in Vienna. He chose not to make his career in the family's industry, but instead studied chemistry. He subsequently chose to do research at the Cavendish laboratory in Cambridge where a few years prior, Desmond Bernal and a co-worker had obtained the first good x-ray diffraction images of a protein. This work paved the way for determining the structure of proteins at the atomic level. Although this work was significant, this was the late 1930s and war threatened. Bernal became involved in politics and then moved to London in 1938 [12]. Perutz remained at the Cavendish facility and selected x-ray diffraction analysis of hemoglobin for his PhD thesis research. It was known that hemoglobin changed its crystal structure when going from the oxygenated to deoxygenated state. This was 1938 and Hitler's annexation of Austria caused Perutz to immigrate. He was able to go to the Cavendish laboratory, and was supported by a grant until the end of World War II. Perutz also was able to bring his Jewish parents to Britain, which may have been life-saving; however, the family business was lost. Perutz submitted his doctoral dissertation on hemoglobin in the spring of 1940 [12]. There were fears in Britain of a German invasion and Perutz was interred as an enemy alien. Eventually, after several transfers to various camps he was sent to Canada. While he was at sea, another troopship with other internees was sunk by a German U-boat and many lives were lost. Perutz organized a camp "university" and his staff included Klaus Fuchs who later was arrested as an infamous Communist "atomic spy" [12]. Perutz subsequently was allowed to return to England where he resumed his research at the Cavendish lab. He became involved in wartime effort that included building an airplane base made of reinforced ice in the Atlantic. This required him to study the crystalline structure of ice. The team eventually developed a mixture of ice and wood pulp (called pykrete) that was said to be as strong as concrete [12]. (This became the subject of a myth on the TV show "Mythbusters", who confirmed this as "plausible"). Pycrete was demonstrated to Churchill and Roosevelt. Based on this success, the team including Perutz was sent to Washington for a time and then returned to the Cavendish lab. Perutz wrote humorously about his adventures[2].

Fig. (9). Max Ferdinand Perutz, awarded the Nobel Prize for determining the molecular structure of hemoglobin.

His work on hemoglobin was relevant to the war effort because of the need for blood components. After the war, a grant from the Medical Research Council funded the establishment of the Unit for the Study of Molecular Structure of Biological Systems with Perutz as director. His first model of hemoglobin was strongly criticized by Crick (later of DNA structure fame). Perutz, to his credit, accepted this criticism and gained from it. The major breakthrough in discerning the correct structure of hemoglobin came in 1953 when he adopted methods developed by others to calculate the distances between atoms from x-ray diffraction patterns and he also found a way to introduce the heavy atom, mercury, into hemoglobin (by what is called "isomorphous replacement") as a replacement atom for iron which was helpful in the analysis. This required great laboratory skill and acumen. Perutz was successful in producing a model of hemoglobin with 5.5 Ängström resolution which uncovered subsequently the mysteries of its cooperative binding of oxygen. One Ängström is one hundred millionths of a centimeter; a typical atom is approximately 1 to 5 Ängströms in diameter. His associate, John C. Kendrew, named previously, was actually the first to fully utilize the methods they developed together and working with the smaller myoglobin molecule, he was the first to publish a model of a globular protein with its structure derived by direct means. As stated, Kendrew shared the Noble Prize in Chemistry in 1962 with Perutz.

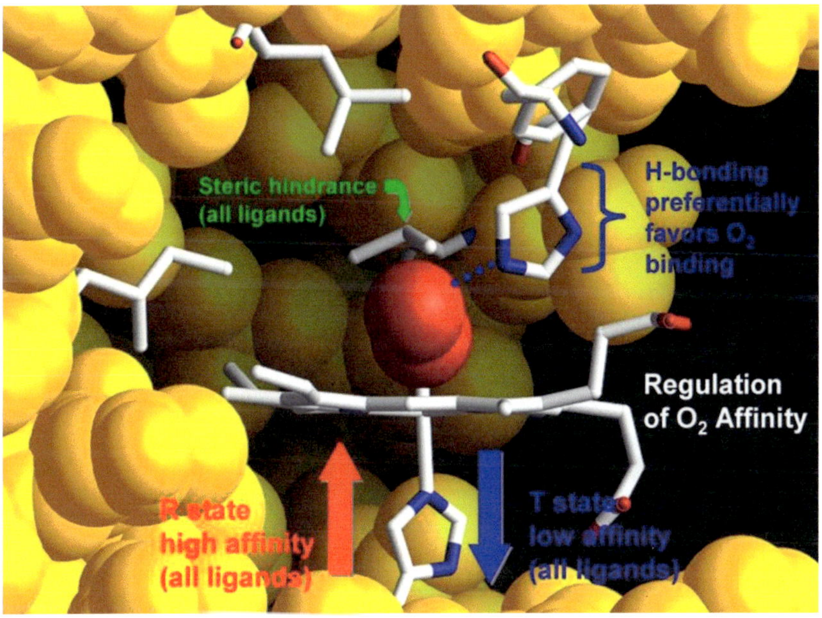

Fig. (10). Molecular model with diatomic oxygen (red) and interactions that alter oxygen binding affinity.

Oxygen Binding by Hemoglobin

Oxygen bonding in hemoglobin involves a key amino acid, histidine, which also is involved in binding iron [13]. When oxygen is bound to the iron near the histidine, the histidine is pulled at the molecular level, and subsequently a chain of events leads to a "shift" of the entire protein configuration (Fig. **10**). These molecular movements and shape changes alter the binding strength of oxygen by the iron [13].

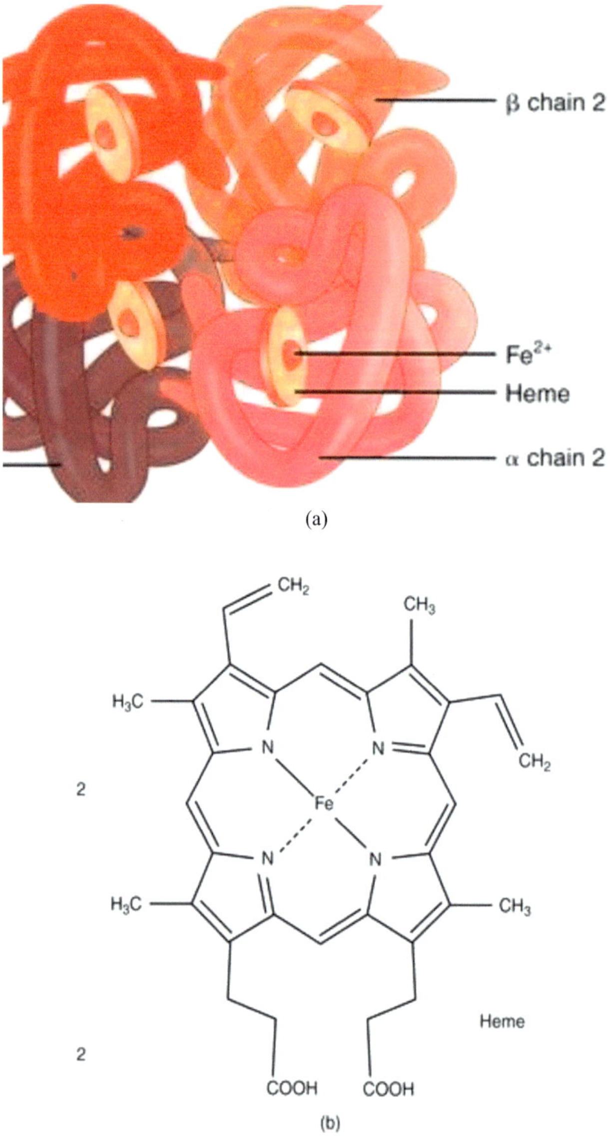

Fig. (11). Relationships of iron, α- and β-chains of hemoglobin (**a**) and porphyrin that binds iron (**b**).

Each hemoglobin molecule has four oxygen-binding sites, each with a porphyrin ring (Fig. **11**). It can be calculated that 1 gram of hemoglobin theoretically can bind 1.39 ml of oxygen. However, in the body a small amount of hemoglobin is either methemoglobin or has carbon monoxide bound tightly, and the actual amount is slightly less (1.34 ml) bound oxygen per gram of hemoglobin. Therefore, in 5 liters of blood (a nominal amount for a human), there are about 750 grams of hemoglobin which can carry approximately 1000 ml of bound oxygen [14]. Hemoglobin can be oxidized to methemoglobin by having the iron atom in the heme group lose an electron (Fe +2 to Fe +3). In the Fe +3 state it does not bind oxygen; normally only about 1.0% of the hemoglobin in the red blood cells is methemoglobin (oxidized).

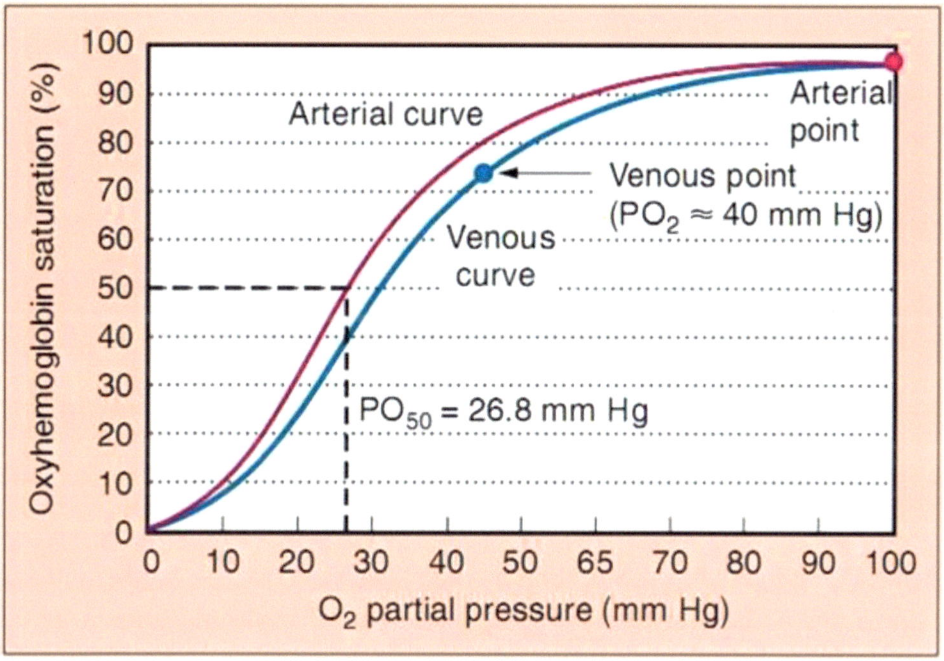

Fig. (12). Graph showing arterial and venous oxyhemoglobin saturation *versus* oxygen partial pressure.

The binding efficiency (tightness with which it is held) of oxygen by hemoglobin iron at the four sites becomes progressively greater as one, then two, then 3, and finally all four sites are occupied. This is called cooperative binding. Oxygen binding under different conditions can be expressed by a graph (plot) and the "curve" that is produced is non-linear (Fig. **12**). Actually, it is sigmoidal: with a partial pressure of oxygen of 40 to 100 mm Hg (in the mid-region of the typical plotted curve at 20% to 80% oxygen saturation) the rate of change is greater than at the lower or higher ends of the plot [9, 15]. This results in oxygen binding to change in exactly the way most beneficial to the body's needs when loading and unloading oxygen in lung and metabolizing tissue cells.

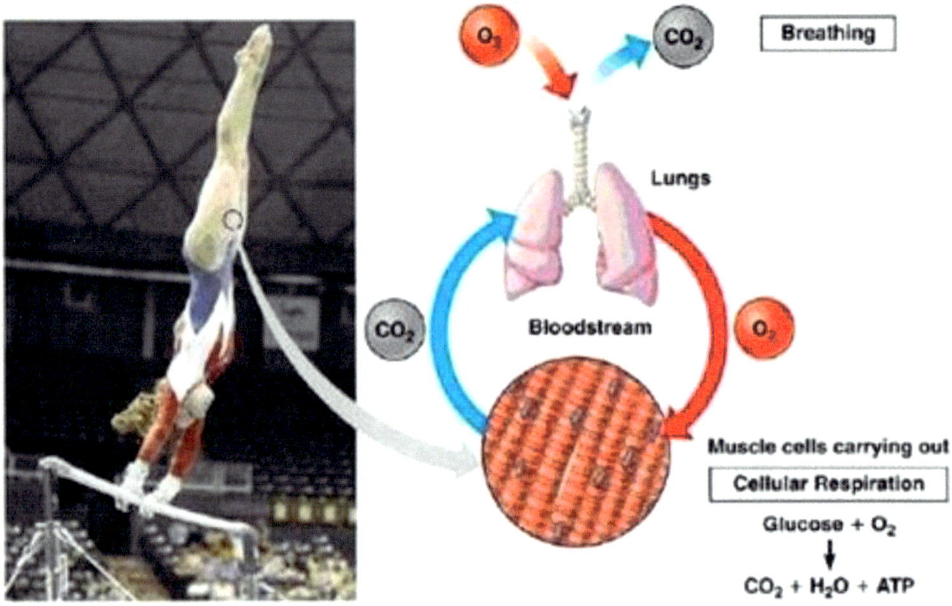

Fig. (13). Exercise makes extreme demands on the systems that must supply oxygen efficiently.

The affinity of hemoglobin for binding oxygen is affected by pH, temperature and partial pressure of carbon dioxide, and by the small molecule known as 2,3-diphosphoglycerate [14]. Increasing the temperature lowers the affinity for oxygen and shifts the oxygen dissociation curve "to the right" as usually graphed. This has relevance to physical exercise where the muscle temperature may be higher than 37 °C, and oxygen is unloaded more easily to muscle cells from red blood cell carriers [14]. Increased hydrogen ions (acidity) also lower the affinity of hemoglobin for oxygen. Exercising muscle generates acid and oxygen is off-loaded more readily [14]. Also, 2, 3-diphosphoglycerate is an intermediate in anaerobic metabolism of glucose (by the Embden-Meyerhoff pathway) and it is produced in red blood cells. Exercise causes an increase in 2,3-diphosphoglycerate which is a charged molecule that does not easily pass through the red cell membrane and, thus, the oxygen dissociation curve is shifted to the right (lower affinity for oxygen and, hence, oxygen is released more readily as is needed by exercising muscle) [14, 16] (Fig. **13**).

Red cells are the most numerous type of cell in the body. Red cells are made from stem cells in the bone marrow at the rate of approximately two million cells per second. They mature in about a week and circulate for a total of about 120 days before being removed from the circulation. During maturation the red cell of mammals loses its nucleus and also loses its mitochondria, the organelles where metabolism and oxidative phosphorylation occur and bioenergetics produces ATP with electrons derived from oxygen. Mature red cells have a simpler energy converting system sufficient for their needs. The loss of mitochondria allows red cells to be small (they are smaller than most cells in the body) and to be very flexible with a biconcave disk shape (somewhat like a deflated basketball). This allows red cells to move in very small capillaries. They eventually may flow one-by-one in small capillaries with close contact with cells where oxygen is released from hemoglobin. Each red cell has about 270 million molecules of hemoglobin, each with 4 binding sites for oxygen. Thus, each red blood cell when fully loaded is capable of carrying more than one billion oxygen molecules. This system provides for preferential, controlled loading and unloading, and protection during transport. It is analogous to a rapid transit system for citizens in a large city. Red cells and hemoglobin are optimally designed to load, transport safely, and unload a large amount of oxygen. If the hemoglobin were in simple solution, not housed within red cells, the viscosity of blood would be too high to be circulated efficiently. The red cell shape allows it to flex, bend and even squeeze its way through the smallest of capillaries.

An average human has a total of approximately 25 trillion red cells in the total blood volume in which there are a total of 6.8 sextillion (6.8×10^{21}) hemoglobin molecules. Thus, 4 times this number of oxygen molecules potentially can be on board at one time; approximately 27×10^{21}. Of course, no one really knows, but one estimate is that there are 1×10^{24} stars in the known Universe. Thus, amazingly, there would be only about 37 stars for each hemoglobin molecule in one person. Another way of comparing is that the number of red cells (not hemoglobin molecules) in one human is more than 4,000 times the number of humans on earth today. A little less than 200 billion red cells are made each day by one person; about 200 billion grow old, are recognized as worn out, and are removed from circulation and eliminated. Without the efficiency provided by red cells in ferrying oxygen, the size of animals would be greatly limited and no large animals, including humans would be possible. Serious diseases result when there are too many or too few red cells. This is further evidence (proof, for me) that oxygen is central to life and that its chemical and biological characteristics in many ways are unique.

Ever wonder why blood in surface veins appears blue? Blue wavelengths of light are reflected preferentially by the skin's surface layers; this is the light that reaches one's eyes. Red wavelengths of light penetrate deeper and blood in the veins absorb most of the red light, plus any blue light that manages to penetrate that far. Therefore, what our eyes detect is the reflected blue light from the skin surface layers– minus the red wavelengths.

Hemoglobinopathies

The genes that code for the protein chains of hemoglobin differ slightly among humans and the amino acid sequences in the proteins of hemoglobins are slightly different [13]. The small changes are not associated with any pathology. However, there are hemoglobinopathies in which significant changes in hemoglobin are produced by gene mutations. One such case is sickle cell anemia. As the name implies there is a change in the shape of red blood cells and a premature destruction of these cells creates anemia. By a mutation in the gene code, the amino acid valine is substituted for the amino acid, glutamate number 6, in the β-chain of the heme protein. When deoxygenated, the hemoglobin of sickle cell anemia deforms and its three-dimensional shape is changed. Red cells with this deformed hemoglobin change their shape from biconcave disks to sickle-shaped cells which are also labile and their early removal leaves the patient anemic. Sickle cells also are able to transport less oxygen. Another genetically caused pathology, thalassemia, results when there is a mutation causing a miss-match in production of α- and β-chains [13].

Binding of Carbon Monoxide and Nitric Oxide to Hemoglobin

Hemoglobin also can bind molecules of nitric oxide and carbon monoxide. Nitric oxide binds to specific cysteine residues in hemoglobin. Subsequently, the nitric oxide can relax smooth muscle of blood vessels and blood pressure is reduced. Thus, hemoglobin regulates blood pressure to a minor extent [13]. Carbon monoxide is toxic; it displaces oxygen and also forms a stable complex with hemoglobin iron. Blood carbon monoxide concentrations from inhaling very low atmospheric amounts can be increased significantly (up to 20%) by cigarette smoking and can cause a significant loss in oxygen transport [7].

The affinity of carbon monoxide for hemoglobin is 200- to 300-times that of oxygen [5]. When carbon monoxide binds to one of the four binding sites (Fe +2) on hemoglobin, the resulting increased affinity of the other binding sites for oxygen reduces "unloading" of oxygen at tissue sites. Also, carbon monoxide binds very tightly and remains at the site. Even very low tissue concentrations of carbon monoxide (1 mm) can effectively block most of the heme sites that otherwise would bind oxygen [5].

Unusual Hemoglobins

Crabs have hemoglobin with copper instead of iron and it is blue in color. Legumes (plants that fix nitrogen such as beans, peas and alfalfa) *via* a symbiotic relationship with particular species of bacteria are able to fix nitrogen (convert N_2 from the air into NH_3). A particular form of hemoglobin (leghemoglobin) is involved in this process. The root system of the legume is invaded by the symbiotic bacterium and each benefits from the process. Bacterial heme synthesis is required for leghemoglobin expression, but the heme moiety is not essential for apoleghemoglobin synthesis by the plant. By symbiosis both contribute to the complete leghemoglobin molecule which is red, similar to blood [17].

All mammals and nearly all vertebrates have hemoglobin; however, the Antarctic "ice fish" lacks hemoglobin. A few invertebrates also have hemoglobin including *Annelids* (worms) which do not have red blood cells, but have the hemoglobin simply dissolved in plasma. Also, the larval stages of a few insect species have hemoglobin [7]. Interesting details about hemoglobin in various species were published by Ross Hardison in 1996 that described the connection between metals, porphyrin rings, cytochromes, chlorophylls, and hemoglobins [18]. A specific structure, protoporphyrin IX, has an iron atom bound in the center of a flat, planar structure in the hemoglobin molecule and also in other proteins including cytochromes and ligninases. The iron can participate in oxidation-reduction cycling (+2 « » +3). Some hemoproteins allow oxygen binding without changing the oxidation state of iron. The amino acid sequences of α- and β- chains of hemoglobin are approximately 50% identical in vertebrate species [18]. Invertebrates including flies, (arthropods) earthworms (annelids and nematodes) including your fishing worms as you may have noted, have hemoglobins [18]. Plant hemoglobins participate in photosynthesis, in the electron transport chain of plant mitochondria, and they also are found in root nodules of legumes as previously stated. The amino acid sequences of leghemoglobin differ from vertebrate hemoglobins at approximately 80% of the sites, but the leghemoglobin of these root nodules fold into the same 3-dimensional structures as animal hemoglobins.

Artificial Blood

There is interest in creating artificial oxygen transporters. To date, none have the versatility and efficiency provided by hemoglobin carried within red blood cells. One approach has been to use outdated human whole blood intended for blood transfusion as a source of hemoglobin. A principle difficulty has been keeping together the 4 protein chains of hemoglobin [19, 20]. Attempts at extracting hemoglobin from such blood or even from animal blood have had limited success. In blood the packaging of the hemoglobin in red cells provides great benefits compared to hemoglobin in solution. Another approach which has had limited success, is the adoption of perflurocarbons as oxygen carriers [21]. Perfluorocarbon emulsions, as blood substitutes, were "first-generation products". Perfluorocarbons are biologically inert (a good thing); oxygen is highly soluble in them (a good thing); however, they are cleared from the blood by the body with a half-life of about 2 to 4 hours (a bad thing). They do not bind oxygen (a potentially bad thing) because there is little means (a bad thing) to exert the wonderful control that hemoglobin has for coordinating loading and unloading as required and so efficiently is regulated by hemoglobin.

In a 2015 publication, Sara J. Grethlein and coauthors [22] stated that work to develop a "viable" blood substitute has spanned 7 decades. I have seen little progress since the 1970s. Most research has been focused on hemoglobin-based oxygen carriers [23]. Concern about transmission of infectious disease and antigenic reactions are driving forces for blood substitutes. However, as of 2015 no oxygen-carrying blood substitutes were approved for medical use by the U.S. Food and Drug Administration [22]. An artificial blood substitute must compete with the (approximately) 1.39 ml of oxygen carried by 1 gram of hemoglobin and hemoglobin's unique ability to cooperatively bind and release oxygen with regulation by temperature, pH and 2,3-diphosphoglycerate, as is described in this Chapter. An "ideal" blood substitute would have similar characteristics. In reality, there have been a number of adverse effects associated with artificial blood substitutes [22]. Adverse effects were described as primarily transient and sometimes clinically minor, and included: hypertension, abdominal pain, skin rash, diarrhea, jaundice, hemoglobinuria, oliguria, fever, stroke, and clinical blood anomalies such as elevated lipase activity [22]. An interesting older review, published in 1997 [21] and previously mentioned in connection with perfluorocarbons, states that clinical trials of both hemoglobin-based and perfluorocarbon-based oxygen carriers were nearing completion. Current applications are limited.

Although they do not technically have a unique hemoglobin, five families of notothenoid fish that inhibit the Southern Ocean (the frigid waters that surround Antarctica), have a protein "antifreeze" in their blood [24] (Fig. **14**). Arthur Devries is credited with discovering antifreeze proteins in Antarctic notothenoid fish in the late 1960's. I recall when this discovery was announced and it became a subject of a seminar at the Dalton Cardiovascular Research Center where I was a research professor/investigator. This extraordinary design allows them to compete so successfully that 90% of the fish biomass in this cold region may be notothenoid fish. Unexpectedly, the protein-bound ice crystals are resistant to melting. The authors claim this may be the first example in nature of superheated ice [24].

Fig. (14). A notothenoid fish from Antarctic waters that has an unusual protein “antifreeze” in its blood.

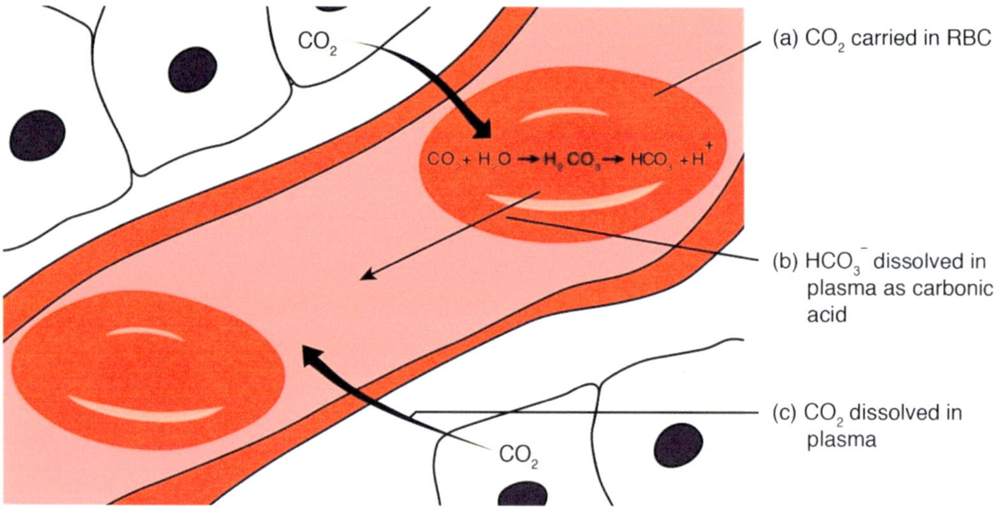

Fig. (15). Red blood cell hemoglobin does double duty and transport carbon dioxide as well as oxygen.

Carbon Dioxide Transport by Hemoglobin

Hemoglobin does double duty: in addition to transporting oxygen efficiently from the lung to cell sites, it also efficiently ferries carbon dioxide, a waste product of cells (Figs. **15**, **16**). The red blood cell enzyme, carbonic anhydrase, is essential for this function. It converts carbon dioxide into bicarbonate at the rate of 600,000 molecules per second. This is 1.7 millionth of a second for each conversion by each molecule of enzyme. Its efficiency is so great that the transformations are limited by the rate at which carbon dioxide molecules reach the enzyme active site by diffusion. In the lung where carbon dioxide concentration is low, the bicarbonate is converted back into carbon dioxide gas which is easily exhaled. Both functions are carried out efficiently by the same enzyme.

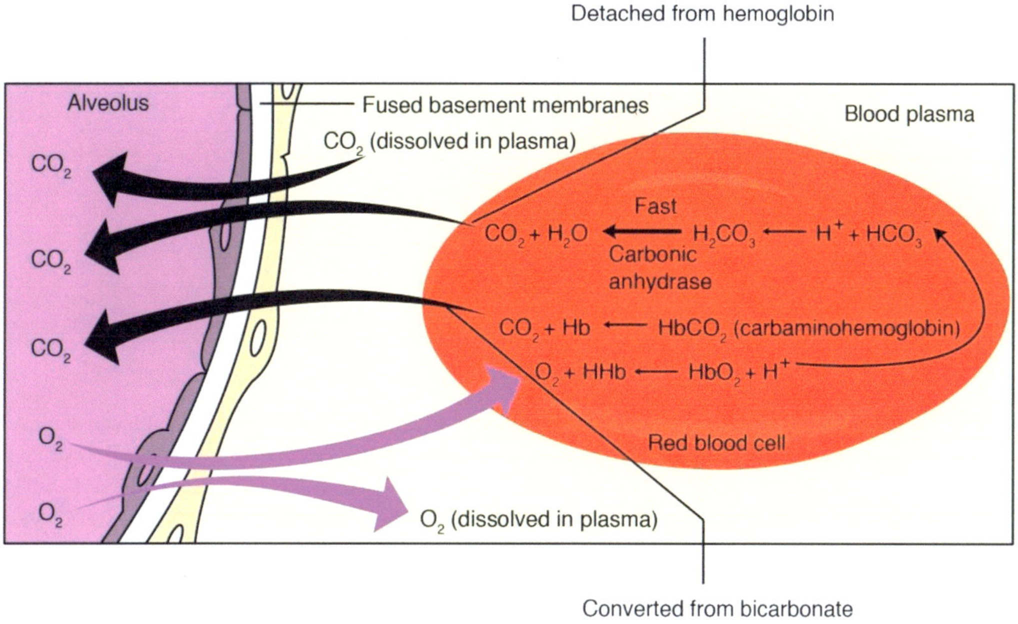

Fig. (16). Details of delivery of carbon dioxide to lung alveolus, and oxygen to the red cell.

OXYGEN IN MITOCHONDRIA

It is well known that oxidative phosphorylation in the electron transport chain of mitochondria is the step at which oxygen is required for catabolism to generate ATP [25] (see also Chapters 2 and 7). The oxygen simply diffuses into the mitochondrion from sites on hemoglobin. Electrons derived from carbohydrates are transferred to oxygen molecules to produce water. These electrons have been passed along an energy gradient and are now at their lowest level after having given up energy at sites of ATP generation. No carrier molecules are needed for oxygen beyond hemoglobin. Oxygen passively diffuses through mitochondrial membrane pores to meet the terminal chtochrome with its oxidase. This also means there is no mechanism for storing oxygen at the use-level in mitochondria. It is readily calculated that an adult human requires approximately 600 to 800 grams of moleculer oxygen per day which equates to approximamtely 420 to 560 liters. Thus, delivery must be continuous and reliable.

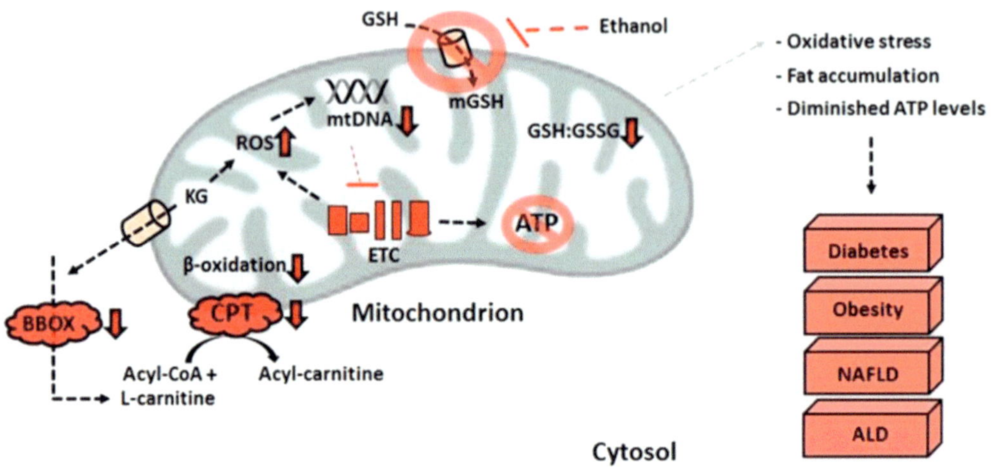

Fig. (17). Mitochondrial oxidant stress increases and become damaging in several medical conditions.

As stated above, the oxygen that is transported to cells, ends up being used by cytochorme oxidase, the terminal enzyme in the electron transport chain. We have previously described this process, including that, under conditions that are poorly understood, reactive oxygen species can be diverted to damage mitochondria, as described in Chapter 7. This process of oxidative formation and damage to mitochondrial oxidative phosporylation is associated with various pathologies (Fig. **17**). Hemoglobin delivers and releases oxygen to the mitochondria. J. B Wittenberg and B. A. Wittenberg have studied this process using mitochondria isolated from pigeon hearts [26]. *In vitro*, they added hemoglobin from various sources including horse, soy bean, and molluscs. These hemoglobins were selected because they have very different binding kinetics for oxygen. Despite these differences, the pigeon heart mitochondria were able to take-up oxygen at the same rate [26]. They also measured the desaturation rates for the various hemoglobins. When oxygen uptake was at half-maximum, they found that all the hemoglobins produced very similar partial pressures at the mitochondrial site. The authors interpreted these results to indicate that the mitochondria were able to obtain sufficient oxygen from any of these hemoglobins to generate the ATP needed at maximum work load [26]. Thus, oxygen diffuses across mitochondrial membranes efficiently without any special transport mechanisms which is a tribute to the uniqueness of oxygen.

T. Koyama, *et al.* [27] have studied the transport kinetics of oxygen through mitochondrial membranes. They assumed that the terminal cystochrome was distributed only on the inner surface of the inner mitochondrial membrane. Using the average oxygen consumption rate of 10 ml of oxygen per 100 grams of tissue per second in beating heart muscle, they compared oxygen transport with and without treatment with a phospholipase known to be activated in ischemic hearts to destroy the microstructure of myocardial cells. The reduced transport indicated that, hypoxia impairs the delivery of oxygen to heart muscle [27].

OXYGEN IN THE BRAIN

It is commonly said that a human can live for a month or more without food, a few days without water, but only a matter of minutes without oxygen. Indeed, the most sensitive organ to oxygen deprivation is the brain. Not only does it fail to function but it suffers irreparable damage from oxygen deprivation. Nerve cells require oxygen to make the ATP required for nerve cells to "fire" and transmit the electrical signals that are their distinguishing characteristic (Fig. **18**). Oxygen must be delivered safely (it has a high oxidizing potential) to all parts of the body; after it does the work of "cold" combustion, the carbon dioxide generated must be efficiently eliminated; and, it must get inside cells and perform the specific functions required for oxidative phosphorylation to produce ATP. The brain has only about 2% of the body weight (normal weight individuals) but it receives approximately 15% of the cardiac output, 20% of the total body oxygen consumption and 25% of the total body glucose utilization, depending on exercise activity state. The energy consumption of the brain (glucose utilization) is approximately 0.1 calories per minute to function, but increases to at least 1.5 with mental activity [28].

In 2014, a new system was reported that combined brain scanning, brain recording and virtual reality to "allow a user to journey through a person's brain in real-time" [29]. This system uses a skull cap with electroencephalogram (EEG) electrodes that measure differences in electric potential to record brain activity. This was combined with a previously scanned magnetic resonance image of the brain. During the demonstration of the system, an individual wore a virtual reality headset that allowed him to explore the brain in 3D while light flashes marked brain activity from the EEG. A website [30] shows this imaging in action and the image is called the 3D 'glass brain' (Fig. **19**). The electrical activity, driven by oxygen, is testament to the amazing brain activity.

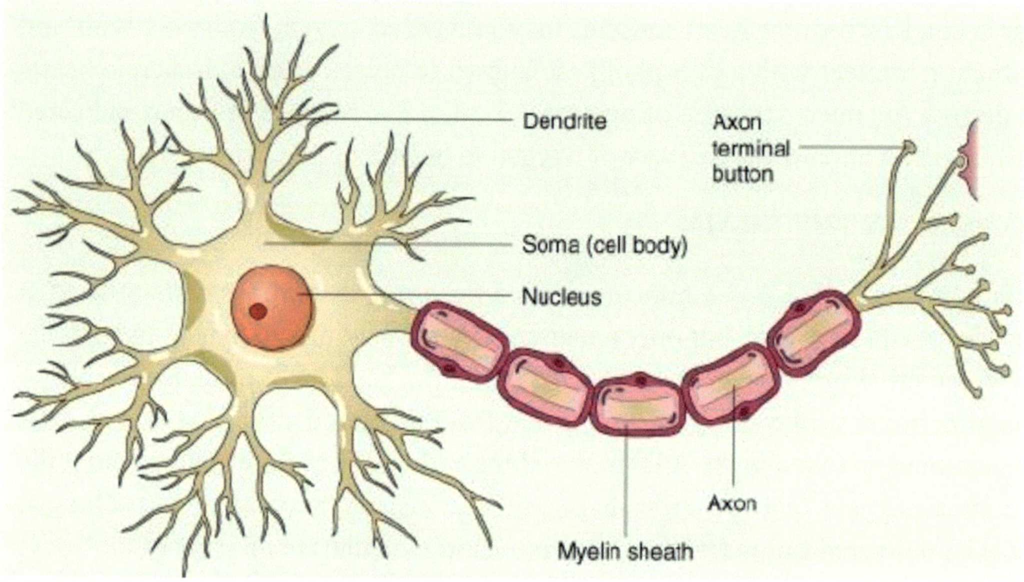

Fig. (18). Brain neuron with cell body containing organelles, including mitochondria, axons, and dendrites that connect *via* synapses to other cells (not shown).

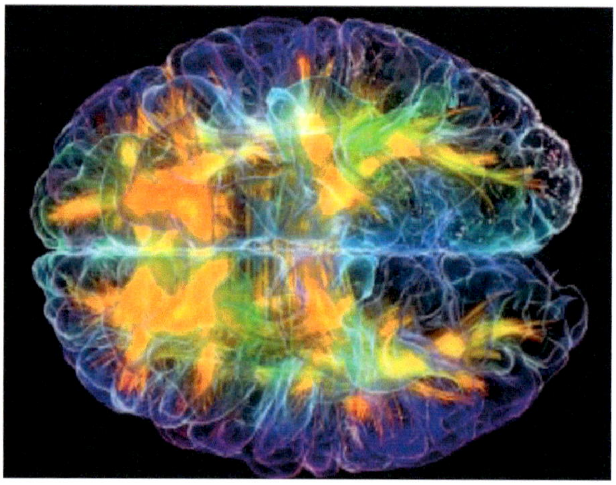

Fig. (19). The 'glass brain'.

NOTES

[1] Emile Dickinson (1830-1886), *"The Brain –is wider than the sky"*. 1862.

[2]*Enemy Alien, the New Yorke,* August 12, 1985, p 35.

REFERENCES

[1] Ikegami M, Weaver TE, Grant SN, Whitsett JA. Pulmonary surfactant surface tension influences alveolar capillary shape and oxygenation. Am J Respir Cell Mol Biol 2009; 41(4): 433-9. [http://dx.doi.org/10.1165/rcmb.2008-0359OC]

[2] Akei H, Whitsett JA, Buroker M, *et al.* Surface tension influences cell shape and phagocytosis in alveolar macrophages 2006; 291(9): L572-9. [http://dx.doi.org/10.1152/ajplung.00060.2006]

[3] The human respiration process Available from: http://www.oxygen-review.com/respiration.html. [cited 2016 Jul 11].

[4] Fairfax RE. Standard Interpretations Available from: https://www.osha.gov/pls/oshaweb/owadisp.show_document?p_table=INTERPRETATIONS&p_id=28604 [cited 2016 Jul 11].

[5] Pittman RN. Regulation of tissue oxygenation. Morgan & Claypool Life Sciences 2011.

[6] Krogh A. The rate of diffusion of gases through animal tissues, with some remarks on the coefficient of invasion. J Physiol 1919; 52(6): 391-408. [http://dx.doi.org/10.1113/jphysiol.1919.sp001838] [PMID: 16993404]

[7] Hemoglobin [Internet]. New World Encyclopedia. Available from: http://newworldencyclopedia.org

[8] Geis I Hemoglobin. Menlo Park, CA: Benjamin/Cummings Pub. Co. 1983.

[9] Garby L. The Respiratory Functions of Blood. New York: Springer 1977. [http://dx.doi.org/10.1007/978-1-4684-2313-6]

[10] Bolton W, Perutz MF. Three dimensional fourier synthesis of horse deoxyhaemoglobin at 2.8 Angstrom units resolution. Nature 1970; 228(5271): 551-2. [http://dx.doi.org/10.1038/228551a0] [PMID: 5472471]

[11] Max F. Perutz- Biographical Available from: http://www.nobelprize.org/nobel_prizes/chemistry/laureates/1962/perutz-bio.html[cited 2016 Jul 11].

[12] Perutz MF. 2008.Complete Dictionary of Scientific Biography 2008. Encyclopedia.com. Available from: http://www.encyclopedia.com/topic/Max_Ferdinand_Perutz.aspx [cited 2016 Jul 12]

[13] Duttta S, Goodsell D. 2003.Hemoglobin. RCSB Protein Data Bank 2003. Available from: http://pdb101.rcsb.org/motm/41 [cited 2016 Jul 7]

[14] Pittman RN. 2011.Oxygen Transport Morgan Claypool Life Sciences Available from: http://www.ncbi.nlm.nih.gov/books/NBK54103/ [cited 2016 May 29].

[15] Barcroft J, Hill AV. The nature of oxyhaemoglobin, with a note on its molecular weight. J Physiol 1910; 39(6): 411-28. Available from: http://doi.wiley.com/10.1113/jphysiol.1910.sp001350

[16] Guyton AC, Widdicombe J. 1974. Respiratory Physiology. Baltimore: University Park Press

[17] O'Brian MR, Kirshbom PM, Maier RJ. Bacterial heme synthesis is required for expression of the leghemoglobin holoprotein but not the apoprotein in soybean root nodules 1987; 84(23): 8390-3. [http://dx.doi.org/10.1073/pnas.84.23.8390]

[18] Hardison RC. Commentary A brief history of hemoglobins: Plant, animal, protist, and bacteria 1996; 93(12): 5675-9.

[19] Chang TM. Future prospects for artificial blood. Trends Biotechnol 1999; 17(2): 61-7. [http://dx.doi.org/10.1016/S0167-7799(98)01242-6] [PMID: 10087605]

[20] Creteur J, Sibbald W, Vincent JL. Hemoglobin solutionsnot just red blood cell substitutes. Crit Care Med 2000; 28(8): 3025-34. [http://dx.doi.org/10.1097/00003246-200008000-00058] [PMID: 10966290]

[21] Scott MG, Kucik DF, Goodnough LT, Monk TG. Blood substitutes: evolution and future applications. Clin Chem Am Assoc Clin Chem 1997; 43(9): 1724-31. Available from: http://www.ncbi.nlm.nih.gov/pubmed/9299967 [cited 2016 Jul 10]

[22] Grethlein SJ. BBlood Substitutes B Available from: http://emedicine.medscape.com/ article/207801-overview#a3 [cited 2016 Jul 10]

[23] Chen JY, Scerbo M, Kramer G. A review of blood substitutes: examining the history, clinical trial results, and ethics of hemoglobin-based oxygen carriers. Clinics (Sao Paulo) 2009; 64(8): 803-13. [San Paluo]. [http://dx.doi.org/10.1590/S1807-59322009000800016] [PMID: 19690667]

[24] Antifreeze proteins in Antarctic fish prevent both freezing and melting Available from: http://www.nsf.gov/news/news_summ.jsp?cntn_id=132798 [cited 2016 Jul 12]

[25] Buzby J. Re: How is oxygen transported to Mitochondria after it crossed the membrane? Available from: http://www.madsci.org/posts/archives/2008-09/ 1220890706.Cb.r.html [cited 2016 Jul 9].

[26] Wittenberg JB, Wittenberg BA. Myoglobin-enhanced oxygen delivery to isolated cardiac mitochondria. J Exp Biol 2007; 210(Pt 12): 2082-90. [http://dx.doi.org/10.1242/jeb.003947] [PMID: 17562881]

[27] Koyama T, Kinjo M, Araiso T. Oxygen diffusion through mitochondrial membranes. Adv Exp Med Biol 1989; 248: 763-7. [http://dx.doi.org/10.1007/978-1-4684-5643-1_86] [PMID: 2551143]

[28] The human body is about two-thirds oxygen Available from: http://www.oxygen-review. com/human-body.html

[29] Lewis T. Virtual reality system lets you explore your brain in real-time Available from: http://www.livescience.com/44000-virtual-reality-system-reveals-brain.html

[30] The dynamic mind: Stunning 3D "glass brain" shows neurons firing off in real-time. Available from: http://www.dailymail.co.uk/sciencetech/article-2581184/The-dynamic-mind-Stunning-3D-glass-brain-shows-neurons-firing-real-time.html

CHAPTER 9

Oxygen: Parkinson's, Alzheimer's, Huntington's

Abstract: Parkinson's, Alzheimer's, and Huntington's are degenerative brain diseases with some similarities in symptoms. Oxidative stress has been linked to these diseases, but causation is unproven. Dementia is a central feature but is not diagnostic. Based on historical, clinical descriptions the order of discovery as a disease is: Parkinson's (1817), Huntington's (1842) although it probably has been known since the Middle Ages and was called "chorea", and Alzheimer's (1906). Alzheimer's is the most common; Parkinson's is second; and Huntington's is the third most common neuronal, degenerative disease. None of these diseases can be cured, and there is generally a long, severe decline in functional ability that is tragic for the individual, the family and friends. Although the causes of these diseases are unknown, there are genetically inherited risks for each. The pathophysiology, brain sites affected, cellular and sub-cellular mechanisms, and genetics for each disease is complex and oxidant stress has been incriminated for some aspects of these diseases. Parkinson's results in a progressive loss of dopaminergic neurons in the substantia nigra pars compacta and the metabolism of dopamine itself generates reactive oxygen species. Therapy with the dopamine precursor levo-dopa (dopamine does not pass the blood- brain barrier) does not provide lasting benefits. Speculation that Parkinson's disease is caused by reactive oxygen species generated from pesticide exposure has not been proven. Discoveries by co-workers in my laboratory pointing to special sensitivity from oxidant stress for certain iron-containing enzymes in amino acid metabolism are promising links to oxidant stress causation. Generation of oxidant stress *via* aberrant mitochondrial oxidative metabolism is a viable thesis but a cure for these diseases is not in sight and is more likely to result from future discoveries using stem cells and perhaps gene therapy.

Keywords: Age-related neurodegeneration, α-synuclein, Alzheimer's, β-amyloid deposits, Cellular redox balance, Dopamine, Gene mutation, Huntingtin gene, Huntingtin protein, Huntington's, Kynurenic acid, Kynurenine pathway, Levo-dopa, Lewy bodies, Miss-folded proteins, Oxidant stress, Parkinson's, Protein phosphorylation, Quinolinic acid, ROS, Tau, Tryptophan metabolism.

Olen R. Brown

INTRODUCTION

Age-related brain degeneration occurs in Parkinson's, Alzheimer's, and Huntington's diseases and there are links to genetic defects, most importantly for the latter. There is evidence that oxidative stress mechanisms are involved in the brain degeneration that is characteristic of each disease. The hereditary basis of Huntington's disease was discovered in 1993, and its hallmark is a mutated huntingtin protein, that is toxic to brain cells with damage most evident in the striatum early in the disease which then spreads to other areas which are conspicuously affected. Alzheimer's disease is characterized by shrinkage of the total brain, and by the formation of twisted aggregates of the tau protein to form neurofibrillary tangles within neurons, and the deposition of another protein, β-amyloid, in spaces between brain cells. Parkinson's disease results primarily from death of dopamine-producing cells selectively in the substantia nigra pars compacta. These diseases progress in severity; they are deeply tragic for the patient, family members and friends; and current therapies are limited and no cures are available.

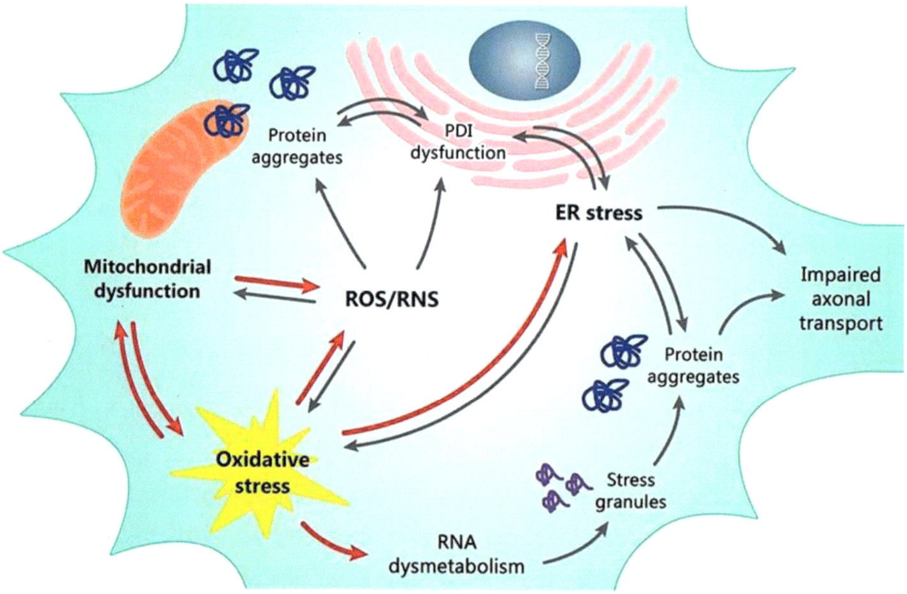

Fig. (1). Oxidative stress and brain dysfunction.

OXIDANT STRESS

Excessive oxidative stress can result from dysfunctions in mitochondrial oxygen utilization, redox cycling of certain xenobiotics and some therapeutic drugs, and other mechanisms that generate a variety of redox-active species. Fig. (**1**) depicts the complexity of proposed interactions that connect generation of oxidative stress and damage to the brain, with a focus on Parkinson's disease.

PARKINSON'S DISEASE

Introduction

Parkinson's disease results from selective death of neuronal cells in an area of the mid-brain called the substantia nigra pars compacta. The neurons that die contain the neurotransmitter dopamine and the neuronal processes extend into the striatum of the brain. The affected neurons control voluntary movement and their degeneration leads to the four cardinal debilitating symptoms of Parkinson's disease: resting tremor, muscular rigidity, bradykinesia, and postural imbalance [1]. In over 90% of cases the cause of the disease is not known, but aging is a strongly-associated factor and the incidence of Parkinson's disease increases exponentially with age after 65 [1, 2]. Genetics has a role in the disease and mutations in a number of genes have been identified. In both idiopathic and genetically-related Parkinson's disease, oxidant stress has been implicated [1 - 6].

James Parkinson[2] (1755-1824) (Fig. **2**) first described in 1817 the basic, clinical symptoms that characterize Parkinson's disease. More than a century passed before the basic pathology was determined to be the loss of neurons in the substantia nigra pars compacta. Progress into the nature of the disease accelerated when Arvid Carlsson (1923-) discovered dopamine in the brain, and this was followed by the anatomical elucidation of the nigrostriatal dopaminergic pathway and discovery that loss of neurons in this region led to striatal dopamine deficiency (responsible for the major symptoms of the disease) and subsequently that oral administration of the dopamine precursor levodopa alleviated symptoms [6]. Carlsson was awarded the Nobel Prize in Physiology or Medicine (shared with two others) "*... for their discoveries concerning signal transduction in the nervous system*"[3].

Fig. (2). James Parkinson.

Disease Manifestations

Fig. (**3**) shows a healthy substantia nigra (center panel), compared to a diseased substantia nigra that is characteristic of Parkinson's disease (right panel), taken from the midbrain (left panel). The principal damage site in the brain is in the pars compacta of the substantia nigra.

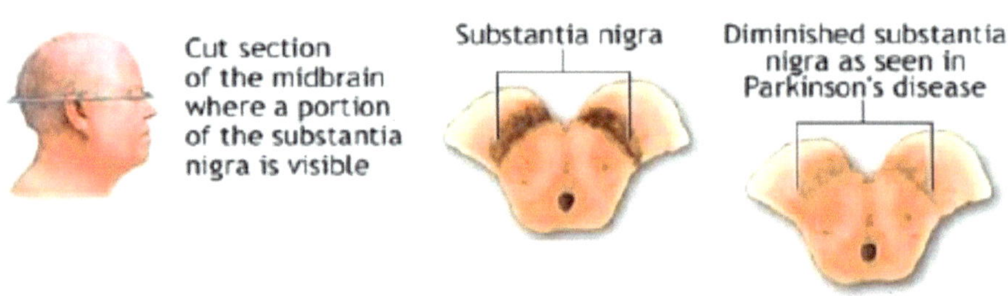

Fig. (3). Location of damaged brain site in Parkinson's disease.

A section through the brain (Fig. **4**) shows details of brain structures, including the substantia nigra, which will be referred to in discussing the pathophysiology of this disease.

Failure of the brain in Parkinson's disease to adequately produce dopamine is central to the pathology of the disease. Dopamine is a neurotransmitter. In the disease there is a selective loss of neurons in the midbrain that produce dopamine. These neurons connect through nerve fibers to the striatum. These neurons control voluntary movements and their degeneration leads to the cardinal signs of the disease. Dopamine, itself, is a source of potential oxidative stress. Although dopamine is usually stored in vesicles, cytosolic dopamine can be accessible to oxidation, spontaneously and enzymatically, to a quinone. Dopamine quinone is reactive and capable of covalent chemistry to modify essential cellular nucleophilic compounds including reduced sulfhydryl groups of glutathione and cysteine residues in proteins [1]. Of special note, dopamine quinone is associated with modification of proteins involved in the pathophysiology of Parkinson's disease including modifying α-synuclein monomer and its conversion to cytotoxic proto fibrils [7]. The complex interactions and associations of oxidant stress with dopamine and brain dysfunction in Parkinson's disease is shown in Fig. (**5**).

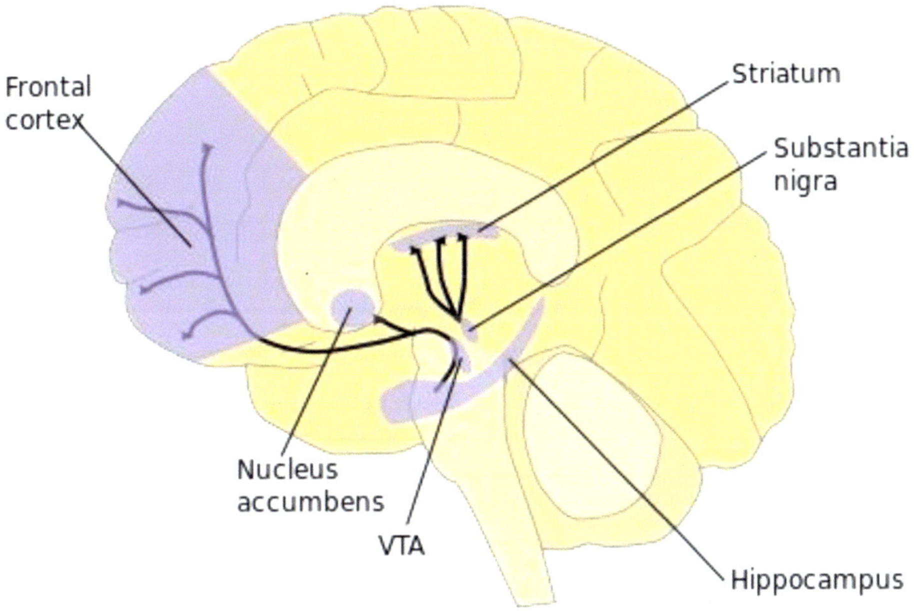

Fig. (4). Mid-brain cross section showing brain regions and relationships to Parkinson's disease.

Levo-dopa Therapy

The first therapy, which is only temporarily effective, for Parkinson's disease was levo-dopa (Fig. **6**).

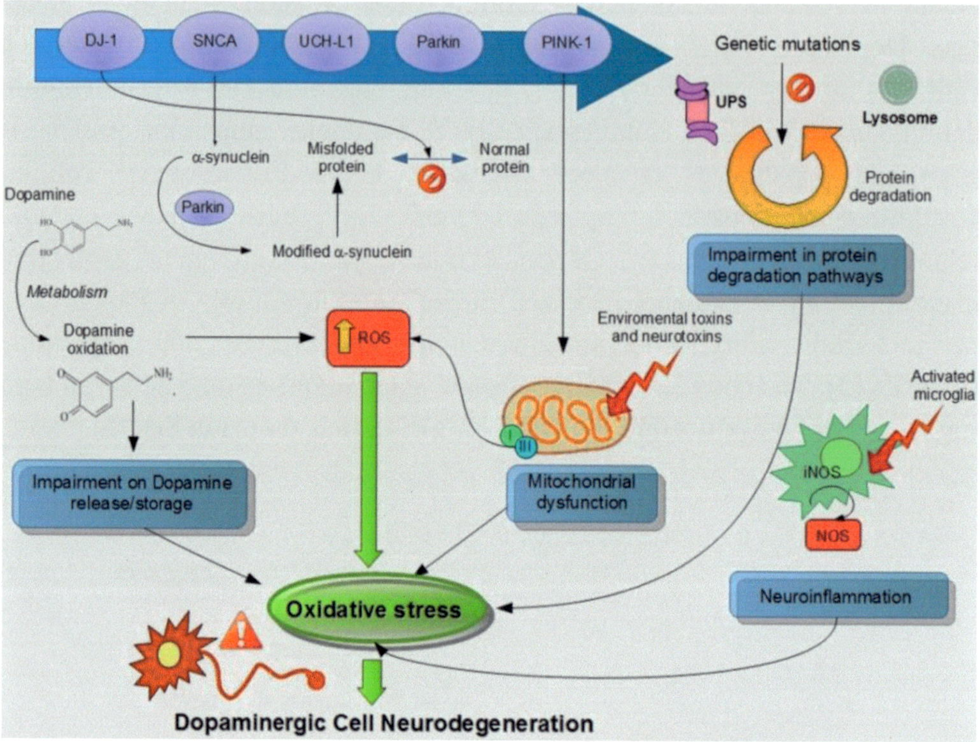

Fig. (5). Interrelationships among dopamine oxidation, mitochondrial dysfunction, oxidative stress, modification of α-synuclein to toxic aggregates (oligomers), and dopaminergic cell neurodegeneration.

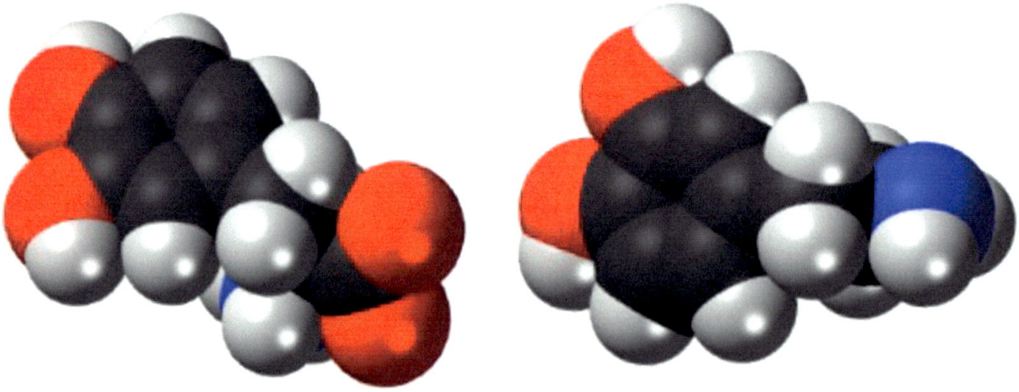

Fig. (6). L-dopa (levo-dopa) (left panel) is converted in the brain to dopamine (right panel) by removal of a carboxyl group (-COOH). Carbon is black, hydrogen is white, oxygen is red, and nitrogen is blue. Due to differences in orientation, the nitrogen atom is barely visible, but is present in L-dopa also.

Dopamine, the brain chemical that is diminished in Parkinson's disease, does not penetrate the blood-brain barrier so it is not a candidate for therapy. Fortunately, the molecule levo-dopa can pass the blood-brain barrier and the brain is capable of converting it to dopamine by removal of a carboxyl (-COOH) group (Fig. **7**). Unfortunately, the initial benefits of levodopa are not sustained with continued therapy and the search for a cure continues.

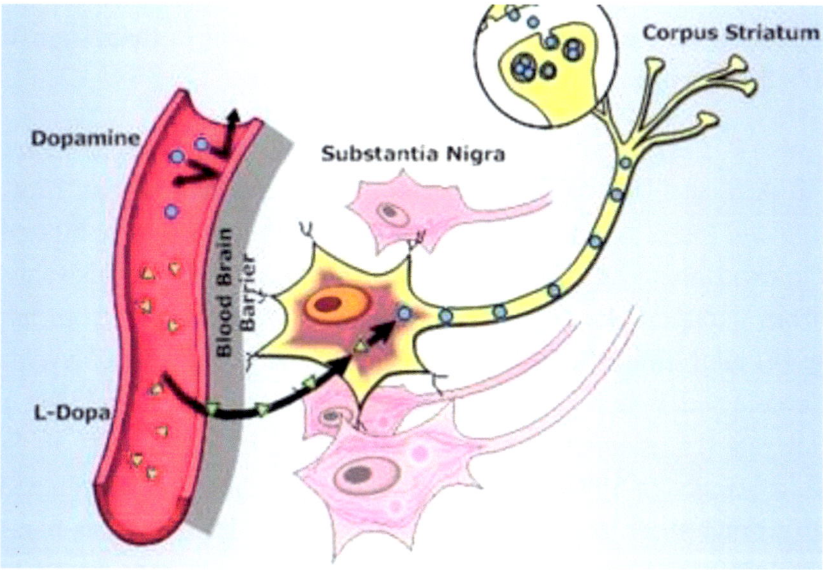

Fig. (7). Dopamine, the neurotransmitter that is deficient in Parkinson's disease, if given therapeutically would not penetrate into the brain; however, L-dopa can pass the blood-brain "barrier" and is then metabolized to the effective transmitter, dopamine.

Sources of Oxidant Stress

Oxidative stress results when there is an imbalance between production of reactive oxygen species and cellular antioxidant activity. A possible source of reactive oxygen species related to Parkinson's disease is from the normal metabolism of dopamine itself as described by G. Cohen [8], see Fig. (**5**). Additionally, it was reported that Parkinson's disease patients who had polymorphisms in specific genes coding for cytochrome P450 enzyme (CYP2D6), which is responsible for detoxifying certain xenobiotics, appeared to be at greater risk for developing the early-onset form of Parkinson's disease [9]. Fig. (**1**) traces a pathway environmental toxins and neurotoxins for the redox-active herbicide paraquat to penetrate into the brain through a membrane carrier and interact with dopamine transporter to redox-cycle with generation of superoxide which can react with α-synuclein to induce aggregation, and Lewy bodies can form. The herbicide dieldrin can cause formation of reactive oxygen species by aberrant oxygen metabolism in mitochondria with damage to cytochrome C, caspases activation, and apoptosis. Paraquat (as one example) can react to destroy dopamine by redox chemistry. The mechanisms shown in Fig. (**5**) should be considered potentially valid but speculative with respect to their significance in Parkinson's disease in any given individual, in my opinion.

Research with P. L. Gilliland from my laboratory and with the other colleagues A. C. Lees, J. A. Fee, Pl. L. Block, and F. C. Neidhardt as early as 1980 [10] in *Escherichia coli* (used as a cellular model) showed that paraquat inhibited growth similarly to exposure to hyperoxia. Research from my laboratory also showed that toxicity from paraquat shared other mechanisms with toxicity from hyperoxia, including: (a) inhibition of NAD biosynthesis by paraquat in *Escherichia coli* in research conducted with M. Heitkamp [11]; (b) that niacin reduced paraquat toxicity for rats (in research conducted with M. Heitkamp and C. S. Song and published in Science in 1981) [12]; (c) with R. L Seither that oxygen and certain redox-active drugs share toxicity sites [13]; (d) with R. L Seither that paraquat and nitrofurantoin (redox-active compounds) induced stringency in *Escherichia coli* [14]; (e) with H. S. Amash, W. E. Perkins, and R. L. Schroder that doxorubicin (a redox-active cancer chemotherapeutic with the side-effect of heart damage) decreased cardiac pyrimidine nucleotide coenzymes in the Bertazzoli mouse model [15]; (f) with R. L Seither that paraquat inhibits NAD biosynthesis at the quinolinic acid synthetase site by inducing stringency [14]; and (g) with G. Siwecki that overproduction of superoxide dismutase does not protect *Escherichia coli* from stringency-induced growth inhibition by paraquat [16]. Overall, these studies support the thesis that redox-active chemicals and exposure to elevated concentrations of oxygen share the toxic mechanism of excess production of reactive oxygen species, but that more than the superoxide radical is involved.

Parkinson's Pathophysiology

Details of the mechanisms and sites of toxicity responsible for the pathophysiology of Parkinson's disease are shown in Fig. (**8**). The α-synuclein aggregation pathway is depicted in the (A) panel; factors that modify α-synuclein are shown on the left; and sites affected by toxic species are shown on the right. The α-synuclein disassociates from the membrane site as a natively unfolded monomer and "aggregation" factors (A) modify it to become tangles of microscopic fibrils with formation of Lewy bodies. The mechanisms resulting in aggregation of proteins to produce clumps and eventually Lewy bodies (panel A) are characteristic of the disease. Toxicity also involves apoptosis, impaired energy production, decreased synaptic release, and other effects (Fig. (**8**), panel B).

(A)

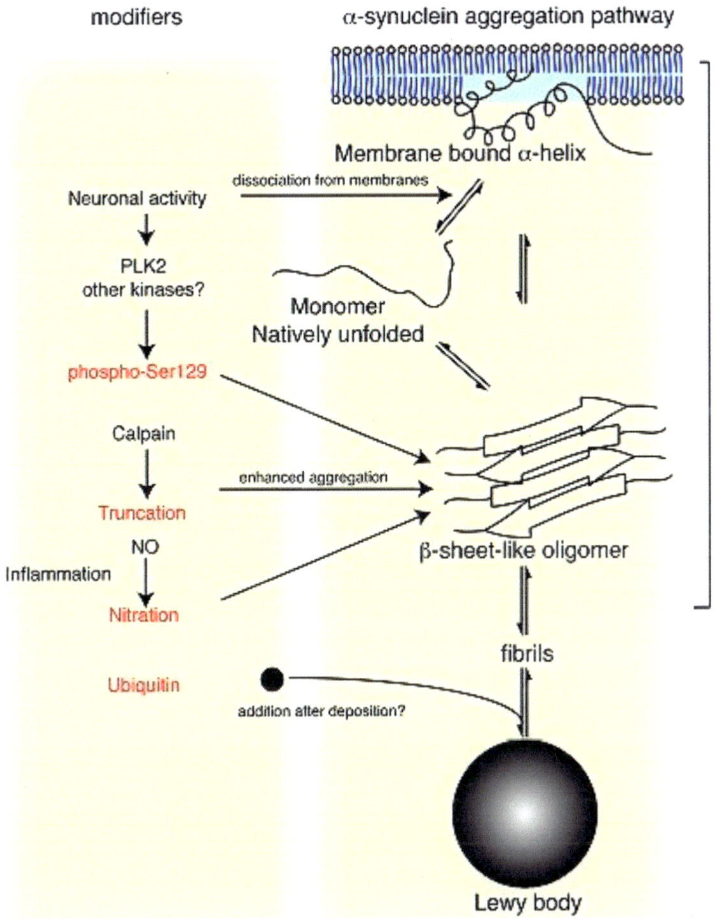

Fig. 8 contd.....

(B)

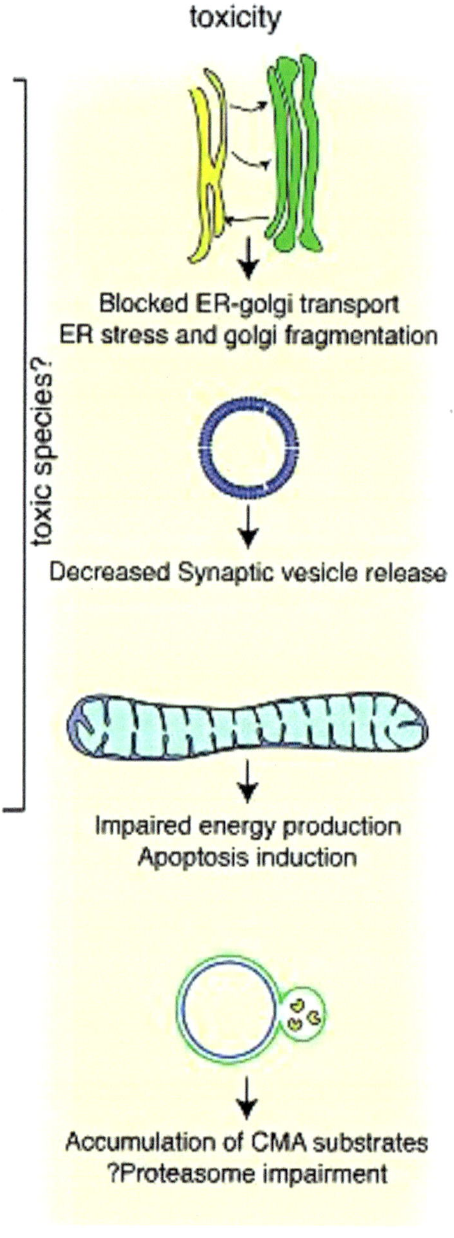

Fig. (8). Sites (**A**) and mechanisms of toxicity (**B**) in Parkinson's disease.

Efforts to clear the brain of the aggregated proteins (α-synuclein) are likely to be ineffective, in my opinion. Devising a means to dissolve, disentangle or otherwise remove these structures from the delicate brain tissue do not seem currently hopeful. A better hope is to find ways of preventing their formation; *i.e.*, more hope for finding preventatives than for finding cures. It does seem clear that oxidant stress is involved, but unclear what causes it to become out-of-control and excessive. Apoptosis is a "normal" process, but can result in destruction of healthy cells when it is not controlled.

Because of the central role of α-synuclein in the neurodegenerative process that is Parkinson's disease, an additional visual aid is provided (Fig. **9**). Perhaps a description of the disease process that relates to the complex technical terms and description of this diagram, but which uses simpler language will be useful.

The top portion of Fig (**9**) shows interactions among three nerve cell bodies in the brain. The disease begins and progresses with the formation of clumps and tangles of protein (shown as packages of arranged fibers shown in green) in cells and that these clumps can be exported to, and taken in by, other brain cells. It also shows that these clumps and tangles can grow in size. Exosomes can transport bundles of α-amyloid fibers, or they can be transported directly into nerve cells and packaged into Lewy bodies, the hallmark of Parkinson's disease.

The bottom portion of Fig. (**9**) is a more detailed (exploded) view of what is occurring in one of the nerve cell bodies (left part of upper diagram). The details are complex and some definition of terms will be helpful for understanding the overall process. Oligomers are collections of protein monomers. They can aggregate into ring-like structures that can leave the cell through pores (shown at the upper left within the box designated by the dotted line). Unfolded monomers and dimers are apparently normal, but can become "propagating" dimers, which means that they began an abnormal aggregation and can receive protein monomers from outside the cell, or from protein monomers made within the cell (shown in the diagram as being cleaved from the cell membrane). The disease results with the aggregation of these monomer proteins to propagating dimers, and larger aggregates: oligomers, small amyloid fibrils (fibrils and small fibers), amyloid fibrils, and Lewy bodies. This sequence is shown along the bottom from left to right within the dotted-line box.

An active field of research is focused on identifying the toxicities of the various forms of α-synuclein and defining their relationships to Lewy body formation, pathology, and symptoms, and devising new strategies for therapy.

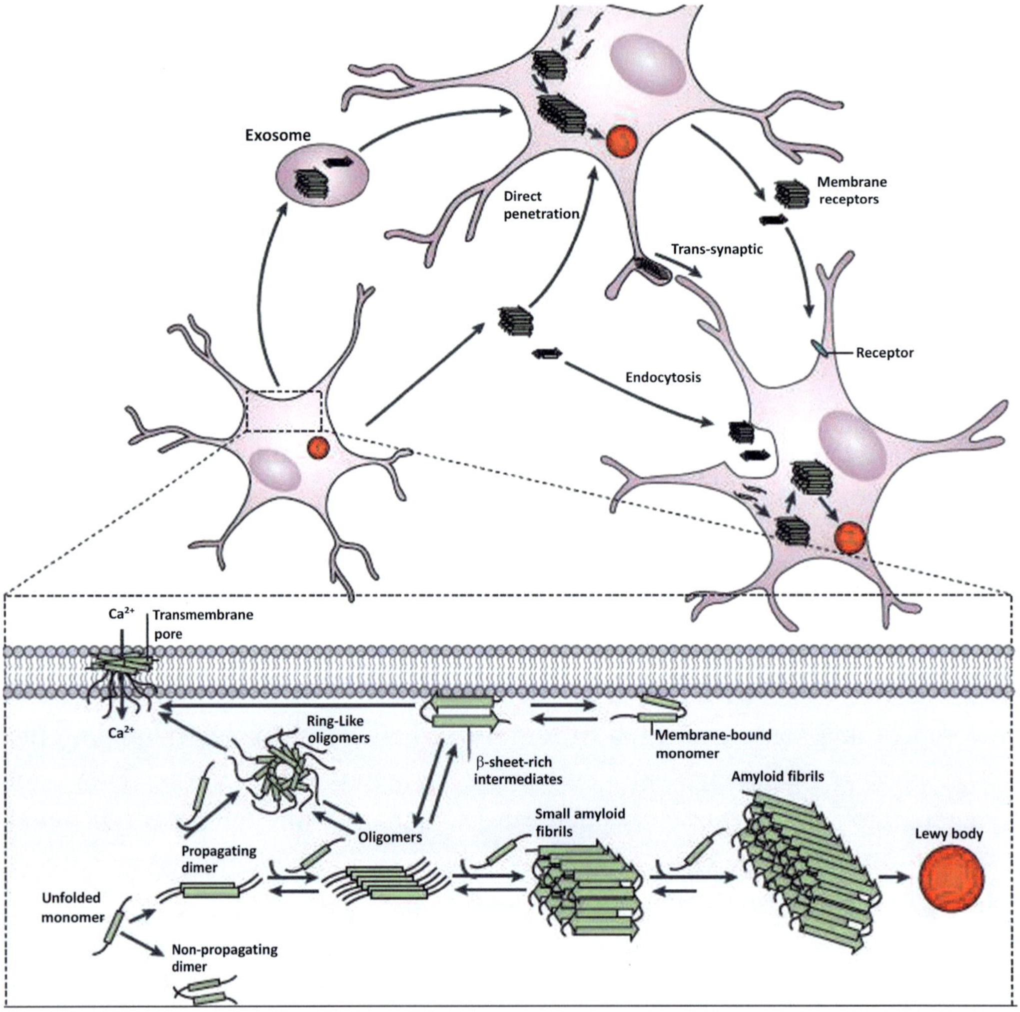

Fig. (9). Interrelationships within a brain cell among the complex mechanisms involving calcium, α-synuclein monomers and oligomers, amyloid fibrils and Lewy body formation.

In a recent 2015 review in the Journal of Neurochemistry, Yan Xu, Yulin Deng, and Hong Qing report [17] that although:

> "... the exact mechanism of PD (Parkinson's disease) is still unclear, α-synuclein is one of the key proteins in PD pathogenesis as it's the main component of the PD hallmark Lewy bodies (LBs)... the study of α-synuclein phosphorylation mechanism related to the PD pathology has become a research hotspot, given that 90% of α-synuclein deposition in LBs is phosphorylated at Ser129, whereas in normal brains, only 4% or less of α-synuclein is phosphorylated at the residue... The phosphorylation of α-synuclein might prove a breakthrough for PD mechanism study and even supply a new therapeutic strategy."

Fig. (**10**) shows anatomical relationships among damage sites in the brain (right of diagram), compared to normal (left of diagram) in Parkinson's disease. In Parkinson's disease the damage is in the substantia nigra (SN, bottom left of diagram). The "normal" connections to the cerebral cortex (shown on the left) and greatly reduced in the disease (shown on the right).

Lewy bodies are the hallmark of Parkinson's disease (Fig. **11**). Intraneuronal accumulation of abnormal α-synuclein fibrils results in Lewy bodies [18]. There are two types of Lewy bodies: brainstem (classical), as shown in Fig. (**11**), and cortical. Both contain α-synuclein: brainstem type Lewy bodies have a dense core of fibrillar and granualr material surrounded by a halo created by fibrills that are radially-oriented. Cortical-type Lewy bodies lack a distinctive core and halo [18]. Lewy body dementia is an umbrella term encompassing two diagnoses: Parkinson's disease and dementia with Lewy bodies. The initial presenting symptoms are somewhat different, but the biological pathologies are the same. Indeed, Lewy bodies can be found also, sometimes, in Alzheimer's disease.

Fig. (**12**) diagrammatically shows positional relationships of the substantia nigra in the midbrain to the forebrain, including various brain areas, in a slice through the brain (left panel). The substantia nigra, the source of Parkinson's pathology, is at the bottom of the diagram. Fig. (**12**) (right panel) depicts pathways to show how damage originating in the substantia nigra results in impairment in motor functions seen in Parkinson's disease. The substantia nigra has nerve connections to putamen and from it to various areas of the brain including for control of motor functions.

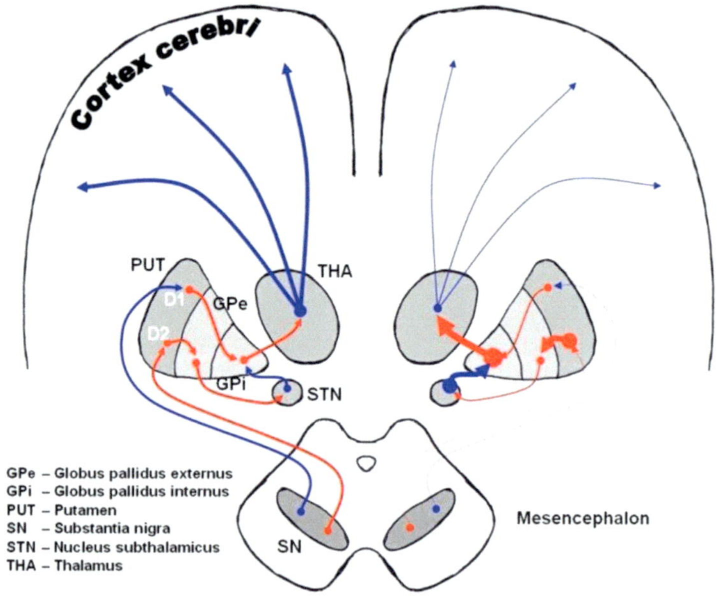

Fig. (10). Sites of brain damage from Parkinson's disease with loss in signalling to the cortex (see text).

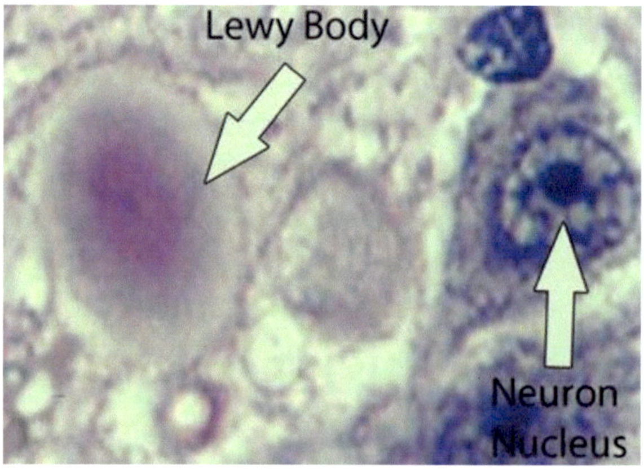

Fig. (11). Classical (brainstem) type Lewy body with dense fibrillar core surrounded by halo.

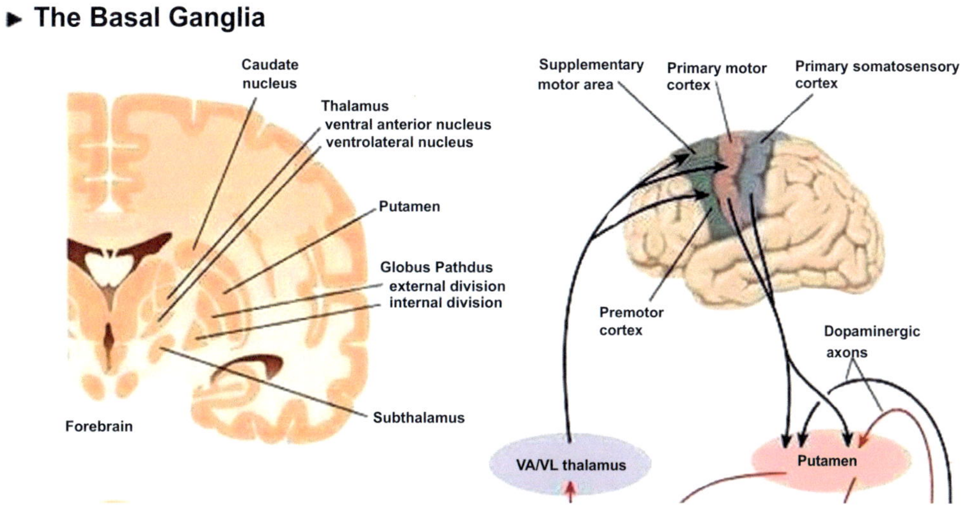

Fig. (12). Anatomical depiction of brain damage sites in Parkinson's disease: forebrain showing the location of the putamen (left panel), and associations of the putamen and various brain regions (right panel).

Diagnosis of Parkinson's Disease

There are no specific tests for diagnosing Parkinson's disease [19]. A neurologist makes a diagnosis of Parkinson's based on the patient's medical history, review of signs and symptoms, and a physical and neurological examination. Blood tests and imaging tests may be required, primarily to rule out other medical conditions.

The usefulness of brain imaging is diminished by the fact that early stages often do not show significant changes from the normal brain. A relatively high dose of levodopa may be given; significant improvement of signs and symptoms confirms the diagnosis of Parkinson's disease [19].

ALZHEIMER'S DISEASE

Introduction

Alzheimer's disease is an irreversible brain disorder. It is the most common form of dementia. It is estimated that 5 million Americans had Alzheimer's disease in 2015 [20] and a report in 2016 placed the number at 5.3 million; 5.1 million of these are age 65 and older; and almost two thirds of Americans with Alzheimer's disease are women [21]. Dementia is defined as a loss of mental ability associated with gradual death of brain cells [21]. It is the 6th leading cause of death among U. S. adults, and this rate is on the rise [20]. Symptoms most often first appear when people are in their mid-60s [20]. An early-onset form of Alzheimer's is known and occurs in people age 30 to 60 [22]. It slowly impairs memory and reasoning and eventually the ability to carry out even simple tasks, including self-care. The disease can be classified as mild, moderate or severe, based on symptoms.

Diagnosis of Alzheimer's

Alzheimer's is not a result of normal aging. The CDC reports [20] that according to the National Institute on Aging, a person with Alzheimer's disease, in addition to memory loss, will experience or present with one or more of the following signs: "gets lost, has trouble handling money and paying bills, repeats questions, takes longer to complete normal daily tasks, displays poor judgment, loses things or misplacing (sic) them in odd places, displays mood and personality changes". These changes are devastating for both the patient and for relatives and loved ones. For a physician's diagnosis, the doctor must be satisfied that there is dementia identified according to guidelines and consisting of cognitive and behavioral symptoms that show a decline from previous levels of functioning and performing and that interfere with ability to function at work or at usual activities [23]. The cognitive decline must be in at least two of five symptom areas specified by the guidelines jointly established by the National Institute on Aging and the Alzheimer's Association, as follows [23]:

1. Worsened ability to take in and remember new information, for example:

- Repetitive questions or conversations
- Misplacing personal belongings
- Forgetting events or appointments
- Getting lost on a familiar route

2. Impairments to reasoning, complex tasking, exercising judgment:

- Poor understanding of safety risks
- Inability to manage finances

- Poor decision-making ability
- Inability to plan complex or sequential activities

3. Impaired visuospatial abilities (but not, for example, due to eye sight problems):

- Inability to recognize faces or common objects or to find objects in direct view
- Inability to operate simple implements, or orient clothing to the body

4. Impaired speaking, reading and writing.

- Difficulty thinking of common words while speaking, hesitations
- Speech, spelling, and writing errors

5. Changes in personality and behavior, for example:

- Out-of-character mood changes, including agitation; less interest, motivation or initiative; apathy; social withdrawal
- Loss of empathy
- Compulsive, obsessive or socially unacceptable behavior

Once the number and severity of these example symptoms confirm dementia, the best certainty that they are because of Alzheimer's disease is given by: "A gradual onset 'Over months to years' rather than hours or days (the case with some other problems) and... marked worsening of the individual person's normal level of cognition in particular areas".

There are additional evaluations and considerations that a physician may consider [23, 24] before confirming whether mental signs and symptoms are severe enough to diagnose.

Three types of brain scans are available for diagnosing Alzheimer's disease. They are used, primarily, for ruling out other diseases. Computerized tomography (CT) scans, computerized axial tomography (CAT) scans, and magnetic resonance imaging (MRI) scans are used for detecting structural brain changes. Single photon emission computerized tomography (SPECT) and positron emission tomography (PET) scans are used less commonly and primarily to detect changes in brain function (blood flow or metabolism). CT and MRI can identify conditions with symptoms similar to dementia including brain bleed, tumor, or fluid build-up. It can be noted that a scan that shows no unexpected changes in the brain does not rule out Alzheimer's disease since in the early stages, brains of Alzheimer's disease patients are quite similar to the normal brain [25].

In the severe stage, there is significant shrinkage of the brain [24] with fewer brain cells and fewer synaptic connections (Fig. **13**). There are other types of dementia and patients can have mixed dementias such as Alzheimer's and vascular dementia. In common with all dementia, Alzheimer's disease is caused by death of brain cells [24].

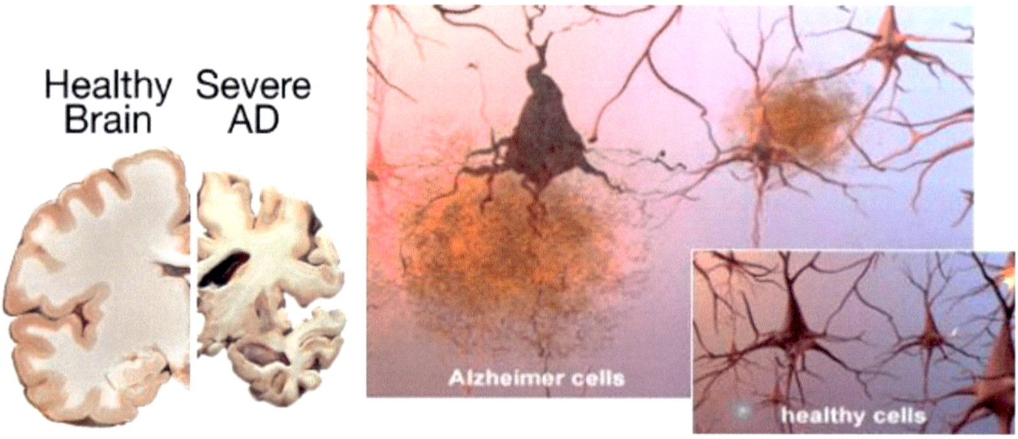

Fig. (13). The brain shrinks in Alzheimer's disease (left panel); and microscopic tangles of tau protein are present in the brain at autopsy (right panel).

Pathophysiology of Alzheimer's

The disease was named for Dr. Alois Alzheimer (1864-1915) who in 1906 observed the symptoms we now ascribe to Alzheimer's in a patient and observed changes in the brain at autopsy [22]. The autopsy changes he observed are now called amyloid plaques and tangled bundles of fibers– tau tangles. These microscopic changes cannot be seen in the living brain (Fig. **13**). The changes are found between dying cells and result from the build-up of a protein, β-amyloid, and tangles within the brain neurons from tau protein [24]. These abnormal inclusions of clumped protein are always present in brain tissue in Alzheimer's disease; however, whether they are cause or consequence of something else at work, is not currently known, and perhaps significantly, similar changes are seen in other brain disorders [24]. There is evidence that brain changes begin pre-clinically a decade or more before memory and cognitive changes are significant [20, 22]. Changes can sometimes be seen by brain imaging (Fig. **14**).

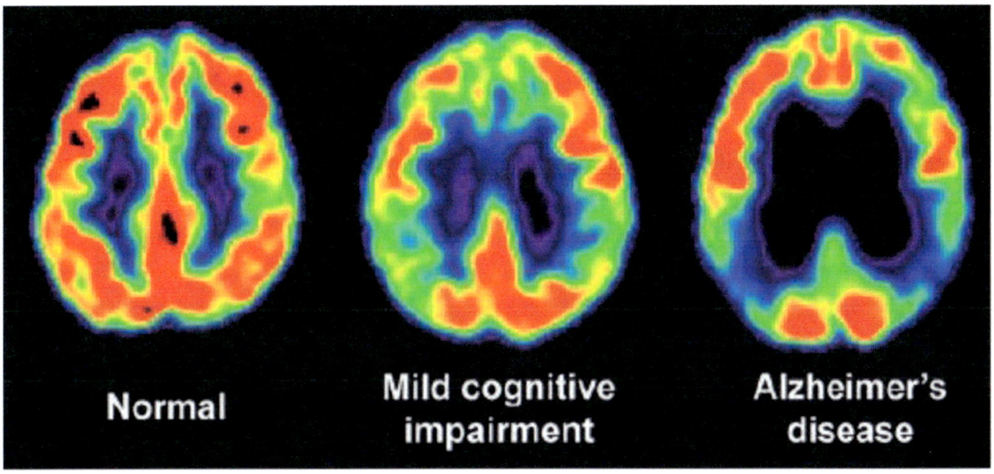

Fig. (14). Brain scan, Alzheimer's.

The cause of Alzheimer's disease in not known; however, genetics is strongly involved. In early-onset Alzheimer's, a genetic mutation is usually present [22]. In most people who develop the late-onset form of Alzheimer's, a particular form of a gene called apolipoprotein E (APOE) is present [22]. The APOE gene is complex and has several different forms. The APOE ε4 gene, for example increases the risk of Alzheimer's but its presence does not mean that a person will develop the disease; also, some people develop the disease but do not have the APOE ε4 gene [22]. Other genes have been identified as associated with an increased risk of Alzheimer's disease, and most people with Down syndrome will develop Alzheimer's. People with down syndrome have an extra copy of chromosome 21 [22].

There is a considerable body of evidence linking Alzheimer's disease with oxidative stress as a mechanism: however there is no proof of causation. β-amyloid, previously mentioned, is a peptide proteolytically cleaved from a trans-membrane protein (amyloid precursor protein) that is generated in brain cells by two membrane-bound protease enzymes, β-secretase and γ-secretase

[26 - 28]. These secretases, by cleavage within the trans-membrane domain, generate several peptides of varying lengths (β-amyloid peptides) with varying neuro-toxic properties [29, 30]. A third cleavage enzyme, α-secretase, prevents production of toxic peptides; consequently increased production or decreased clearance of β-amyloid protein can result. Accumulation of β-amyloid appears to stimulate multiple cell-signaling pathways resulting in degeneration of synapses, death of neurons, and declined cognitive function [27, 29 - 33].

Oxidative Stress-Induced β-amyloid Cyto-toxicity

There is a body of research that strongly implicates excess oxidative stress in neurodegeneration and neurotoxicity from β-amyloid peptide [34]. For example, β-amyloid peptide treatment of cell cultures increased the concentrations of hydrogen peroxide and lipid peroxidases [35]. Using transgenic mouse models, increased hydrogen peroxide and nitric oxide production, and oxidative modification of proteins and lipids were correlated with age-associated β-amyloid accumulation which confirmed in these studies that oxidative stress and β-amyloid are associated [36 - 40]. A role has been suggested for reactive oxygen species toxicity in hippocampal neuronal cell cultures *via* activation of N-methyl-D-aspartate receptors, suggesting a proximal role for soluble β-amyloid oligomers in synaptic impairment and neuronal cell death [41].

Some research, using systems defective in oxidant stress defenses, has indicated that β-amyloid can be promoted by excess oxidant stress [42, 43], and that dietary antioxidants can lower the otherwise increased concentrations of oxidized proteins and lower the burden of plaques [44].

Over-expression of the manganese form of superoxide dismutase (See Chapter 7) in a transgenic mouse model of Alzheimer's disease resulted in reduction of oxidative stress, amyloid deposition, and memory defect [45]. K. Murakami *et al.* reported that in a mouse model of Alzheimer's disease (that over-expresses amyloid precursor protein), deletion of the copper-zinc form of superoxide dismutase (see Chapter 7) resulted in increased β-amyloid oligomerization and losses in spatial learning and memory, suggesting that oxidative stress is involved [46].

The mechanisms by which oxidative stress causes the pathology seen in Alzheimer's disease are not proven and are complex. It is difficult to prove cause and effect but the evidence incriminating oxidant stress is considerable. β-amyloid production is enhanced by increased oxidative stress which decreases the activity of the enzyme α-secretase (to which we previously referred), and promotes expression and activation of γ-secretase (to which we also previously referred). Together, these two enzymes act to enhance formation of β-amyloid peptide from amyloid precursor protein [47 - 51]. These processes involve induction/activation of the major cell-signaling "cascade" (c-Jun N-terminal kinase) which is known to be stimulated by oxidative stress [52, 53]. The terminal kinase enzyme (c-Jun N-terminal kinase) has also been linked to apoptosis [54]. Apoptosis (the Greek meaning is "falling off") is a programmed cell death. Thus, there is a reasonable linkage of molecular mechanisms to pathology and symptoms.

Alternative explanations are also logical. The enhanced production of β-amyloid peptide production by oxidative stress could be compensatory for oxidative stress. Research has shown that oxidative damage to neurons was enhanced in Alzheimer's patients with lesser quantities of β-amyloid peptide deposition and also in patients with shorter disease duration [55, 56]. It also was reported [57] that there was an inverse relationship between the extent of oxidative damage to RNA in neurons and their content of one type of β-amyloid peptide. Nunomura and co-workers [58] have hypothesized that β-amyloid peptide may provide protection against excessive oxidant stress. Also, evidence indicates that low concentrations (nanomolar to picomolar) of β-amyloid peptide is functionally neuroprotective [59, 60] and that physiologic concentrations of β-amyloid peptide efficiently inhibited autoxidation of lipoproteins in cerebrospinal fluid and in plasma while high nanomolar concentrations of β-amyloid peptide resulted in the known pathology characteristic of Alzheimer's disease [61]. These opposite effects of β-amyloid peptide were shown to depend on the state of aggregation of β-amyloid peptide [60, 62].

In summary, excess oxidative stress is best characterized as closely linked to the pathogenesis of neurodegeneration in Alzheimer's disease, and oligomers of β-amyloid cause cognitive impairment and dysfunctional synapses, but cause and effect has not been proven and relationships to disease progression is unclear.

Oxidative Stress and Tau Pathology

In addition to β-amyloid, tau protein is involved in the brain degeneration that occurs with Alzheimer's disease. Tau is a hyperphosphorylated protein that is characteristic of Alzheimer's disease and it correlates with neurodegeneration and with cognitive decline [27]. Tau hyperphosphorylation is important because it impairs the binding with tubulin and interferes with microtubule assembly which causes self-aggregation to form filaments [63]. Tau deserves further explanation.

In Alzheimer's disease and related pathologies, a specific type of slow, progressive neurodegeneration occurs. This degeneration involves an abnormal type of hyperphosphorylation of a particular type of protein called microtubule-associated protein (MAP) tau. This results in neurofibrillary degeneration and neurofibrillary tangles, neuropil threads, and dystrophic neuritis which appears to be required for the clinical expression of Alzheimer's disease [63]. Several phosphatases (enzymes) regulate the extent of phosphorylation of tau, and protein phosphatase-2A is down-regulated in the brain of people with Alzheimer's disease [63]. There is evidence that excess oxidative stress is linked to tau pathology. In particular, it was shown that cells over-expressing tau protein were more susceptible to oxidative stress, perhaps due to depletion of peroxisomes which makes cells vulnerable to oxidative stress and also leads to their degeneration [64]. Among the evidence for a role for oxidative stress were the findings of M. Dumont *et al.*, experimenting with transgenic mice, who reported that the brain of these mice exhibited high levels of oxidative stress including increased protein carbonyl concentrations in cortex mitochondria, alterations in activity and amount of mitochondrial enzymes involved in reactive oxygen species formation suggestive that oxidative stress and mitochondrial dysfunction were involved in tau pathology [65]. These transgenic mice over-expressed the human tau mutated gene. They developed synaptic deficits and microglial activation prior to any neurodegeneration and tangles formation. Changes in mitochondrial proteome appeared prior to tau hyperphosphorylation and tangle formation.

Research indicates that factors other than genetic defects and oxidant stress have some role in Alzheimer's. Relationships are being actively researched between cognitive decline and various vascular conditions including heart disease, stroke, and high blood pressure, and there are associations with certain metabolic diseases including diabetes and obesity [22]. It is commonly said that Alzheimer's probably is caused by a "combination of genetic, environmental, and lifestyle factors" [22] which is a rather non-helpful generalization.

Prospects for Therapy

Changes in the brain in Alzheimer's disease are complex and, in this brief review, not all the mechanisms of the disease and the systems that are affected at the sub-cellular level have been addressed. Fig. (**15**). shows relationships among "activated" microglia, signalling molecules, inflammatory stress, and neuronal death. It includes arrows showing relationships, including those where positive therapy might be effective, and yellow lines where "blocking" therapy might be effective. Therapy directed to anti-oxidant mechanisms are only a small part of potential approaches to therapy.

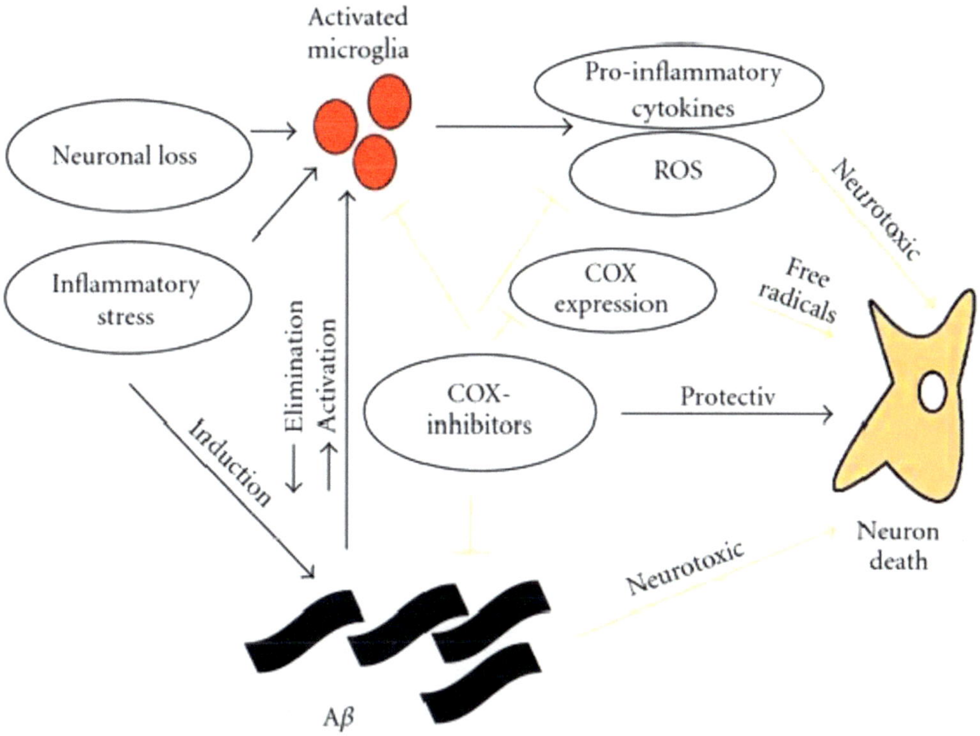

Fig. (15). Alzheimer's disease pathophysiology.

HUNTINGTON'S

Introduction

Huntington's is a rare, inherited disease caused by an autosomal dominant mutant in either of an individual's two copies of a gene called *huntingtin*. Thus, the child of a parent with Huntington's disease has a 50% chance of inheriting the disease. In the example (Fig. **16**), a mother carrying the gene for Huntington's disease and a father without the trait, can produce male and female children as shown in the first line of descent that will be: affected female, affected male or normal male, and normal female. In subsequent lines of descent, the possibilities also are shown. The disease does not "skip" a generation.

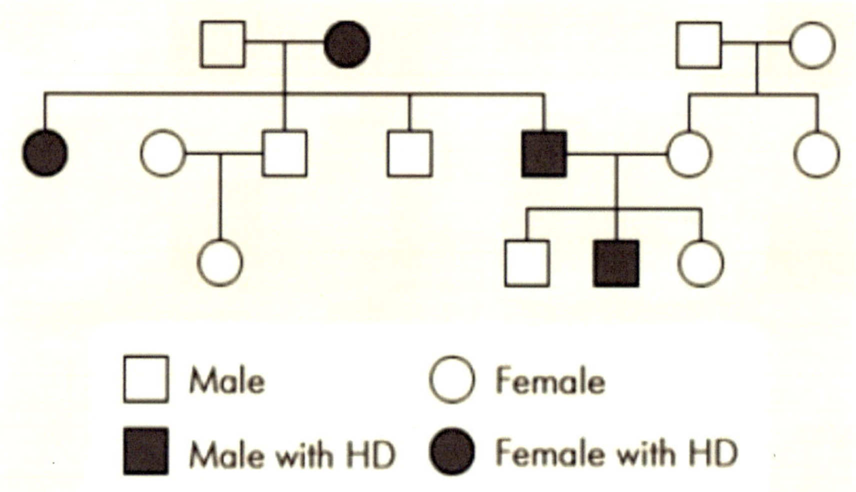

Fig. (16). Huntington's is an inherited trait.

Genetic testing, even before symptoms occur, can detect the disease. This has caused ethical considerations. Some people are more stressed knowing, than not knowing, that they have the mutant gene. The disease is autosomal dominant. This is a pattern of inheritance where the simplest case is that an individual has one copy of a mutant gene (in this case, the *huntingtin* gene) and one copy of the normal gene for this trait. As stated, this produces a 50-50 chance that offspring will have the mutant gene. If both parents have the mutant gene, the risk is 75%, and when either parent has two copies, 100% of children will be affected.

Oxidant Stress

Because Huntington's disease is a genetically inherited disease, justification for including it in this Chapter is required. The precise mechanisms of neuronal death are not known, and evidence from many sources suggest that free radicals may be involved including energy metabolism dysfunction and increased oxidative stress may be involved [66]. In studies reported during the 5 years after the elucidation of the genetic mechanism of the disease, oxidative damage indications included malondialdehyde, heme oxygenase and other products in areas of brain where degeneration occurs [66]. Although the abnormal protein found in Huntington's disease is expressed throughout various organs, cell degeneration occurs primarily in the brain and evidence is suggestive that overproduction of reactive oxygen species with an imbalance of pro-oxidant and antioxidant systems results in oxidative damage to proteins, lipids and DNA [67]. In that review, the evidence for the use of antioxidants such as essential fatty acids, coenzyme Q10, and creatine as potential therapies was also reviewed. A balanced view of the potential of antioxidants as therapy [68] states:

> "Free radicals contribute to chronic diseases from cancer to heart disease and Alzheimer's disease to vision loss. This doesn't automatically mean that substances with antioxidant properties will fix the problem..."

The concept of "oxidative stress" was addressed succinctly and authoritatively by McCord and coworkers [69] and we shall review it in the separate section "Oxidant Stress" in this Chapter.

Research with coworkers in my laboratory was the first to report evidence of oxygen toxicity for amino acid biosynthesis [70], the kynurenine pathway [71], tryptophan metabolism [72], comparative effects on enzymes in the kynurenine pathway [73], effects on 3-hydroxyanthranilate oxidase [74], and protein A of quinolinate synthetase [75]. This was described in Chapter 7. Subsequently, Stoy, *et al.* [76] reported that abnormalities in the kynurenine pathway, related to oxidative stress may have a role in Huntington's disease.

Pathogenesis and Symptoms

Huntington's causes neurodegeneration with loss of muscle coordination, mental decline and behavioral symptoms. The disease usually progresses slowly and predictably with earliest symptoms of subtle changes in mood or cognition [77]. As the disease worsens, there is loss of coordination and gait problems that progress to uncoordinated, jerky body movements [77]. The condition is sometimes known as Huntington's chorea because of the involuntary movements. Physical symptoms do not usually begin until age 35-44; however, the disease has variants named: juvenile, akinetic-rigid, and Westphal variant HD. Woody Guthrie, a song writer and performer recognized for his music in the depression era is perhaps the best know person to have had the disease.

Fig. (17). George Huntington as a young man; he was only 22 when he described the disease Huntington.

George Huntington (1850-1916) (Fig. **17**) was the first to thoroughly describe the condition, although what is likely the same condition had been recognized as a pathology since the Middle Ages [78]. Huntington (in 1872), when he was 22 years old, in one of only two papers he ever published [78], described the heritable nature of the disease as follows:

> "When either or both the parents have shown manifestations of the

disease..., one or more of the offspring almost invariably suffer from the disease... But if by any chance these children go through life without it, the thread is broken and the grandchildren and great-grandchildren of the original shakers may rest assured that they are free from the disease."

Huntington's Genetics

Everyone has two copies of the *huntingtin* gene (*HTT*). This gene codes for the huntingtin protein (HTT) (Fig. (**18**), left panel). The huntingtin mutation causes the length of a repeated section of a gene (the *HTT* gene) on chromosome 4 to have a number of repeats that exceed the normal range of repeats (generally less than 36 repeats) [79]. The repeats code for the amino acid glutamine and result in translation into a protein that is the "mutant" huntingtin protein which damages certain types of neurons in the brain (Fig. (**18**), right panel) although the protein is expressed in cells throughout the body where it can also cause pathological expressions of abnormality including muscle atrophy, cardiac failure, impaired glucose tolerance, weight loss, and osteoporosis, for example.

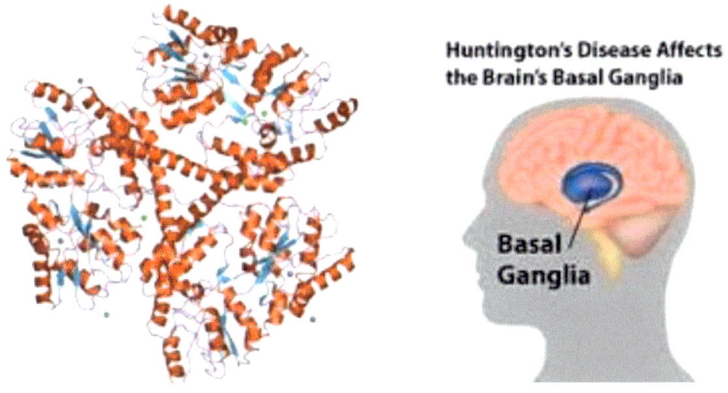

Fig. (18). Ribbon structure of Huntingtin protein (left); site of Brain damage in Huntington's disease (right).

The genetics of Huntington's disease is somewhat complex. Everyone has two copies of the *huntingtin* gene which encodes the genetic information for making the huntingtin protein (note the spelling difference for the man Huntington). The nucleotide triplicate CAG, the part of the gene that codes for amino acid glutamine, may be repeated in normal genomes, but the number of repeats is increased in people with Huntington's disease (Fig. **19**).

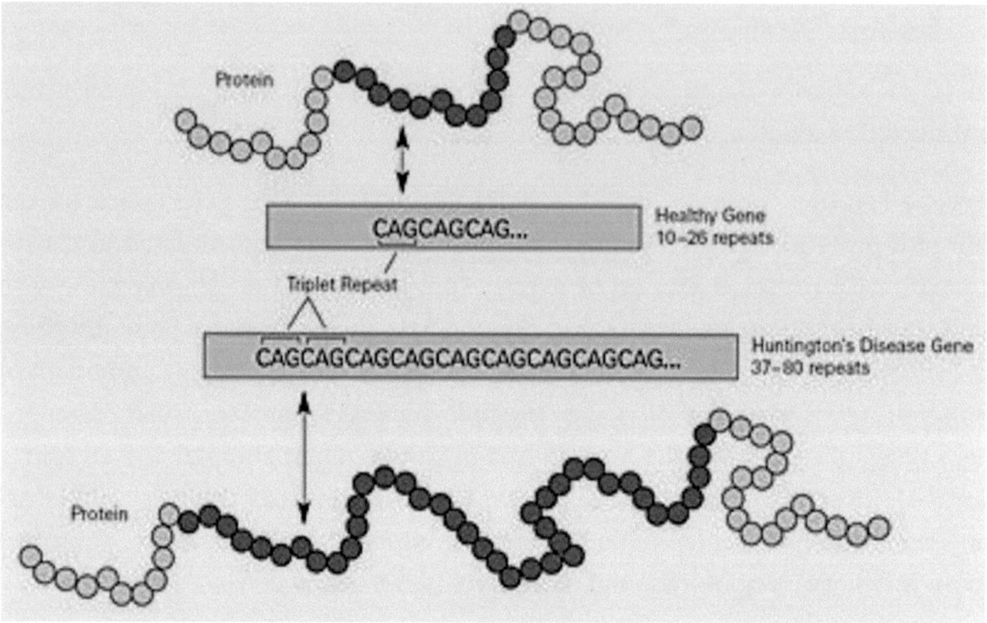

Fig. (19). Mutated *huntingtin* gene has excessive repeats of CAG that results in toxic huntingtin protein.

When the number (and length) of this repeated information in the gene code reaches approximately 36 (or more) glutamines, an altered protein results (Fig. **19**). These altered protein molecules are the cause of the pathology seen in the disease. A significant pathology is that it increases the decay rate of certain types of neurons, which are present in different regions to different extents, and the basal ganglia are severely impaired early in the disease [79, 80], but other brain cells are also affected [81]. The basal ganglia function in movement and behavior control but their role is incompletely understood [81].

Future Therapy Prospects

Currently, research is focused toward identifying the functions of the protein huntingtin (HTT). The mutated *huntingtin* (mHTT) differs from or interferes with the native protein, and is related to brain pathology [82]. Transgenic animal models have been developed including: nematodes, fruit flies, mice, rats, sheep, pigs and monkeys that express mutant huntingtin and develop progressive signs that mimic the disease [82]. Therapeutic strategies include: reducing the amount of the mutant huntingtin protein (including by gene silencing), improving neuronal survival by reducing the damage to specific pathways that occurs *via* the mutant protein, and replacing lost neurons [83].

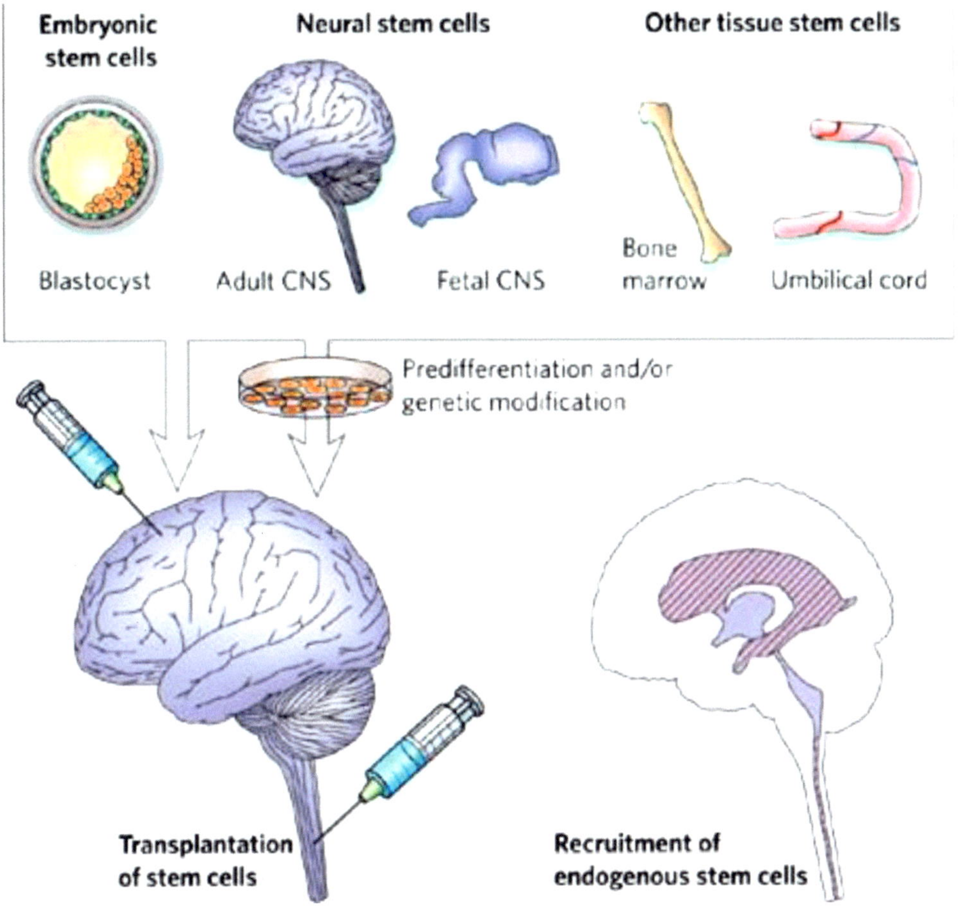

Fig. (20). Stem cell therapy briefly explained.

Genetic silencing is directed toward reducing the production of mutant protein, based on the fact that a single dominant gene causes the production of the toxic protein characteristic of the disease, and gene suppression has already shown positive results [84]; safety has been demonstrated in primates [85]; and trials with human patients began in 2015 [86].

Research approaches to improve cell survival include: correcting transcriptional regulation using inhibitors of histone deacylase, modulation of huntingtin aggregation, restoring damage to mitochondrial metabolism, and repairing synapses to restore function [84].

Stem cell therapy is an exciting approach with promise that lost neurons can be replaced (Fig. **20**). Stem cell research has already been used to study the disease process [87], and there have been experiments to replace cells in afflicted brain regions with animal models of Huntington's disease and even preliminary human clinical trials with mixed results [88]. As of this writing, several clinical trials are planned and/or "in the pipeline [83].

Several chemicals have failed as therapy or prevention (including remacemide, coenzyme Q10, riluzole, creatine, minocycline, ethyl-EPA, phenylbutyrate and dimebon) [89], and I am not aware of any exciting progress on the horizon.

NOTES

[1] Seneca the Younger (5? B.C.- A. D. 65). "Of Peace of Mind" (5), *Minor Dialogues,* tr. Aubrey Stewart, 1889.

[2] James Parkinson was the first to publish a description of the disease that bears his name in his classic 1817 monograph "*Essay on the Shaking Palsy"*.

[3] With Paul Greengard and Eric R. Kandel. The Nobel Prize in Physiology or Medicine 2000, *Nobelprize.org*.

REFERENCES

[1] Hwang O. Role of oxidative stress in Parkinson's disease 2013. [http://dx.doi.org/10.5607/en.2013.22.1.11]

[2] Surmeier DJ, Guzman JN, Sanchez-Padilla J, Goldberg JA. The origins of oxidant stress in Parkinson's disease and therapeutic strategies 2011; 14(7): 1289-301. [http://dx.doi.org/10.1089/ars.2010.3521]

[3] Hsu LJ, Sagara Y, Arroyo A, *et al.* Alpha-synuclein promotes mitochondrial deficit and oxidative stress. Am J Pathol 2000; 157(2): 401-10. Available from: http://www.ncbi. nlm.nih.gov/ pubmed/ 10934145 [cited 2016 Jun 25]

[4] Jenner P, Olanow CW. Oxidative stress and the pathogenesis of Parkinson's disease 1996; 47(6(suppl 3)): 1615-705.
[http://dx.doi.org/10.1212/WNL.47.6_Suppl_3.161S]

[5] Henchcliffe C, Beal MF. Mitochondrial biology and oxidative stress in Parkinson disease pathogenesis 2008; 4(11): 600-9.
[http://dx.doi.org/10.1038/ncpneuro0924]

[6] Dauer W, Przedborski S. Parkinsons disease: mechanisms and models. Neuron 2003; 39(6): 889-909.
[http://dx.doi.org/10.1016/S0896-6273(03)00568-3] [PMID: 12971891]

[7] Conway KA, Rochet JC, Bieganski RM, Lansbury PT. Kinetic stabilization of the alpha-synuclein protofibril by a dopamine-alpha-synuclein adduct 2001; 299(5545): 1346-9.
[http://dx.doi.org/10.1126/science.1063522]

[8] Cohen G. Oxy-radical toxicity in catecholamine neurons. Neurotoxicology 1984; 5(1): 77-82.
[PMID: 6326007]

[9] Sandy MS, Armstrong M, Tanner CM, *et al.* CYP2D6 allelic frequencies in young-onset Parkinsons disease. Neurology 1996; 47(1): 225-30.
[http://dx.doi.org/10.1212/WNL.47.1.225] [PMID: 8710083]

[10] Brown OR, Gilliland PR, Lees AC, Fee JA, Block PL, Neidhardt FC. oxygen stasis of bacterial growth: analogy between the stasis of E. coli by hyperbaric oxygen and by aerobic paraquat. Biochem Int 1980; 1: 304-11.

[11] Heitkamp M, Brown OR. Inhibition of NAD biosynthesis by paraquat in *Escherichia coli.* 1981; 676(3): 345-9.
[http://dx.doi.org/10.1016/0304-4165(81)90169-0]

[12] Brown OR, Heitkamp M, Song CS. Niacin Reduces Paraquat Toxicity in Rats 1981; 212(4502): 1510-2.
[http://dx.doi.org/10.1126/science.7233236]

[13] Brown OR, Seither RL. 1983. Oxygen and redox-active drugs: Shared toxicity sites. Toxicol Sci. 1983 ;3(4):209–14. Available from: http://www.ncbi.nlm. nih.gov/ pubmed/6195038

[14] Seither RL, Brown OR. 1984. Paraquat and nitrofurantoin inhibit growth of Escherichia coli by inducing stringency. J Toxicol Environ Health. 1984; 14(5-6):763–71. Available from: http://www.ncbi.nlm.nih.gov/pubmed/6394769

[15] Brown OR, Amash HS, Perkins WE, Schroeder RL. Doxorubicin decreases cardiac pyridine nucleotide coenzymes in the Bertazzoli mouse model. Med Sci Res 1987; 15: 497-8.

[16] Siwecki G, Brown OR. 1990. Overproduction of superoxide dismutase does not protect Escherichia coli from stringency-induced growth inhibition by 1mM paraquat. Biochem Int. 1990; 20(1):191–9. Available from: http://www.ncbi.nlm.nih.gov/pubmed/2183802

[17] Xu Y, Deng Y, Qing H. 2015. The phosphorylation of α-synuclein: development and implication for

the mechanism and therapy of the Parkinson's disease. J Neurochem. 2015; 135(1):4–18. Available from: http://doi.wiley.com/10.1111/jnc.13234

[18] Kalia LV, Kalia SK. 2015. α-Synuclein and Lewy pathology in Parkinson's disease. Curr Opin Neurol. 2015; 28(4): 375–81. Available from: http://content.wkhealth.com

[19] Parkinson's disease, Tests and diagnosis. Available from: http://www.mayoclinic.org/ diseases-conditions/ parkinsons-disease/ basics/tests-diagnosis/CON-20028488 [cited 2016 Jun 28]

[20] Alzheimer's disease. Available from: http://www.cdc.gov/ aging/ aginginfo/alzheimers.htm [cited 2016 Jun 21]

[21] MacGill M. Alzheimer's disease: causes, symptoms, and treatments. Available from: http://www.medicalnewstoday.com/articles/159442.php [cited 2016 Jun 21].

[22] Alzheimer's Disease Fact Sheet. Available from: https://www.nia.nih.gov/ alzheimers/ publication/ alzheimers-disease-fact-sheet [cited 2016 Jun 21].

[23] McKhann GM, Knopman DS, Chertkow H, *et al.* The diagnosis of dementia due to Alzheimers disease: recommendations from the National Institute on Aging-Alzheimers Association workgroups on diagnostic guidelines for Alzheimers disease. Alzheimers Dement 2011; 7(3): 263-9. [http://dx.doi.org/10.1016/j.jalz.2011.03.005] [PMID: 21514250]

[24] Dementia: Hope Through Research. Available from: http://www.ninds.nih.gov/ disorders/ dementias/ detail_dementia.htm [cited 2016 Jun 21].

[25] Assessment and Diagnosis. Alzheimer's Society. Available from: https://www.alzheimers.org. uk/site/ scripts/ documents_info.php?documentID=260 [cited 2016 Jun 25].

[26] Querfurth HW, LaFerla FM. 2010. Alzheimer's Disease. N Engl J Med 2010; 362(4):329–44. Available from: http://www.nejm.org/doi/abs/10.1056/NEJMra0909142 [cited 2016 Jun 25]

[27] Goedert M, Spillantini MG. A century of Alzheimer's disease 2006. Science. American Association for the Advancement of Science; 2006; 314(5800):777–81. Available from: http://www.ncbi.nlm.nih.gov/pubmed/17082447 [cited 2016 Jun 25] [http://dx.doi.org/10.1126/science.1132814]

[28] Steiner H. Uncovering gamma-secretase 2004. Curr Alzheimer Res. 2004; 1(3):175–81. Available from: http://www.ncbi.nlm.nih.gov/pubmed/15975065 [cited 2016 Jun 25]

[29] Walsh DM, Selkoe DJ. 2007. A beta- Oligomers - a decade of discovery. J Neurochem. Blackwell Publishing Ltd; 2007; 101(5):1172–84. Available from: http://doi.wiley.com/ 10.1111/ j.1471-4159. 2006.04426.x [cited 2016 Jun 25]

[30] Yankner BA, Lu T. 2009. Amyloid beta-protein toxicity and the pathogenesis of Alzheimer disease. J Biol Chem. 2009; 284(8):4755–9. Available from: http://www.ncbi.nlm.nih.gov/pubmed/18957434 [cited 2016 Jun 25]

[31] Roberson ED, Mucke L. 2006. 100 years and counting: prospects for defeating Alzheimer's disease. Science. American Association for the Advancement of Science; 2006; 314(5800):781–4. Available from: http://www.ncbi.nlm.nih.gov/pubmed/17082448 [cited 2016 Jun 25]

[32] Maccioni RB, Muñoz JP, Barbeito L. The molecular bases of Alzheimer's disease and other neurodegenerative disorders; 32(5): 367-81.

[http://dx.doi.org/10.1016/S0188-4409(01)00316-2]

[33] Hsiao K, Chapman P, Nilsen S, *et al.* 1996. Correlative memory eficits, a elevation, and amyloid plaques in ransgenic mice. Science (80-). 1996; 274(5284):99–103. Available from: http://www.sciencemag.org/ cgi/ doi/ 10.1126/ science.274.5284.99 [cited 2016 Jun 25]

[34] Mattson MP. Cellular actions of beta-amyloid precursor protein and its soluble and fibrillogenic derivatives; 77(4): 1081-32 1997.

[35] Behl C, Davis JB, Lesley R, Schubert D. 1994. Hydrogen peroxide mediates amyloid β protein toxicity. Cell. Cell Press; 1994;77(6):817–27. Available from: http://linkinghub. elsevier. com/ retrieve/ pii/ 0092867494901317

[36] Matsuoka Y, Picciano M, La Francois J, Duff K. Fibrillar beta-amyloid evokes oxidative damage in a transgenic mouse model of Alzheimer's disease; 104(3):609-13 2001.

[37] Smith MA, Hirai K, Hsiao K, *et al.* 1998. Amyloid-beta deposition in Alzheimer transgenic mice is associated with oxidative stress. J Neurochem. 1998; 70(5):2212–5. Available from: http://www.ncbi.nlm.nih.gov/ pubmed/ 9572310

[38] Mohmmad Abdul H, Sultana R, Keller JN, St. Clair DK, Markesbery WR, Butterfield DA. Mutations in amyloid precursor protein and presenilin-1 genes increase the basal oxidative stress in murine neuronal cells and lead to increased sensitivity to oxidative stress mediated by amyloid β-peptide (1-42), H2O2 and kainic acid: implications for A. J Neurochem. 2006; 96(5):1322–35. Available from: http://doi.wiley.com/10.1111/j.1471-4159.2005.03647.x

[39] Manczak M, Anekonda TS, Henson E, Park BS, Quinn J, Reddy PH. Mitochondria are a direct site of A beta accumulation in Alzheimer's disease neurons: implications for free radical generation and oxidative damage in disease progression 2006; 15(9): 1437-49

[40] Apelt J, Bigl M, Wunderlich P, Schliebs R. Aging-related increase in oxidative stress correlates with developmental pattern of beta-secretase activity and beta-amyloid plaque formation in transgenic Tg2576 mice with Alzheimer-like pathology. Int J Dev Neurosci 2004; 22(7): 475-84. [http://dx.doi.org/10.1016/j.ijdevneu.2004.07.006] [PMID: 15465277]

[41] De Felice FG, Velasco PT, Lambert MP, *et al.* 2007. Abeta oligomers induce neuronal oxidative stress through an N-Methyl-D-aspartate receptor-dependent mechanism that Is blocked by the alzheimer drug memantine. J Biol Chem. 2007 Apr 13; 282(15):11590–601. Available from: http://www.jbc.org/cgi/doi/ 10.1074/ jbc.M607483200

[42] Calingasan NY, Yu F, Mauck WM, *et al.* Increased plaque burden in brains of APP mutant MnSOD heterozygous knockout mice. J Neurochem [Internet]. 2004 Jun [cited 2016 Jun 25];89(5):1308–12. Available from: http://www.ncbi.nlm.nih.gov/pubmed/15147524

[43] Nishida Y, Yokota T, Takahashi T, Uchihara T, Jishage K, Mizusawa H. Deletion of vitamin E enhances phenotype of Alzheimer disease model mouse 2006; 350(3): 530-36 2006. [http://dx.doi.org/10.1016/j.bbrc.2006.09.083]

[44] Lim GP, Chu T, Yang F, Beech W, Frautschy SA, Cole GM. The curry spice curcumin reduces oxidative damage and amyloid pathology in an Alzheimer transgenic mouse. J Neurosci. 2001; 21(21):8370–7. Available from: http://www.ncbi. nlm.nih.gov/ pubmed/ 11606625

[45] Dumont M, Wille E, Stack C, Calingasan NY, Beal MF, Lin MT. Reduction of oxidative stress,

amyloid deposition, and memory deficit by manganese superoxide dismutase overexpression in a transgenic mouse model of Alzheimer's disease 2009. [http://dx.doi.org/10.1096/fj.09-132928]

[46] Murakami K, Murata N, Noda Y, *et al.* 2011. SOD1 (copper/zinc superoxide dismutase) deficiency drives amyloid β protein oligomerization and memory loss in mouse model of Alzheimer disease. J Biol Chem. 2011; 286(52):44557–68. Available from: http://www. ncbi. nlm.nih. gov/pubmed/ 22072713

[47] Tamagno E, Bardini P, Obbili A, *et al.* Oxidative stress increases expression and activity of BACE in NT2 neurons. Neurobiol Dis 2002; 10(3): 279-88. [http://dx.doi.org/10.1006/nbdi.2002.0515] [PMID: 12270690]

[48] Oda A, Tamaoka A, Araki W. Oxidative stress up-regulates presenilin 1 in lipid rafts in neuronal cells. J Neurosci Res. 2010; 88(5):1137–45. Available from: http://www.ncbi. nlm.nih.gov/ pubmed/ 19885829

[49] Yoo M-H, Gu X, Xu X-M, *et al.* Delineating the role of glutathione peroxidase 4 in protecting cells against lipid hydroperoxide damage and in Alzheimer's disease. 2010; 12(7): 819-27. [http://dx.doi.org/10.1089/ars.2009.2891]

[50] Chen L, Na R, Gu M, Richardson A, Ran Q. Lipid peroxidation up-regulates BACE1 expression in vivo: a possible early event of amyloidogenesis in Alzheimer's disease. J Neurochem. 2008; 107(1):197–207. Available from: http://www.ncbi. nlm.nih.gov/ pubmed/ 18680556

[51] Quiroz-Baez R, Rojas E, Arias C. Oxidative stress promotes JNK-dependent amyloidogenic processing of normally expressed human APP by differential modification of α-, β- and γ-secretase expression. Neurochem Int. Elsevier; 2009; 55(7):662–70. Available from: http://linkinghub. elsevier.com/ retrieve/ pii/S019701860900196X

[52] Tamagno E, Guglielmotto M, Aragno M, *et al.* Oxidative stress activates a positive feedback between the γ- and β-secretase cleavages of the β-amyloid precursor protein. J Neurochem. Blackwell Publishing Ltd; 2007; 0(0):071115163316002 –??? Available from: http://doi. wiley. com/ 10.1111/ j.1471-4159.2007.05072.

[53] Shen C, Chen Y, Liu H, *et al.* Hydrogen Peroxide Promotes A Production through JNK-dependent Activation of -Secretase. J Biol Chem. American Society for Biochemistry and Molecular Biology; 2008; 283(25):17721–30. Available from: http://www. jbc.org/ cgi/ doi/ 10.1074/ jbc.M800013200

[54] Yao M, Nguyen T-V, Pike CJ. Beta-amyloid-induced neuronal apoptosis involves c-Jun N-terminal kinase-dependent downregulation of Bcl-w. J Neurosci. 2005; 25(5):1149–58. Available from: http://www.ncbi.nlm.nih.gov/pubmed/15689551

[55] Nunomura A, Perry G, Aliev G, *et al.* Oxidative damage is the earliest event in Alzheimer disease. J Neuropathol Exp Neurol. 2001; 60(8):759–67. Available from: http://www.ncbi. nlm.nih.gov/ pubmed/ 11487050

[56] Nunomura A, Chiba S, Lippa CF, *et al.* Neuronal RNA oxidation is a prominent feature of familial Alzheimers disease. Neurobiol Dis 2004; 17(1): 108-13. [http://dx.doi.org/10.1016/j.nbd.2004.06.003] [PMID: 15350971]

[57] Nunomura A, Tamaoki T, Tanaka K, *et al.* Intraneuronal amyloid β accumulation and oxidative

damage to nucleic acids in Alzheimer disease. Neurobiol Dis. Elsevier; 2010; 37(3):731–7. Available from: http://linkinghub.elsevier.com/retrieve/pii/S0969996109003672

[58] Nunomura A, Castellani RJ, Zhu X, Moreira PI, Perry G, Smith MA. Involvement of oxidative stress in Alzheimer disease. J Neuropathol Exp Neurol. 2006; 65(7):631–41. Available from: http://www.ncbi.nlm.nih.gov/pubmed/16825950

[59] Luo Y, Sunderland T, Roth GS, Wolozin B. Physiological levels of beta-amyloid peptide promote PC12 cell proliferation 1996; 217(2-3): 125-8

[60] Plant LD, Boyle JP, Smith IF, Peers C, Pearson HA. The production of amyloid beta peptide is a critical requirement for the viability of central neurons. J Neurosci. Society for Neuroscience; 2003; 23(13):5531–5. Available from: http://www.ncbi.nlm.nih.gov/pubmed/12843253

[61] Kontush A, Berndt C, Weber W, *et al.* Amyloid-β is an antioxidant for lipoproteins in cerebrospinal fluid and plasma. Free Radic Biol Med 2001; 30(1): 119-28. [http://dx.doi.org/10.1016/S0891-5849(00)00458-5] [PMID: 11134902]

[62] Zou K, Gong J-S, Yanagisawa K, Michikawa M. Novel function of monomeric amyloid beta-protein serving as an antioxidant molecule against metal-induced oxidative damage. J Neurosci. 2002; 22(12):4833–41. Available from: http://www.ncbi.nlm.nih.gov/pubmed/12077180

[63] Iqbal K, Liu F, Gong C-X, Alonso AD, Grundke-Iqbal I. Mechanisms of tau-induced neurodegeneration. 2009; 118(1): 53-69. [http://dx.doi.org/10.1007/s00401-009-0486-3]

[64] Stamer K, Vogel R, Thies E, Mandelkow E, Mandelkow E-M. Tau blocks traffic of organelles, neurofilaments, and APP vesicles in neurons and enhances oxidative stress. J Cell Biol. Rockefeller University Press; 2002; 156(6):1051–63. Available from: http://www.jcb.org/ lookup/ doi/ 10.1083/ jcb.200108057

[65] Dumont M, Stack C, Elipenahli C, *et al.* Behavioral deficit, oxidative stress, and mitochondrial dysfunction precede tau pathology in P301S transgenic mice. FASEB J. 2011; 25(11):4063–72. Available from: http://www.ncbi.nlm.nih.gov/pubmed/21825035

[66] Browne SE, Ferrante RJ, Beal MF. Oxidative Stress in Huntington's Disease Brain Pathol. Blackwell Publishing Ltd; 2006; 9(1):147–63. Available from: http://doi.wiley.com/ 10.1111/ j.1750-3639. 1999.tb00216.x

[67] Gil-Mohapel J, Brocardo PS, Christie BR. The role of oxidative stress in Huntingtons disease: are antioxidants good therapeutic candidates? Curr Drug Targets 2014; 15(4): 454-68. [http://dx.doi.org/10.2174/13894501156661401151113734] [PMID: 24428525]

[68] Kalin S, Phares E, Skerrett P, Dart H. Antioxidants: Beyond the Hype. Available from: https://www.hsph.harvard.edu/nutritionsource/antioxidants

[69] Hybertson BM, Gao B, Bose SK, McCord JM. Oxidative stress in health and disease: the therapeutic potential of Nrf2 activation. Mol Aspects Med 2011; 32(4-6): 234-46. [http://dx.doi.org/10.1016/j.mam.2011.10.006] [PMID: 22020111]

[70] Boehm DE, Vincent K, Brown OR. Oxygen and toxicity inhibition of amino acid biosynthesis. Nature 1976; 262(5567): 418-20. [http://dx.doi.org/10.1038/262418a0] [PMID: 785275]

[71] Dale WE, Dang Y, Amiridze N, Brown OR. Evidence that kynurenine pathway metabolites mediate hyperbaric oxygen-induced convulsions. Toxicol Lett 2000; 117(1-2): 37-43. [http://dx.doi.org/10.1016/S0378-4274(00)00232-0] [PMID: 11033231]

[72] Dale WE, Dang Y, Brown OR. Tryptophan metabolism through the kynurenine pathway in rat brain and liver slices. Free Radic Biol Med 2000; 29(2): 191-8. [http://dx.doi.org/10.1016/S0891-5849(00)00341-5] [PMID: 10980407]

[73] Dang Y, Dale WE, Brown OR. Comparative effects of oxygen on indoleamine 2,3-dioxygenase and tryptophan 2,3-dioxygenase of the kynurenine pathway. Free Radic Biol Med 2000; 28(4): 615-24. [http://dx.doi.org/10.1016/S0891-5849(99)00272-5] [PMID: 10719243]

[74] Dang Y, Xia C, Brown OR. Effects of oxygen on 3-hydroxyanthranilate oxidase of the kynurenine pathway. Free Radic Biol Med 1998; 25(9): 1033-43. [http://dx.doi.org/10.1016/S0891-5849(98)00136-1] [PMID: 9870556]

[75] Dracyznska-Lusiak B, Brown OR. Protein a of quinolinate synthetase is the site of oxygen poisoning of pyridine nucleotide coenzyme synthesis in Escherichia coli 1992. [http://dx.doi.org/10.1016/0891-5849(92)90042-F]

[76] Stoy N, Mackay GM, Forrest CM, *et al.* Tryptophan metabolism and oxidative stress in patients with Huntington's disease. J Neurochem. Blackwell Science Ltd; 2005; 93(3):611–23. Available from: http://doi.wiley.com/10.1111/j.1471-4159.2005.03070.x

[77] Liou S. The behavioral symptoms of Huntington's disease. Huntington's Outreach Project for Education, at Stanford. 2010. Available from: http://web. stanford. edu/ group/hopes/cgi-bin/hopes_test/the-behavioral-symptoms-of-huntingtons-disease

[78] Bellenir K. Huntington Disease, Genetic Disorders Sourcebook. 3rd ed. Bellenir K, editor. Detroit: Omnigraphics; 2004. 159-179 p.

[79] Walker FO. Huntingtons disease. Lancet 2007; 369(9557): 218-28. [http://dx.doi.org/10.1016/S0140-6736(07)60111-1] [PMID: 17240289]

[80] Montoya A, Price BH, Menear M, Lepage M. Brain imaging and cognitive dysfunctions in Huntingtons disease. J Psychiatry Neurosci 2006; 31(1): 21-9. [PMID: 16496032]

[81] Lobsiger CS, Cleveland DW. Glial cells as intrinsic components of non-cell-autonomous neurodegenerative disease. Nat Neurosci 2007; 10(11): 1355-60. [http://dx.doi.org/10.1038/nn1988] [PMID: 17965655]

[82] Ross CA, Tabrizi SJ, Group HC, Langbehn D, Brinkman R, Falush D, *et al.* Huntington's disease: from molecular pathogenesis to clinical treatment. 2011; 10(1): 83-98. [http://dx.doi.org/10.1016/S1474-4422(10)70245-3]

[83] Wild EJ, Tabrizi SJ. Targets for future clinical trials in Huntington's disease: what's in the pipeline? 2014. [http://dx.doi.org/10.1002/mds.26007]

[84] Munoz-Sanjuan I, Bates GP. The importance of integrating basic and clinical research toward the development of new therapies for Huntington disease. J Clin Invest. American Society for Clinical Investigation; 2011; 121(2):476–83. Available from: http://www.ncbi.nlm.nih.gov/pubmed/21285520

[85] Kordasiewicz HB, Stanek LM, Wancewicz EV, *et al.* Sustained therapeutic reversal of Huntington's disease by transient repression of huntingtin synthesis 2012; 74(6): 1031-44 [http://dx.doi.org/10.1016/j.neuron.2012.05.009]

[86] Gallagher J. Landmark Huntington's trial starts. Available from: http://www. bbc.com/ news/ health-34552041

[87] Cundiff PE, Anderson SA. Impact of induced pluripotent stem cells on the study of central nervous system disease. Curr Opin Genet Dev 2011; 21(3): 354-61. [http://dx.doi.org/10.1016/j.gde.2011.01.008] [PMID: 21277194]

[88] Clelland CD, Barker RA, Watts C. Cell therapy in Huntington disease. 2008; 24(3-4): E9. [http://dx.doi.org/10.3171/FOC/2008/24/3-4/E8]

[89] HSG Research Studies Trials in Progress. Available from: http://huntingtonstudygroup.org/ current-clinical- trials

CHAPTER 10

Oxidant Stress: Cardiovascular and Lung Disease, Exercise, and Aging

Abstract: Oxygen utilization by essential metabolic processes inevitably has the potential to damage or destroy essential cell components by oxidative mechanisms. This negative effect of oxygen is called oxidant stress. Over a lifetime oxidant stress contributes to the decreased functioning observed by aging; it occurs during exercise with both detrimental and adaptive potential; and it is associated with or causative for certain lung and cardiovascular diseases. Oxygen uptake and utilization becomes maximal during athletic performances with high workloads for extended time periods. The associated oxidative stress is positive in outcome *via* enhanced antioxidant defenses. There is sufficient evidence that the primary source of excess oxidant stress is at the site of oxygen utilization at the inner membrane of mitochondria. Normally, a small fraction of mitochondrial use of oxygen is diverted from the secure transfer, one at a time, of electrons *via* the cytochrome system to generate water and ATP in a complex, carefully orchestrated process. Four electrons are required to reduce diatomic oxygen completely and partially reduced intermediates, including peroxides, and the free radicals superoxide and hydroxyl radical can form. Oxidative damage occurs, and damaged mitochondria can produce even more free radicals. There are also other biochemical sources and mechanisms that generate oxidative radicals and create oxidant stress. The lung, which is exposed to the highest concentration of oxygen, is especially vulnerable to oxidant stress when hyperoxia is used therapeutically. Lipid membrane damage involving the alveoli, and potentially fibrosis, is a consequence for premature infants and for chronic obstructive pulmonary disease and other lung pathologies in adults. Oxidation of LDL-cholesterol by oxygen free radicals during excess oxidant stress is the initial event in a sequence leading to atherosclerosis and potentially to cardiovascular events and strokes. A natural balance of antioxidant defenses is protective for oxidant stress and is enhanced by antioxidants in the diet and as a positive benefit from exercise. However, there is evidence that, over time, the net effect of oxidant stress contributes to detrimental deteriorations associated with the aging process in the brain and throughout the body.

Olen R. Brown

Keywords: Aerobic metabolism, Aging, Anaerobic metabolism, Atelectasis, COPD, Environmental factors, Exercise, Fibrosis, Foam cells, Genetics, LDL-cholesterol, Longevity, Macrophage, Maximum oxygen uptake, Monocyte, Oxidant stress, Plaque, Prematurity, Second messenger, Supplements.

INTRODUCTION

Oxidant stress, resulting from the necessary biochemistry that occurs in the body is the common thread that connects the subjects of exercise physiology, aging, and certain pathologies in the lung and cardiovascular system described in this chapter. Of course, oxidant stress does not fully explain all the pathology seen in all the medical conditions addressed in this chapter. It would also be incorrect to infer that oxidant stress, however defined, is the singular cause for each of these conditions. However, oxidant stress is an inevitable result of the body's use of oxygen: its transport; oxidative phosphorylation to produce ATP at the mitochondrial level; and redox-reactions that can damage essential molecules and even complex structures in cells. The latter includes: genes, chromosomes, DNA, and associated regulatory factors; proteins functioning as enzymes and essential structural components; the broad group of cellular components that have lipids as components; and other molecules that do not fit easily into defined categories biochemically. Obviously, the topics chosen for this chapter are vast in scope and complexity. Focusing on the role of oxidant stress is not intended to be a clarion call that all answers lie at the doorstep of oxygen.

Some level of oxidant stress in the body is appropriate. A medical dictionary definition is useful as a beginning. Oxidant stress has been defined [1]: "Physiological stress on the body that is caused by the cumulative damage done by free radicals inadequately neutralized by antioxidants and that is held to be associated with aging". The emphasis on aging is appropriate; it indicates that oxidant stress is cumulative and may increase sequentially. In the context of the four areas of this chapter, the stated definition obviously requires broadening.

Normal Level of Oxidative Stress

It is reasonable, given the chemical nature of oxygen and the labile nature of the biological components of living systems that some untoward consequences will occur. One important site (Fig. **1**) is the mitochondrion where oxygen is utilized to bioenergetically produce ATP (oxidative phosphorylation). This site functions in nearly all cells in the body. There are other sources of oxidant stress and the body has an elaborate compensatory system designed to prevent unwanted oxidation and to reverse or restore oxidative damage that does occur.

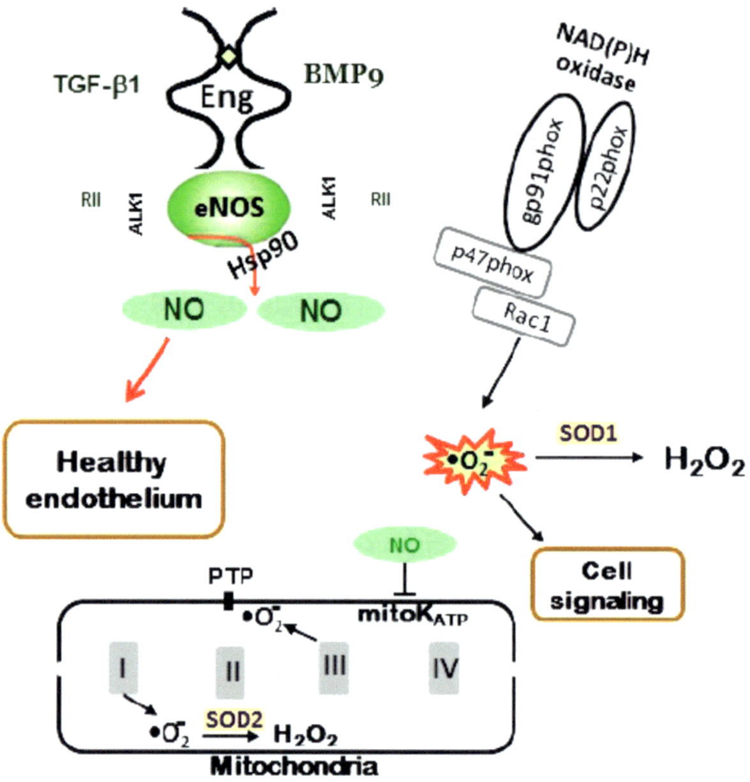

Fig. (1). Sites of oxidant stress at the mitochondrial level.

OXIDATIVE STRESS AND CARDIOVASCULAR DISEASE

A brief overview of biological production of oxygen radicals is useful; for additional details refer to Chapter 7. In the early 1930s Fritz Haber (1868-1934) and Joseph Weiss (1905-1972) proposed that hydroxyl radicals were produced by a reaction of superoxide and hydrogen peroxide. Subsequently this came to be known as the Haber-Weiss reaction. Haber, a German Chemist, was awarded the Nobel prize for Chemistry in 1918 for discoveries related to nitrogen fixation [2].

The hydroxyl radical is extremely reactive and is an extremely strong oxidant–one capable of oxidizing (according to most evidence) any biological molecule, and the rate at which the reaction occurs is so fast that it is diffusion-limited (the rate is as fast as theoretically possible). The reaction is better characterized biologically as the iron-catalyzed Haber-Weiss reaction, or the superoxide-driven Fenton reaction. The latter name is appropriate because more than 100 years ago, Henry Fenton (1854-1929) observed that the reducing agent Fe^{+2} in the presence of hydrogen peroxide was capable of oxidizing various organic compounds (See Chapter 7).

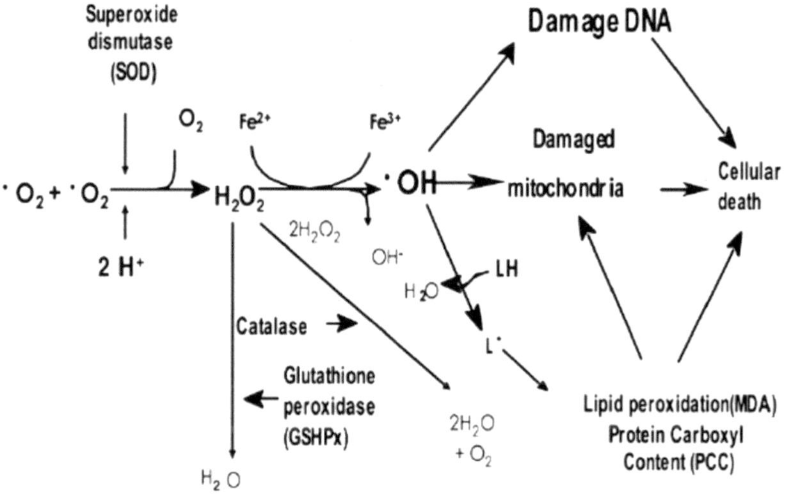

Fig. (2). Overview of cellular sites and mechanisms of oxidant stress.

Some interactions among cell sites and oxidant stress are shown in Fig. (**2**). Attention is called to the heme proteins and metals, particularly to iron. It may be noted that although iron, as Fe^{+2}, is conducive to generation of oxygen free radicals, and potentially damaging in the body, the iron carried on porphyrin rings in hemoglobin is "protected" and safely is transported in biologically very high amounts (See Chapter 8). There are abundant, potential sites of generation of oxygen radicals but the body is in control of and has defenses for these; otherwise, cell death from oxidant stress would be unacceptably high.

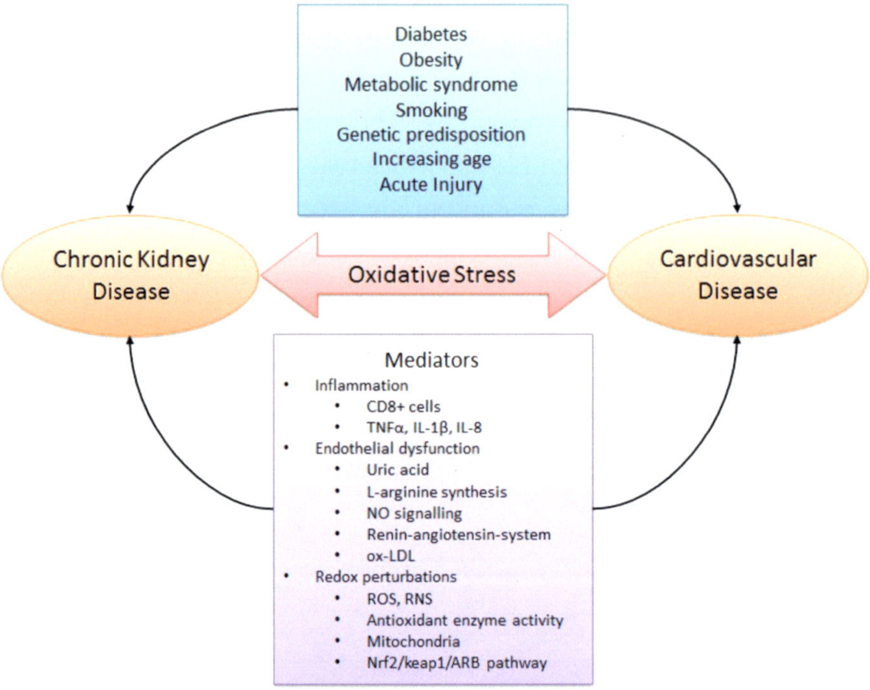

Fig. (3). Relationships of oxidant stress to mediators and diseases, including cardiovascular disease.

There is a level of excessive oxidative stress, stress beyond what the body defenses can manage effectively, that becomes harmful and such stress can be related to diseases including cardiovascular disease (Fig. **3**). What the excess oxidative stress level is, how it can be clinically reduced, and what total impact such management would have on cardiovascular disease is unclear. There are many known mediators of oxidant stress at the cellular level including inflammation, dysfunctions at the endothelial level in blood vessels, and disturbances in particular pathways that generate or modify reactive oxygen species (Fig. **3**). It is beyond the scope here to evaluate the large body of evidences for and against oxidant stress as causation for cardiovascular disease; or the extent to which modification of oxidant stress from these sources would prevent, treat, or modify cardiovascular disease. One thing is clear; cardiovascular disease is multi-factorial. This thought seems to prevent oversimplification of causation and treatment options.

A review article by U. Ravnskov and K. S. McCully [3] states:

> "According to the cholesterol hypothesis, atherosclerosis is initiated by endothelial dysfunction caused by hypercholesterolemia, hyperhomocystemia or other toxic factors. Endothelial dysfunction is said to allow LDL-cholesterol and monocytes to enter the arterial wall, where LDL-cholesterol becomes oxidized and taken up by the macrophages. These events are considered to cause inflammation and to be the starting point of atherosclerosis."

The authors then state six categories of contradictory evidence. However, it is also true that main-stream medicine strongly recommends and prescribes cholesterol-lowering drugs [3]. The evidence on the topic of lipid-lowering drugs and the role of oxidant stress in cardiovascular disease is voluminous. My purpose in this chapter is not to review all this evidence but to present the general line of argument that relates atherosclerotic cardiovascular disease to the subject of this book: oxygen, and specifically to oxygen radicals and oxidant stress. I state clearly that by providing evidence for oxidant stress involvement, I am not denying or even attempting to refute the evidence supporting the current guidelines that advocate and expand the use of statins for prevention and treatment of cardiovascular disease and stroke. No one should modify their personal treatment decisions based on any information in this Chapter or otherwise in this book.

The argument for a role for oxidant stress and oxygen radicals in cardiovascular disease and stroke can be stated as follows.

1. Atherosclerosis has multiple causes (is multi-factorial):
    - The lines of evidence for this are many, including that genetics has a role, life style factors (including diet and exercise) contribute, it is age-related, it is gender-related, population differences are known, lipid-lowering drugs are not proven to <u>prevent</u> a significant percentage of primary cases, associations are known for several medical conditions, there is a relationship to hypertension, there is a relationship to smoking, dyslipidemia, diabetes,

obesity, and there are other factors. These associations are largely uncontested; however, they are associated as risk factors not direct causes.

2. Significant evidence supporting oxidant stress and oxidant radical damage as contributory to atherosclerosis, and cardiovascular disease has been reviewed [4]:
   - LDL-cholesterol has a variety of sizes and densities.
   - Small, dense LDL-cholesterol particles (Fig. **4**) are atherogenic.
   - Small, dense LDL-cholesterol particles are more readily taken up by the arterial endothelial lining than are larger LDL-cholesterol particles.
   - Oxidant susceptibility increases and associated antioxidant concentration decreases with LDL-cholesterol particle size.
   - Interventions to reduce small LDL-cholesterol particles have been shown to be clinically useful.
   - LDL-cholesterol size shows a positive correlation with plasma HDL-cholesterol and a negative correlation with triglyceride concentration; the former is beneficial and the latter negative as a predictor of cardiovascular disease.
3. There is a logical mechanism and sequence for the involvement of oxygen radicals and oxidant stress in atherosclerosis and heart disease as follows:
   - Oxygen, oxygen radicals, leukocytes, macrophages, LDL-cholesterol, foam cells, the blood vessel intima, proteoglycans, chronic inflammation, VCAM-1 and ICAM-1, growth factors, plaque, and thrombus are connected. Monocytes engulf oxidized LDL-cholesterol and start the process of plaque formation (Fig. **5**).
   - Oxidation of LDL-cholesterol is proposed to be the initial step.
   - A search of published articles relating to oxidized low density cholesterol lipoproteins (using several variants as terms) over the years from 1977 through 2013 by Tatsuya Sawamura *et al.* [5], produced 3,357 publications that had been cited in other publications a total of 40,788 times. There were few publications from 1977 through about 1985 and a significant, steady increase after about 1997 with a leveling off about 2010.

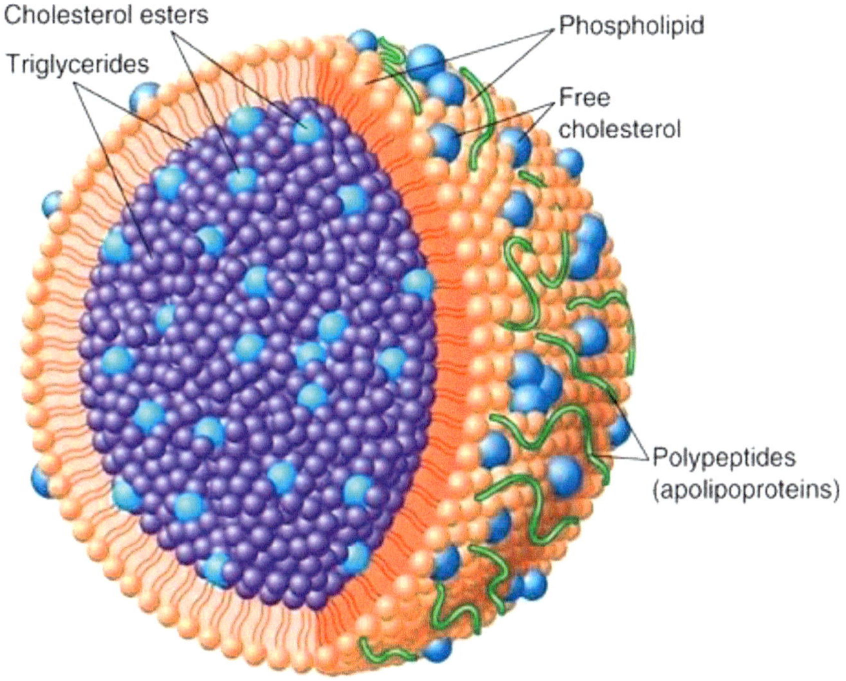

Fig. (4). Low Density Lipoprotein (LDL-cholesterol) cholesterol particle.

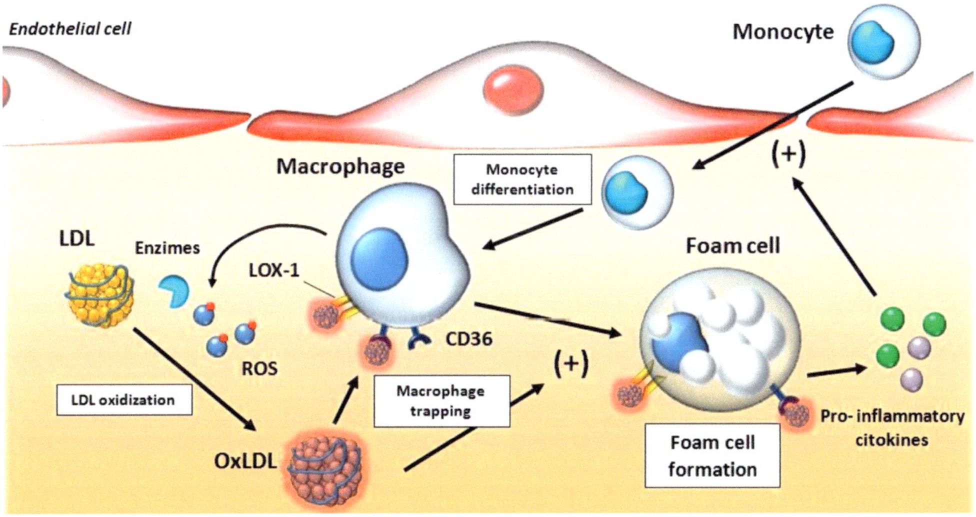

Fig. (5). Details (see text) relating oxidized LDL-cholesterol to foam cell formation.

According to Dayuan Li and Jawahar Mehta [6], it was generally accepted by about 2005 that oxidized LDL-cholesterol is more important than native LDL-cholesterol for atherogenesis, and some of the mechanisms by which oxidized LDL-cholesterol induced endothelial dysfunction were being described as the "earliest" event in atherosclerosis. Included was evidence that Oxidized LDL-cholesterol enhanced expression of pro-inflammatory genes leading to monocyte recruitment into the walls of arteries. Generation of oxygen free radicals is involved and there is impaired nitric oxide gene expression and activity. Inflammatory signaling pathways are activated, including CD40/CD40L; and increased gene expression and activity of metaloproteinases (MMP-1 and -3) occur in endothelial cells.

Atherosclerosis begins with a complex sequence of events initiated by oxidized LDL-cholesterol, a step involving free radicals and oxidant stress. The disease progresses and the principle details have been elucidated and the molecular details are reasonably clear. It is reported that the ratio of oxidized-LDL cholesterol to native LDL-cholesterol is predictive of coronary artery disease and, more specifically, that various calculated ratios involving oxidized LDL-cholesterol and other native cholesterol values are better predictors than measured values of any of the native cholesterols for discrimination between healthy subjects and patients with coronary artery disease [7]. A review published in 2009 by Manfried Rizzo, *et al.* [8] indicates there is continuing support for the concept that oxidized LDL-cholesterol is the significant, early factor, along with dyslipidemia, in the sequence of mechanisms present in atherosclerosis and its associated complications such as frank cardiovascular disease. In 2010, Rizzo, *et al.*, published a paper entitled: "Should we measure routinely oxidized and atherogenic dense low-density lipoproteins in subjects with type 2 diabetes?" They concluded that although routine measurement of particle size oxidized extent was not recommended it was "strongly encouraged" [9].

Dania Mohty, *et al.* in 2007 published research documenting an association between LDL-cholesterol particle size and inflammation in patients with aortic stenosis [10]. In the introduction to the paper, they noted that there was increasing evidence supporting that calcified aortic stenosis was not a passive degenerative disease, but was an active cellular process sharing clinical and histological similarities with atherosclerosis including lipoprotein deposition, chronic inflammation, and tissue remodeling with associated oxidized low-density lipoproteins involvement. As early as 2001, careful work by Akira Kondo, *et al.* [11] led them to conclude that: "The threshold for oxidative susceptibility coincided with the point of LDL size separation between LDL subclass pattern A and B as an atherosclerotic risk". They measured particle size using gradient gel electrophoresis and malondialdehyde-modified LDL measured using ELISA to study the association between triglyceride concentration, LDL particle size and oxidized LDL-cholesterol (identified as "malondialdehyde modified" LDL). They identified a "threshold" for oxidation susceptibility at the LDL size of 25.5 nanometers and a triglyceride concentration of 150 mg/dL.

An editorial [12] published in the International Journal of Clinical Practice in 2007 by A. S. Wierzbicki was entitled; "Quality as well as quantity? Beyond low-density lipoprotein-cholesterol- the role of particle size". This report provides insight into clinical practice at the time. Wierzbicki pointed out that "Hypertriglyceridaemia has always been a controversial risk factor in cardiovascular medicine as epidemiological studies often failed to show an association of triglycerides with CVD risk". It was also pointed out that most cardiovascular risk calculators at that time (it was 2007) used the ratio of total cholesterol to high-density HDL lipoprotein cholesterol because this ratio was the principal lipid determinant of risk and triglycerides correlated so well inversely with HDL-cholesterol that they added little to risk prediction [reference given]".

In a review in 2012, Hui Yang *et al.*, declared oxidized low density lipoprotein (Ox-LDL), to be a risk factor of atherosclerosis; that it facilitated the formation and vulnerability of atherosclerotic plaques and that it contributed to several clinical complications [13]. They then described the status of research directed toward use of stem cells in attempts to repair vascular damage after atherosclerosis. In a section titled "Oxidized low density lipoprotein and atherosclerosis" they point out that: "...the liver loads lipids onto apoB and secretes very-low-density lipoproteins (VLDLs), which undergo lipolysis by lipoprotein lipase to form low-density lipoproteins (LDLs)". Indeed, ApoB is a principal transporter of lipids and is a main component of LDLs. These authors also stated: "LDL is essential for normal cellular function, but in high concentrations, especially after oxidized or carbonylated [two references are provided] it would facilitate atherosclerotic plaque formation".

In a paper published in 2001, A. Mertens and P. Holvoet described the processes and mechanisms by which oxidized LDL-cholesterol is involved in atherothrombosis:

> "Circulating oxidized LDL does not originate from extensive metal ion-induced oxidation in the blood but from mild oxidation in the arterial wall by cell-associated lipoxygenase and/or myeloperoxidase. Oxidized LDL induces atherosclerosis by stimulating monocyte infiltration and smooth muscle cell migration and proliferation. It contributes to atherothrombosis by inducing endothelial cell apoptosis, and thus plaque erosion, by impairing the anticoagulant balance in endothelium, stimulating tissue factor production by smooth muscle cells, and inducing apoptosis in macrophages."

They also noted that HDL-cholesterol levels are inversely related to risk of coronary artery disease (elevated levels are protective) and that HDL "… prevents atherosclerosis by reverting the stimulatory effect of oxidized LDL on monocyte infiltration and that the HDL-associated enzyme paraoxonase inhibits the oxidation of LDL". In my view, this is important because it links an important mechanism of protection (HDL) with the thesis that oxidized LDL-cholesterol is a significant contributor to atherosclerosis. Fig. (**6**) also depicts the sequence of events leading to the development of foam cells and their involvement in plaque formation.

As long ago as 1996 [14] there was published information that myeloperoxidase, a heme protein secreted by phagocytes, is a potent catalyst for LDL-cholesterol oxidation *in vitro*, and noted that the enzyme was present in human atherosclerotic lesions. This report provided new evidence for involvement of myeloperoxidase including finding a "family of chlorinated sterols". Oxidation of LDL-cholesterol by myeloperoxidase required hydrogen peroxide and chloride ion, which they believed indicated that hypochlorous acid was an intermediate in the reaction;

however, they found evidence that Cl_2, derived from HCLO, was the chemical specie responsible for oxidation, or that it may occur in the artery wall.

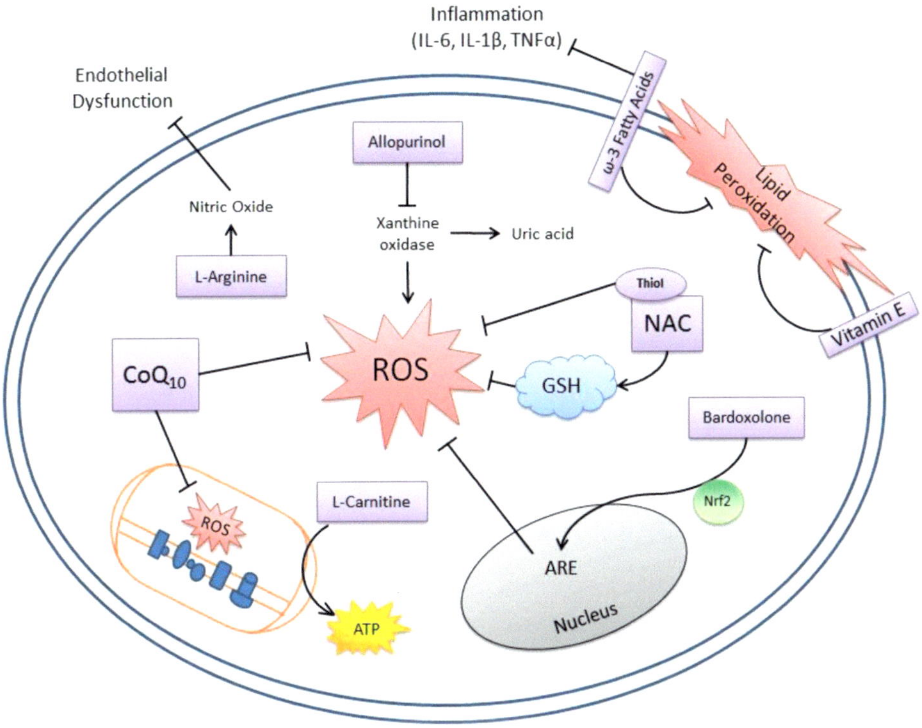

Fig. (6). Relationships among reactive oxygen species, inflammation, molecular damage sites, and defense mechanisms.

Sampath Parthasarathy and co-workers published a report in 2012, entitled "Oxidized Low-Density Lipoprotein" containing 157 cited references [15]. They reported a critical view (somewhat overcritical, in my opinion):

> "Oxidized low-density lipoprotein (Ox-LDL) has been studied for over 25 years. Numerous pro- and anti-atherogenic properties have been attributed to Ox-LDL. Yet, Ox-LDL has neither been defined nor characterized, as its components change depending on its source, method of preparation, storage, and use. It contains unoxidized and oxidized fatty acid derivatives both in the ester and free forms, their

decomposition products, cholesterol and its oxidized products, proteins with oxidized amino acids and cross-links, and polypeptides with varying extents of covalent modifications with lipid oxidation products, and many others. It seems to exist *in vivo* in some form not yet fully characterized. Until its pathological significance, and how it is generated *in vivo* are determined, the nature of its true identity will be only of classical interest."

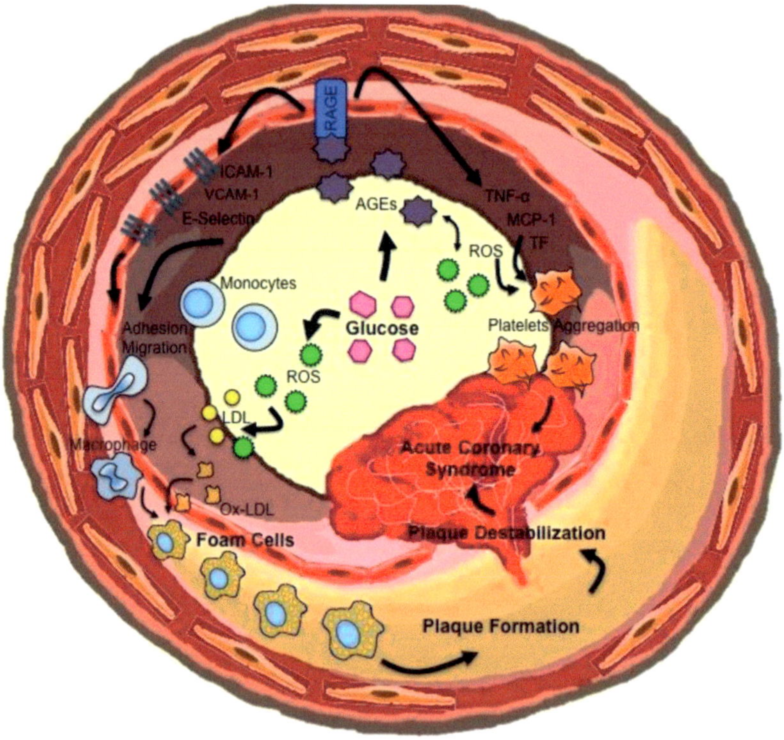

Fig. (7). Depiction of events and factors involved in plaque formation.

The hypothesized sequence of events and processes that link reactive oxygen species with oxidant stress damage and antioxidant protective effects are shown in a general way in Fig. (**6**) Lipid peroxidation, in a cell membrane in this example, is antagonized by vitamin E but can lead to inflammation (cytokines IL-6, IL-1β, and TNFα). Xanthine oxidase can be a source of ROS which is antagonized by Coenzyme Q-10 and reduced glutathione. ROS is generated in mitochondria and a connection to attacking the cell nucleus is shown. The reactive nitrogen species, nitric oxide is shown leading to endothelial dysfunction.

Fig. (7) diagramatically shows the elements that have been described in the sequence of events connecting oxidant stress and formation of plaque and its rupture to cause acute coronary syndrome. Glucose (red hexagon) is shown as a source of ROS (green circle) and this, not shown, may arise from mitochondrial oxidative phosphorylation with the terminal cytochrome reacting with oxygen to produce water and where an unbalanced oxidant state can generate excessive oxygen radicals. LDL (yellow circle) becomes oxidized (brown irregular shaped particle); taken up by macrophage (blue irregular shape); formation of foam cell (irregular tan particle with blue circle); formation of plaque (tan crescent) on arterial wall; platelets aggregate (orange, irregular shape); plaque (red mass) can destabalize and block the corornary artery (large circle).

OXIDANT STRESS AND LUNG DISEASE

Introduction

There is compelling evidence that oxidative damage has a role in several lung pathologies. The lung is exposed to the highest oxygen tensions of any tissue in the body. The interface of oxygen is in the alveoli with delicate membranes composed of lipids that are potentially susceptible to oxidation. Lorrain Smith (1862-1931) near the begnning of the 20th Century described the effects of oxygen on the lung. Chapter 5 has information about Lorrain Smith who was the first to describe the response of the lung to high concentrations of oxygen in the period of 1897-1899. Smith described the acute histolgical findings including atelectasis, inflammation, and vascular congestion that could lead to death. James D. Crapo has provided an excellent description of the morphological changes seen in pulmonary oxygen toxicity [16]. Exposure by breathing high concentrations of oxygen results in alterations throughout the respiratory tract in humans and other animals. Damage occurs to the epithelial lining of the airways, the arterial vascular bed, the pleura, and to the alveoli. Atalectasis, alveolar edema (an edema in the surrounding lung tissue), pleural effusions, and changes in cell structure and function can result.

Acute Lung Injury and Acute Respiratory Distress Syndrome

In a 2010 review article, P. A. Ward [17] described the acute and progressive lung injury that follows significant oxidative stress including acute lung injury (ALI) and induction of the acute respiratory distress syndrome (ARDS). The initial inflammatory response can progress to fibrosis, particularly in premature infants receiving prolonged exposures to high concentrations of oxygen. Oxidant stress-related acute lung injury can result from deposition in airways of lipopolysaccharide or IgG antibody immune complexes. Neutraphils can be recruited and macrophages, resident in alveoli, can release toxic factors and extensive endothelial and epithelial damage can occur that is too great for protective and repair mechanisms and pulmonary fibrosis can result.

Chronic Obstructive Pulmonary Disease

Chronic obstructive pulmonary disease (COPD) is a pathology that induces respiratory muscle dysfunction (Fig. **8**). Resistance to breathing occurs and reactive oxygen species and oxidative stress are produced [18]. Reactive oxygen species, *via* cellular signalling cascades, activate molecules including protein kinases (enzymes) and nuclear factor κB. Cytokines are released and can damage the diaphragm with sarcomer (muscle) dysruption. Respiratory fatigue can ensue resulting in a compensating shift to type I muscle fibers which have increased resistance to hyperoxia (compared to type II fibers). The lung typically hyperinflates and intercostal muscle dysfunction exacerbates the disease symptoms. There currently is no cure, but respiratory muscle training and even lung volume reduction surgery have been considered. As is the case in many oxidant stress-related maladies, the oxidant stress triggers a cascasde of events at the tissue, cellular, and subcellular levels.

In 2004, Gerritsen *et al.* published research that focused on markers of inflammation and oxidate stress in chronic obstructive pulmonary disease patients [19]. They noted that these patients had lung damage that was characterized by injury to the small airways resulting from an inflammatory process plus an imbalance they described betweeen oxidants and antioxidants. The inflammatory process primarily was neutrophilic and involved several cytokines and cell adhesion molecules. They concluded from work with patients that hydrogen peroxide in inhaled air was a marker, and that IL-8 and ICAM in serum were useful markers for monitoring exacerbated chronic obstructive disease. This work again documented that a cascade of cellular responses follow oxidative insult in this disease.

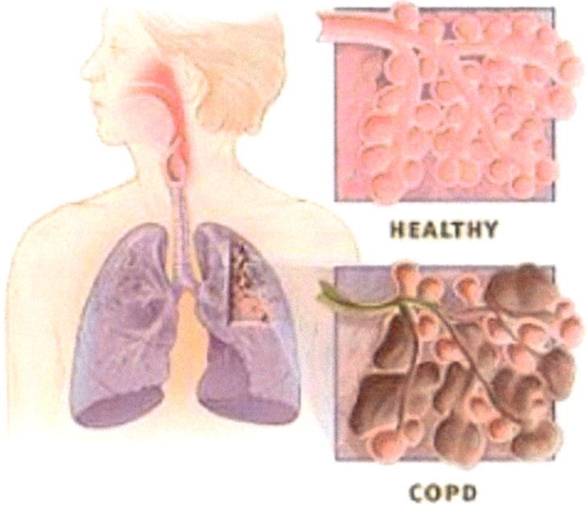

Fig. (8). Depiction of damage to lung in chronic obstructive pulmonary disease.

In a publication in 2011, Ivana Stanojkovic *et al.* [20] reported that chronic obstructive disease is a major worldwide health problem that is increasing in prevalence, morbidity and mortality. In the introduction to the report, the authors note that neutrophils and macrophages may infiltrate the lung tissue and exert the key damage due to the production of reactive oxygen species *via* induction of nitric oxide (NO) synthase expression and production of superoxide radical. The elevated oxidant stress leads to oxidative damage to cell proteins, lipids, DNA and carbohydrates. Irreversable damage may occur to both parenchyma and airway walls, and also activate molecular mechanisms initiating both lung and systemic inflammation and can even lead to atherosclerosis. In new results reported, they found an imbalance between oxidant and antioxidant that was particularly evident in patients with chronic obstructive pulmonary disease that was at least 24 hours post-hospital admission and after patients were clinically stable enough to be discharged. Increased oxidative stress, elevated systemic inflammation, and decreased antioxidant defence were common in end-stage disease and particularly in patients who had also ischemic heart disease [20].

Particulates and Smoking

The toxic effect of a particulate is dependent on the nature of the substance and also on its size. Particulates less than 10 microns in aerodynamic diameter are of concern because they are small enough to be inhaled beyond the nasal passages (Fig. **9**). As shown in this figure, smaller particles can be categorized by the depth to which they penetrate into the lung. Particles as fine as 1 micron (as small as small bacteria) can penetrate into the alveolar spaces, and even smaller particles may get into the blood stream.

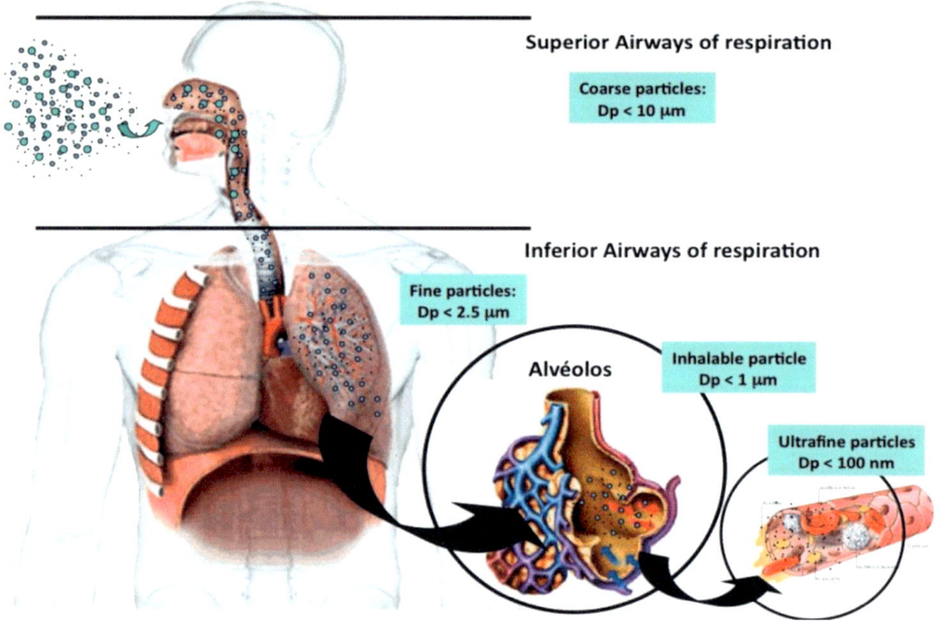

Fig. (9). Based on aerodynamic diameter, small particles differentially penetrate the respiratiory system.

Based on epidemiological data, B. Stringer and L. Kobzik focused on oxidant stress as a key mechanism in lung effects from exposure to particulates in air pollution [21]. They hypothesized that preexisting lung inflammation "primed" lung epithelial cells for enhanced interlukin-8 (IL-8) production from particulates in air pollutants. Using an *in vitro* model, they identified specific paticulates that increased IL-8 production in primed, compared to normal, cells in a concentration-dependent manner over a particle concentration range of 0-200 micrograms per ml. They interpreted their results to indicate that lung epithelial cells primed by inflammatory mediators produced more cytokines after exposure to air pollution prticulates and that oxidant stress was the primary mechanism.

Jonathan Mayo and colleagues [22] reported that iron-mediated oxidative damage was involved in inhalation of airborne pollutants containing iron. Cigarette smoking increases the lung burden of iron. They showed that iron at a physiologial dose stimulates ferritin synthesis without altering the transferrin receptor-mRNA in cells, but acts primarily by reducing a specific iron regulatory protein. Smoke inhalation diminished ferritin levels and increased a particular mRNA in cells. Ferritin serves as a cytoprotective agent against oxidant stress. Thus, exposure of lung cells to low levels of smoke, as present in urban environments, could result in reduced cytoprotection by ferritin with enhanced lung damage by iron-mediated oxidative stress. Oxidative stress is considered to be a common mechanism for various toxicities of cigarette smoking leading to inflammation and lung injury as shown in Fig. (**10**).

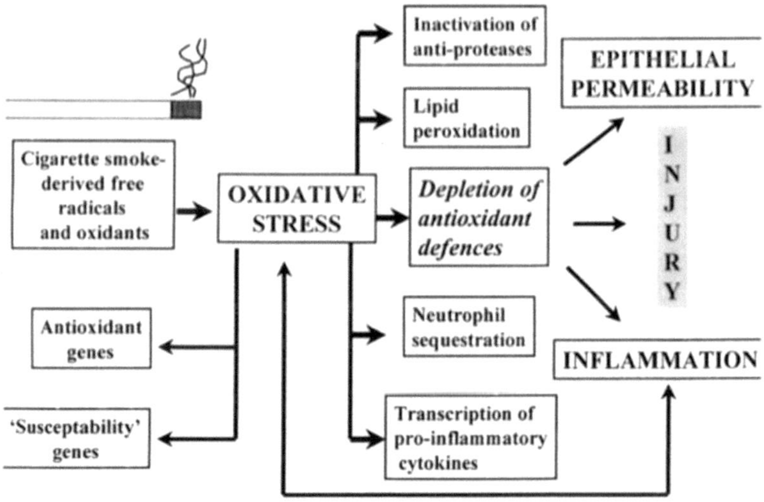

Fig. (10). Mechanisms relating oxidative stress to cigarette smoking and lung damage.

Bronchopulmonary Dysplasia in Premature Infants

Premature infants have severly reduced antioxidant defenses and are particularly sensitive to the effects of hyperoxia. However, because their lungs are insufficiently developed, supplemental oxygen therapy may be required. The chronic lung condition bronchopulmonary dysplasia may result as a consequence [23]. This condition is characterized by dysregulated inflammation amd altered expression of proteases and several growth factors. Fibrosis can result with asymmetric aeration and respiratory insufficiency [23]. Despite its necessity as therapy and its widespread use, optimal strategies for the use of oxygen in preterm infants remains controversial and even today it presents a challenge to treating preterm infants [24].

A report presented by Stephen Welty at the symposium "Pediatric Pulmonary Insufficiency: Nutritional Strategies for Prevention and Treatment" in San Diego in 2000, and published in 2001 [25] gives insight into historical progress with the disease. Welty stated that "... in the last decades, advances in the acute care of premature infants has reduced the reliance on therapy with high concentrations of supplemental oxygen". He also noted that the incidence had not changed significantly; there were qualitative and quantitative differences in oxidation of lipids and proteins compared to infants who did not develop the condition; and that lung injury occurred within hours to days from delivery with "oxidation" a major contributor to the pathological process. He also noted that: "Unfortunately, early attempts at delivery of antioxidants to the lung have not been successful".

A review by A. Phillips in 2013 documented progress made over the past five decades in the use of respiratory support, including supplemental oxygn therapy, for preterm infants [26]. Oxygen toxicity is recognized as one of the main factors contributing to ventilatory-induced lung injury. Protective measures include noninvasive respiratory support, controlling tidal volume during mechanical ventilation, continuous positive airway pressure, antinatal corticosteroids and surfactant replacement. The latter has led to great improvements in care and reducing damage from oxygen therapy. However, lung damage has not been eliminated and a clinical solution may depend on fundamantal new discoveries at the cell mechanism level.

However, a somewhat more optimistic view was expressed in a 2012 paper: "Bronchopulmonary displasia: then and now" [27]. This paper states that "Bronchopulmonary dysplasia was first described in 1967 and the use of assisted ventilation in neonates was in its infancy". High concentrations of oxygen were implicated as causation and the disease was equated to pulmonary oxygen toxicity. Subsequently, causation was found to be related to both oxygen and peak inspiratory pressure as well as their durations. Bronchopulmonary dysplasia was a dreaded complication during the 1970s and the 1980s. These authors state that with the availablilty and therapeutic introduction of exogenous surfactant, "classical" bronchopulmonary dysplasia began to disappear and was replaced by the "new" bronchopulmonary dysplasia. The former was seen in more mature preterm infants and the latter in less mature infants (less than 28 weeks) and was defined as having comparitively mild radiographic abnormalities requiring continued oxygen support at 36 weeks. Pathologically, the "classical" form of bronchopulmonary dysplasia had obstructive bronchiolitis and fibrosis of lung parenchyma while the "new" form had minimal fibrosis [27]. The pathogenesis of bronchopulmonary dysplasia is complex but it primarily is defined [28] as a "heterogeneous group of lung disorders in infants that is commonly associated with prematurity and surfactant deficiendy". Oxygen toxicity from supplemental oxygen administration is recognized as a contributing factor. There may be cycles of lung damage and repair.

OXIDANT STRESS, EXERCISE AND AGING

Introduction

The subjects of exercise and aging are connected. There is evidence that oxidant stress is contributory to aging and there is controversy about whether the net effect of exercise is positive *via* enhancing anti-oxidant mechanisms or negative by increasing reactive oxygen species. Oxygen intake and utilization are certainly increased by exercise and oxygen is primarily metabolized in mitochondria. In the terminal step in oxidative phosphorylation, cytochrome c oxidase receives electrons and converts molecular oxygen to two molecules of water. In the process four protons are tranlocated across the mitochondrial memebrane. The transmembrane potential drives the synthesis of ATP. The process is efficient but, under some conditions, oxygen radicals can be formed and damage can occur. There is controversy, typified by the expression: "You can either rust out or wear out". However, there is abundant evidence about the value of reasonable exercise and proper diet to promote health and longivity, compared to alternatives. D. Harman [29] defined aging rather vaguely as progressive "... accumulation of diverse deleterious changes in cells and tissues with advancing age that increases the risk of disease and death".

Oxygen Radical Formation During Exercise

A review article in 1990 by Bertil Sjodin *et al.* described the biochemical mechanism by which oxygen creates free radicals during exercise [30]. They concluded that evidence had accumulated that during exercise, skeletal muscle damage and inflammation were attributable to oxygen free radicals (Fig. **11**). At the molecular level they concluded that identified sources of oxygen free radicals were semiquinone in mitochondria and xanthine oxidase in capillary endothelial cells. During high intinsity exercise, oxygen utilization increases in skeletal muscles and ATP utilization rate can exceed ATP production rate. During resting or moderate exercise, oxygen free radicals are produced in low amounts and cellular defenses (scavengers and antioxidant systems) are able to destroy them efficiently. Strenuous exercise can exceed the abilities of the body's defense and scavenging systems to remove oxygen free radicals. Accumulated oxygen radicals can attack essential cell components including membranes, proteins and DNA leading to inflammatory responses and ultimately to cell damage that over time can accumulalte with intense and/or exhaustive exercise [30].

A review by Kelsey Fisher-Wellman and Richard J. Bloomer titled: "Acute exercise and oxidative stress: a 30-year history" was published in 2009 [31]. They cited 321 references and nearly 300 of them were published since 1978. They noted that a single session of aerobic or anaerobic exercise was capable of inducing acute oxidative stress, as detected by the presence of "oxidized molecules" in various tissues. They also found that exercise-induced oxidative stress up-regulated (increased) endogenous antioxidant defenses. The authors of this review cite extensive sources that have found strong correlations between strongly oxidized redox-state and diseases including: cardiovascular disease, diabetes, and chronic obstructive pulmonary disease. They also found that the redox-state of cigarette smokers was imbalanced toward oxidation and 16 references were cited. An oxidized redox-state also was associated with exercise but the association was variable and studies differed in many relevant factors including mode, duration, intensity, and state of conditioning. The authors noted that in the past it was common to consider that exercise produced detrimental oxidant stress. However, the authors concluded:

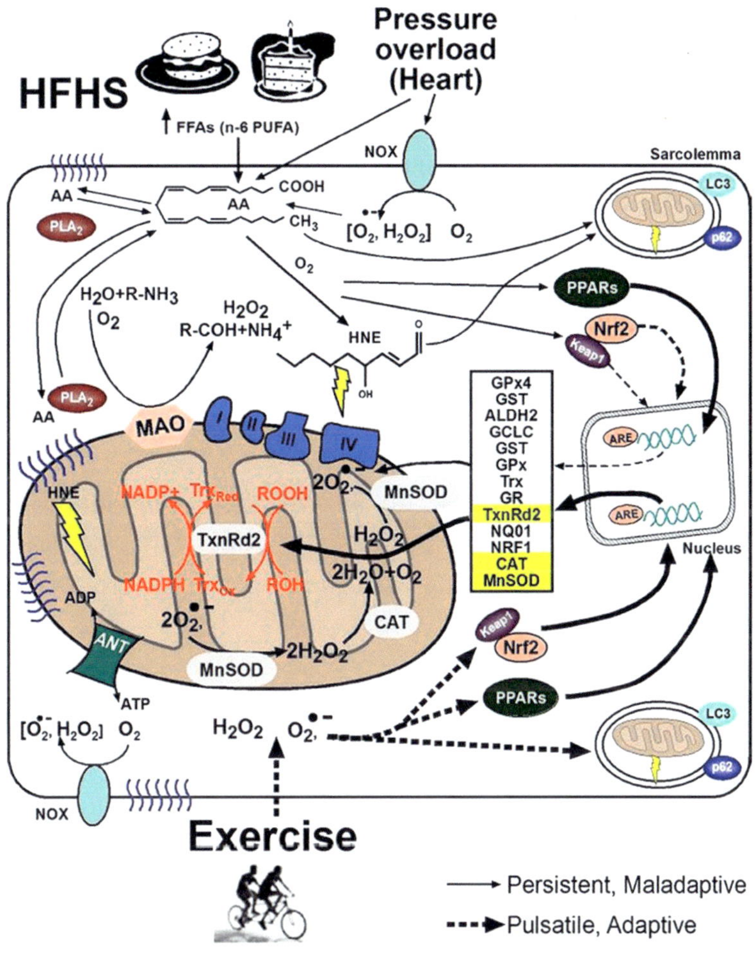

Fig. (11). Complex relationships of exercise to toxic and adaptive mechanisms of oxidant stress.

> "While excessive RONS (reactive oxygen nitrogen species) production and oxidative stress certainly has the ability to result in physiological damage, perhaps leading to the development of ill-health and/or disease over time, an optimal level of prooxidant production may actually serve as the necessary stimulus for the upregulation of antioxidant defenses, thereby providing protection against future RONS attack and disease development. Although the role of oxidative stress in exercise–induced adaptations, as well as in human physiology remains to be completely elucidated, it appears based on the extensive body of literature that a currently undefined optimal level of RONS production may be imperative in order for optimal adaptive potential and physiological function to be achieved".

Barry Halliwell and Carol Cross have made major contributions to the field of oxidant stress and oxygen free radical biology and biochemistry [32]. In this publication, oxidant stress is defined as:

> "… a condition in which the delicate balance existing between prooxidant (free radicals) production and their subsequent anelioration *via* the antioxidant defense system becomes skewed in favor of free radical expression."

Molecular oxygen itself is a "free radical" (a free radical is any species with one or more unpaired electrons). Diatomic oxygen itself can accept 4 electrons. Free radicals include: hydrogen atoms, transition metal ions, carbon-centered radicals, and sulfur-centered radicals. Species derived from oxygen or nitrogen are possibly the most relevant biologically, for original sources see [31]. Such species plus non-radical species created by interactions with free radicals are collectively called reactive oxygen/nitrogen species. Biological defenses against damage from such species include bilirubin, uric acid, speroxide dismutase, catalase, glutathione peroxidase, and dietary factors including carotenoids, tocopherols, ascorbate, bioflavinoids, and other water-soluble and lipid-soluble molecules.

In a review in 2000, R. G. Allen and Maria Tresini indicated that it is generally accepted that a basal level of oxidant stress in the body is continually countered by defenses, and this determines the redox-state and has a significant role in aging and some degenerative diseases [33]. Additionally, changes in oxidative metabolism are known to occur during cell differentiation and development, and experimental manipulation of cellular redox-state has been shown to influence these processes. These authors also state that discovery of specific genes and pathways affected by oxidants led:

> "... to the hypothesis that reactive oxygen species serve as subcellular messengers in gene regulatory and signal transduction pathways... antioxidants can activate numerous genes and pathways." [33].

An article written by Armi Legge in 2013 provides 164 references and assessed the connection asserted by the article title: "The Truth about Extreme Exercise, Oxidative Stress and Your Health" [34]. Evidence was cited relating to issues including: (a) the free radical theory of aging, (b) why oxidatgive stress isn't always bad, (c) when exercise may be too much of a good thing, (d) a definition of extreme exercise, (e) is extreme exercise especially bad, (f) extreme exercise and oxidative stress, (g) exercise and protection against oxidative stress, and (h) how much exercise is too much. Conclusions included: "hard exercise" creates free radicals and causes some oxidative damage that is quickly repaired and also stimulates protection agains free radicals. With continued exercise training, the antioxidant defenses increase; it is possible to "overtrain" to damage the body by factors including oxidant stress; but, overtraining is not the same as high volume training. Armi Legge concluded:

> "Most evidence indicates that if you train in a progressive, intelligent manner, with adequate recovery between workouts, you can build up to extremely high training loads and still be protected against potentially dangerous levels of oxidant stress."

A similar conclusion was stated in a review by Jose Viña *et al.* [35] which cited the classical work of Britton Chance and coworkers (which will be described in the section below: "Oxidant Stress and Aging") who discovered that "state 4" of mitochondrial oxidation resulted in about 2% of oxygen consumption being turned not to energetics but to formation of reactive oxygen species. The work of K. Davies, L. Packer and associates, also was reviewed as one of the earliest reports that free radicals damage from exercise produces damage tissue. These findings and evidence that xanthine oxidase is a source of oxidant radicals is described in Chapter 7.

OXIDANT STRESS AND AGING

Introduction

Aging with deterioration in functioning is obviously an inevitable part of living; it occurs in all multicellular life forms and depending on definition may be an inevitable consequence for all life stemming from the Second Law of Thermodynamics. Some of the evidence connecting oxidant stress with aging is found in the previously cited research on exercise in this chapter. Herein, the focus will be on the hypothesis and the evidence that, over time, oxidant stress contributes to aging (Fig. **12**). In some very general sense, aging in humans begins with conception. Aging can refer to a process of change which can include growth and positive accretion of characteristics. For example, a person may be considered to "come of age", or wine may age; both generally considered positive characteristics. However, in this chapter, aging will be considered in the sense that it is generally thought of as a physiologic state of progressive decline that becomes evident at some point in the life of an individual. Genetics, life style, specific diseases, and factors other than cumulative oxidant damage are important contributors to aging.

In 2006, Devin Kregek and Hanna Zhang wrote an extensive, invited review entitled: "An Integrated view of oxidative stress in aging: basic mechanisms, functional effects, and pathological considerations" [36]. They concluded:

> "The 'oxidative stress theory' holds that a progressive and

irreversible accumulation of oxidative damage caused by ROS impacts on critical aspects of the aging process and contributes to impaired physiological function, increased incidence of disease, and a reduction in life span. While compelling correlative data have been generated to support the oxidative stress theory, a direct cause-an--effect relationship between the accumulation of oxidatively mediated damage and aging has not been strongly established."

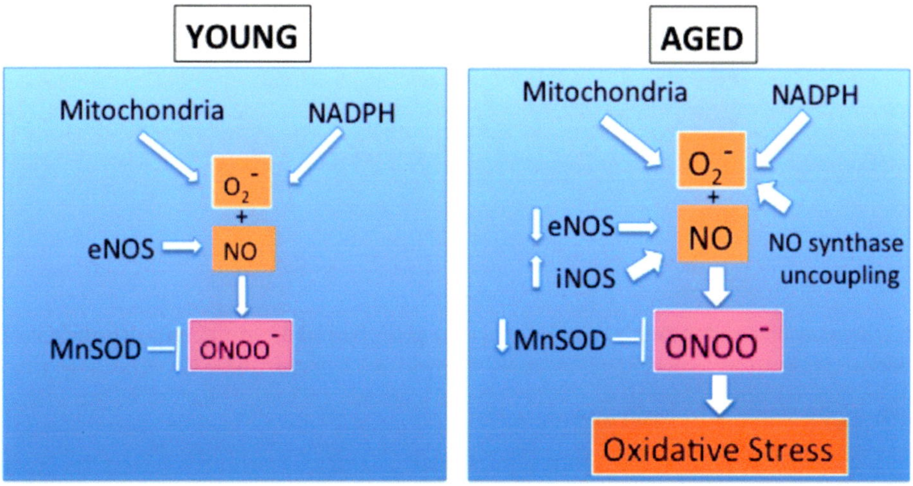

Fig. (12). Excessive oxidant stress over time is proposed to result in significant aspects of aging.

The thesis that oxidant damage was a cause of aging probably can be traced back at least to the "rate of living theory" published by R. Pearl in 1928 [37]. This theory postulates that the lifespan of individuals depends on the rate of energy utilization; *i.e.*, the metabolic rate, and on the genetically-determined quantity of energy consumed during adult life. However, birds have a high metabolic rate and some live longer than many mammals [38]. Despite their high rate of metabolism (oxygen utilization) birds have a low rate of oxygen radical formation and their mitochondria are reported to produce up to 10-times less molecules of reactive oxygen species *in vitro* [39]. A reasonable conclusion is that the mitochondrial rate of free radical generation is more strongly correlated with longevity than is metabolic rate. Indeed, what is known at the "free-radical" theory of aging (Fig. **13**) seems to have originated in the 1950s with a publication by D. Harman: "Aging: a theory based on free radical and radiation chemistry" [40].

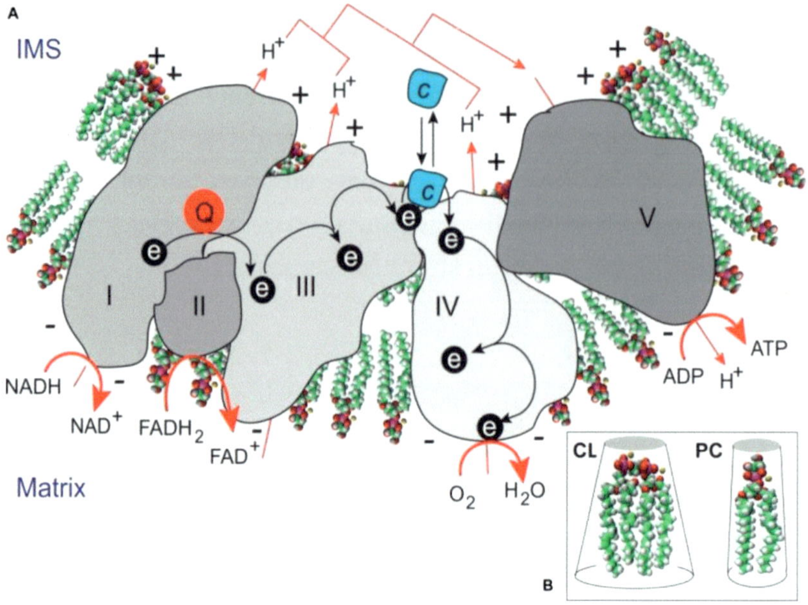

Fig. (13). Diagram of the cytochrome site of oxidative phosphorylation in the inner membrane of the mitochondrion (see text) with transfer of electrons, derived from reduced NAD (NADH) one-at-a-time to oxygen to produce water with formation of ATP; when the process goes awry reactive oxygen species are formed.

The discovery in 1969 of superoxide dismutse (an enzyme whose sole function seems to be removal of superoxide radiacls) sparked research that has generated considerable evidence that free radicals are involved in aging [41]; for more evidence see Chapter 7.

The mechanisms by which oxidant stress is generated by mitochondrial oxidative phosphorylation can be described simply as follows (Fig. **13**). Mitochondria are considered to be the main source of oxidant radical stress in cells [42]. The process occurs in the inner membrane of mitochondria with oxidation of NADH to eventually transfer energy, *via* electron transfer, to generate ATP. The electron transfer process involves transfer of four-electrons to H_2O. In this process, reduced nicotinic acid dinucleotide (NADH) and succinate donate electrons perspectively to complex 1 (of the cytochrome system) *via* NADH dehydrogenase and complex II (*via* succinic dehydrogenase) of the electron transport chain. Coenzyme Q accepts electrons from complexes I and II and transfers electrons to cytochrome b in complex III (*via* ubiquinone-cytoshrome c reductase). Complex III donates electrons to cytochrome c, then to conmplex IV; (*via* cytochrome c oxidase) which reduces H_2O to water. In this process, however, a one-electrton reduction of oxygen creates superoxide radical. *In vitro* studies, by Britton Chance, Helmut Seis, and A. Boveris indicated that up to 2% of oxygen molecules could be diverted into this process that generates the radical, superoxide. I recommend this 78-page paper, written in 1979, for anyone who wants a thoughtful, groundbreaking, comprehensive view of formation of damaging oxidant radicals in cells. Subsequently, other studies under more physiological conditions suggest this value is lower; however, it still is considred to be the major source of mitochondrial oxidant stress [43].

It is estimated that approximately 90% of cellular oxygen is used within mitochondria *via* oxidative phosphorylation (Fig. **13**). Indeed, mitochondria are both sites of, and targets of, reactive oxygen species. Mitochondria, damaged by oxidant effects progressively become less effective, lose functional integrity, and release even more oxidant radicals culminating in an accumulation of dysfunctional mitochondria with age [41].

Studies have shown that reactive oxygen species are important in homeostatic processes that work in cells normally as messengers with positive outcomes. Nitrous oxide (NO) is one example of a cell-signaling molecule. Growth factors also induce production of free radicals that regulate the proliferative response [44]. Humans have very complex and effective antioxidant defenses that include superoxide dismutase, catalase, glutathione peroxidases, and glutathione reductase as described by Carmelina Gemma *et al.* [41]. However, with extended production of reactive oxygen species antioxidants can become depleted leading to cell damage.

There are also non-mitochondrial sources of reactive oxidant species. Peroxisomal β-oxidation of fatty acids generates hydrogen peroxide. Peroxisomes are intracellular organelles that degrade fatty acids and some other molecules. They have a high concentration of catalase and it is still uncertain whether they contribute significant cytosolic oxidant stress. Phagocytosis is another process that generates oxidants during engulfment and destruction of microbes and to oxidize and clear debris by oxidative mechanisms including evolvement of nitric oxide, hydrogen peroxide and superoxide [41]. Cytochrome P450 enzymes (there are many) include CYP2E1 that, in the liver, metabolizes ethanol, carbon tetrachloride, and the drug acetaminophen (Tylenol©) to products with generation of reactive oxygen species. CYP2E1 is induced (is increased) by metabolism of ethanol and its activity increases with increased ethanol intake. It is responsible for metabolism of alcohol by the liver with formation of reactive oxygen species (Fig.**14**). Arthur Cederbaum in a 2011 paper concluded on the basis of research with mice that CYP2EI was a major contributor to fatty liver and oxidant stress from alcohol consumption [45]. In the introduction to this paper Cederbaum stated: "The ability of acute and chronic ethanol treatment to increase production of reactive oxygen species and enhance peroxidation of lipids, proteins, and DNA has been demonstrated in a variety of systems, cells, and species, including humans".

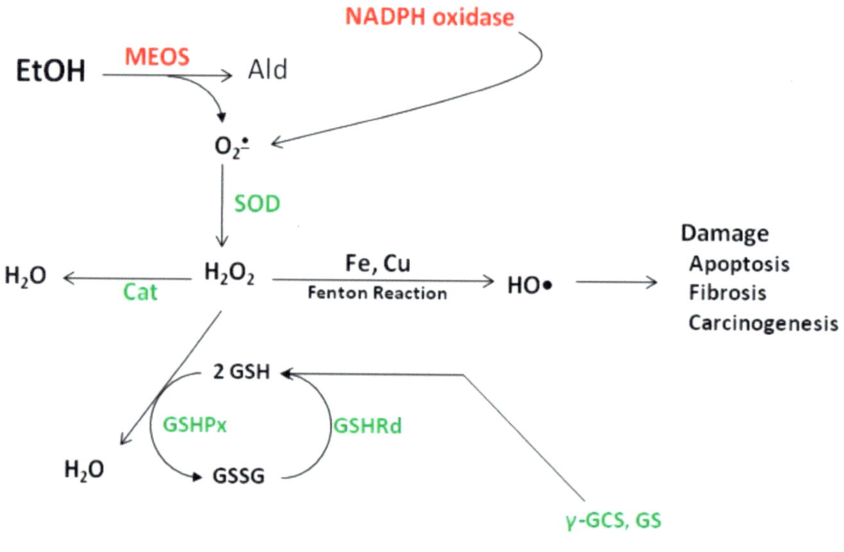

Fig. (14). Ethanol metabolism in liver results in formation of reactive oxygen species.

Carmelina Gemma *et al.* [41] also cite references that document that the aging brain and the brain of Alzheimer's patients, decline in oxidant defenses resulting in increased vulnerability to the deleterious effects of oxidative stress. It is also hypothesized that free radicals of mitochondrial origin are a significant cause of mitochondrial DNA damage seen in postmortem brains of aged subjects. Mitochondrial DNA appears more susceptible to oxidative damage for reasons including that it is near the inner mitochondrial membrane, the site of oxidative phosphorylation by oxygen.

Carmelina Gemma *et al.* [41] also reported that aging is accompanied by damaging changes to cell membranes, primarily decreases in polyunsaturated fatty acids and increases in monounsaturated fatty acids. Polyunsaturated fatty acids are more susceptible to oxidation, and their observed decrease may result from oxidation over time by reactive oxygen species. Polyunsaturated fatty acids, such as arachidonic acid, when peroxidized, form malondialdehyde which induces DNA damage with formation of adducts that disrupt DNA base-pairing. This and other related evidence was stated to support the hypothesis that lipid peroxidation contributes to the deteriorations seen sometimes in aging brains.

There is also evidence that chronic inflammation contributes to aging of the brain and development of age-related degenerative disease. Carmelina Gemma *et al.* also have reviewed this subject [41] and described key cellular signaling events accompanying inflammation in the brain as follows. A significant event was described to be accumulation of "reactive" microglia in degenerative brain areas. One function/response of microglia is to trigger and/or amplify the innate immune response *via* constitutively-expressed cell-surface receptors. These include complement receptors, cytokine receptors, chemokine receptors, and major histocompatibility complex II. When cellular damage occurs they respond by inducing a protective immune response. This immune response usually resolves pathogenic conditions, but when chronic inflammation occurs, prolonged microglia activation can be neurotoxic with production of pro-inflammatory cytokines including interlukin-1 and interleukin-6, and tumor necrosis factor. It is hypothesized that increase in activation of microglia is an early event in oxidative damage because it is reported that microglia-derived radicals appear early in oxidative brain damage. Indeed, activated microglia are said to be the most abundant source of free radicals in the brain and both superoxide and nitric oxide are formed. However, microglia do have high concentrations of reduced glutathione and other antioxidant defense enzymes. It is proposed that with excessive, prolonged oxidant stress these become depleted and damage ensues.

Liang-Jun Yan, in a mini-review in 2014 [46], provided strong evidence that reactive oxygen and reactive nitrogen species are produced as part of normal cellular metabolism (Fig. **15**). Reactive nitrogen species are generally produced by nitric oxide synthase, but reactive oxygen species are generated by a variety of other enzymes and metabolic pathways including: mitochondrial complexes I through III of the electron transport chain (cytochromes), dihydrolipoamide dehydrogenase in the α-keto acid dehydrogenase complexes, NADPH oxidase, xanthine oxidase, monoamine oxidase, and cytochrome P450. Under certain conditions all of these sites can contribute to generating excessive oxidant stress. Basal (normal) levels of reactive nitrogen and reactive oxygen species are essential for redox-signaling and even for functioning and survival of cells; elevated levels become damaging. However, it is recognized that a moderate level of oxidative stress induces adaptive features that are beneficial. Thus, to create positive oxidative conditions, some oxidative stress level is required; however, this author points out that at high levels oxidant stress is harmful. The "best examples" this author gives of positive oxidant stress are exercise-induced ischemic tolerance including pre-conditioning and post-conditioning. Included are: mitochondrial electron transport chain inhibitors, hypoxia, hyperoxia, hyperthermia, hypothermia, and short episodes of ischemia. The author states that positive oxidant stress works because, although the free radicals are extremely short-lived, they relay and extend the positive oxidative effects through "molecular signatures" and modified oxidative products like peroxidized lipids and protein oxidation-adducts. Chemically altered proteins include those modified by S-sulferation, and S-nitrosylation. The author concluded that:

> "... a moderate level of oxidative stress, induced by a variety of stressors, can yield great beneficial effects on adaptive cellular responses to pathological challenges in aging and aging-associated disease tolerance... I term this moderate level of oxidative stress as positive oxidative stress, which usually involves imprinting molecular signatures on lipids and proteins *via* formation of lipid peroxidation by-products and protein oxidation adducts... these molecular

signatures can thus execute the ultimate function of ROS/RNS... positive oxidative stress could be a valuable prophylactic and/or therapeutic approach targeting aging-associated diseases."

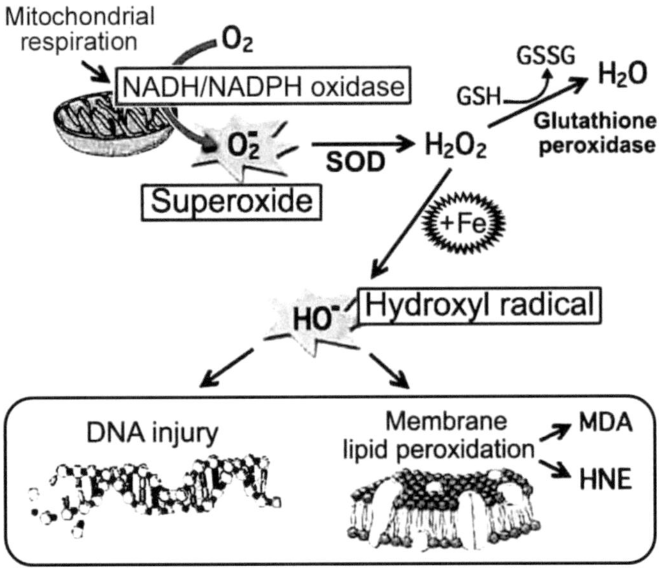

Fig. (15). Oxygen radical formation is linked to mitochondrial respiration, iron catalysis, and oxidative damage to DNA and membranes (proteins can also be damaged (not shown).

NOTES

[1] Quotationary, Leonard Roy Frank, Ed. Random House, New York 2001, p 259; Bob Costa TV interview, NBC, 22 January, 1995 ibid, p. 17.

[2] Musician (1883-1983), remark made on his 100th birthday five days before his death. In *Observer,* British Newspaper, 13 February 1983.

REFERENCES

[1] Medical definition of oxidative stress. Available from: http://www.merriam-webster.com/medical/oxidative stress

[2] Jensen WB. Fritz Haber German Chemist. Available from: https://www.britannica.com/biography/Fritz-Haber

[3] Ravnskov U, McCully KS. Infections may be causal in the pathogenesis of atherosclerosis. Am J Med Sci 2012; 344(5): 391-4.
[http://dx.doi.org/10.1097/MAJ.0b013e31824ba6e0] [PMID: 22495581]

[4] Rizzo M, Berneis K. Low-density lipoprotein size and cardiovascular risk assessment QJM. The Oxford University Press; 2006; 99(1) :1–14. Available from: http://www.ncbi.nlm.nih.gov/pubmed/16371404
[http://dx.doi.org/10.1093/qjmed/hcl146]

[5] Sawamura T, Witztum J, Mehta J, Otabe H, Kuma N. Oxidized low density lipoprotein. Available from: http://academic. research. microsoft.com/ Keyword/ 29792/ Oxidized-Low-Density-Lipoprotein

[6] Li D, Mehta JL. Oxidized LDL, a critical factor in atherogenesis. Cardiovasc Rs. 2005; 68(3): 353-4.
[http://dx.doi.org/10.1016/j.cardiores.2005.09.009]

[7] Huang H, Mai W, Liu D, Hao Y, Tao J, Dong Y. The oxidation ratio of LDL: a predictor for coronary artery disease Dis Markers [Internet]. 2008 [cited 2016 Jul 13]; 24(6): 341–9. Available from: http://www.ncbi.nlm.nih.gov/pubmed/18688083
[http://dx.doi.org/10.1155/2008/371314]

[8] Rizzo M, Kotur-Stevuljevic J, Berneis K, *et al.* Atherogenic dyslipidemia and oxidative stress: a new look. Transl Res 2009; 153(5): 217-23.
[http://dx.doi.org/10.1016/j.trsl.2009.01.008] [PMID: 19375682]

[9] Rizzo M, Berneis K, Koulouris S, *et al.* Should we measure routinely oxidised and atherogenic dense low-density lipoproteins in subjects with type 2 diabetes? Int J Clin Pract. 2010; 64(12): 1632–42. Available from: http://www.ncbi.nlm.nih.gov/pubmed/20831734
[http://dx.doi.org/10.1111/j.1742-1241.2010.02378.x]

[10] Mohty D, Pibarot P, Després J-P, *et al.* Association between plasma LDL particle size, valvular accumulation of oxidized LDL, and inflammation in patients with aortic stenosis 2008. Arterioscler Thromb Vasc Biol. 2008; 28(1): 187–93. Available from: http://www.ncbi.nlm.nih.gov/pubmed/17975118

[11] Kondo A, Muranaka Y, Ohta I, *et al.* Relationship between Triglyceride Concentrations and LDL Size Evaluated by Malondialdehyde- modified LDL. Clin Chem 2001; 47(5): 893-900.

[12] Wierzbicki AS. Quality as well as quantity? Beyond low-density lipoprotein-cholesterol - the role of particle size. Int J Clin Pract. 2007; 61(11): 1780–2. Available from: http://www.ncbi.nlm.nih.gov/pubmed/17935540

[13] Yang H, Mohamed AS, Zhou S, *et al.* Oxidized low density lipoprotein, stem cells, and atherosclerosis Lipids Health Dis. BioMed Central 2012; 11(1):85. Available from: http://lipidworld. biomedcentral.com/ articles/ 10.1186/ 1476-511X-11-85
[http://dx.doi.org/10.1186/1476-511X-11-85]

[14] Hazen SL, Hsu FF, Duffin K, Heinecke JW. Molecular chlorine generated by the myeloperoxidase-hydrogen peroxide-chloride system of phagocytes converts low density lipoprotein cholesterol into a family of chlorinated sterols. J Biol Chem. 1996; 271(38):23080–8. Available from: http://www.ncbi.nlm.nih.gov/pubmed/8798498

[15] Parthasarathy S, Raghavamenon A, Garelnabi MO, Santanam N. Oxidized low-density lipoprotein 2010. Methods Mol Biol. NIH Public Access; 2010; 610: 403–17. Available from: http://www.ncbi.nlm.nih.gov/pubmed/20013192
[http://dx.doi.org/10.1007/978-1-60327-029-8_24]

[16] Crapo JD. Morphologic Changes in Pulmonary Oxygen Toxicity 2003. Annual Reviews 4139 El Camino Way, P.O. Box 10139, Palo Alto, CA 94303-0139, USA

[17] Ward PA. Oxidative stress: acute and progressive lung injury 2010. Ann N Y Acad Sci. 2010; 1203(1): 53–9. Available from: http://doi. wiley.com/ 10.1111/ j.1749-6632.2010.05552.x [http://dx.doi.org/10.1111/j.1749-6632.2010.05552.x]

[18] Zuo L, Hallman AH, Yousif MK, Chien MT. Oxidative stress, respiratory muscle dysfunction, and potential therapeutics in chronic obstructive pulmonary disease 2012. Front Biol (Beijing). SP Higher Education Press; 2012; 7(6): 506–13. Available from: http://link.springer.com/10.1007/s11515- 012-1251-x
[http://dx.doi.org/10.1007/s11515-012-1251-x]

[19] Gerritsen WB, Asin J, Zanen P, *et al.* Markers of inflammation and oxidative stress in exacerbated chronic obstructive pulmonary disease patients 2005. Respir Med. Elsevier; 2005; 99(1): 84–90. Available from: http://www. ncbi. nlm. nih. gov/ pubmed/ 15672854
[http://dx.doi.org/10.1016/j.rmed.2004.04.017]

[20] Stanojkovic I, Kotur-Stevuljevic J, Milenkovic B, *et al.* Pulmonary function, oxidative stress and inflammatory markers in severe COPD exacerbation. Respir Med 2011; 105 (Suppl. 1): S31-7.
[http://dx.doi.org/10.1016/S0954-6111(11)70008-7] [PMID: 22015083]

[21] Stringer B, Kobzik L. 1998. Environmental particulate-mediated cytokine production in lung epithelial cells (A549): role of preexisting inflammation and oxidant stress. J Toxicol Environ Health A. 1998; 55(1): 31–44. Available from: http://www.ncbi. nlm.nih. gov/ pubmed/ 9747602

[22] Mayo JJ, Kohlhepp P, Zhang D, Winzerling JJ. Effects of sham air and cigarette smoke on A549 lung cells: implications for iron-mediated oxidative damage 2004. Am J Physiol Lung Cell Mol Physiol. American Physiological Society; 2004; 286(4): L866–76. Available from: http://www.ncbi.nlm. nih.gov/pubmed/15003939

[23] Weinberger B, Laskin DL, Heck DE, Laskin JD. Oxygen Toxicity in Premature Infants 2002. Toxicol Appl Pharmacol. Academic Press; 2002; 181(1): 60–7. Available from: http://linkinghub.elsevier. com/retrieve/pii/S0041008X02993877
[http://dx.doi.org/10.1006/taap.2002.9387]

[24] Cherian S, Morris I, Evans J, *et al.* Oxygen therapy in preterm infants 2014. Paediatr Respir Rev. Elsevier; 2014; 15(2): 135–41. Available from: http://www.ncbi.nlm.nih.gov/pubmed/23402990
[http://dx.doi.org/10.1016/j.prrv.2012.12.003]

[25] Welty SE. 2001. Is there a role for antioxidant therapy in bronchopulmonary dysplasia? J Nutr. American Society for Nutrition; 2001; 131(3): 947S-950S. Available from: http://www.ncbi.nlm. nih.gov/pubmed/11238792

[26] Berger TM, Fontana M, Stocker M. The journey towards lung protective respiratory support in preterm neonates Neonatology. Karger Publishers; 2013; 104(4): 265–74. Available from: http://www.karger. com?doi=10.1159/000354419
[http://dx.doi.org/10.1159/000354419]

[27] Philip AG. Bronchopulmonary dysplasia: Then and now. Neonatology. Karger Publishers; 2012; 102(1): 1–8. Available from: http://www. karger. com/ doi/ 10.1159/ 000336030

[28] Hayes D Jr, Feola DJ, Murphy BS, Shook LA, Ballard HO. Pathogenesis of bronchopulmonary dysplasia Respiration. Karger Publishers; 2010; 79(5): 425–36. Available from: http://www.karger.com/ doi/ 10.1159/ 000242497

[29] Harman D. Aging: Overview. Ann N Y Acad Sci. Blackwell Publishing Ltd; 2006; 928(1): 1–21. Available from: http://doi. wiley.com/ 10.1111/ j.1749-6632. 2001.tb05631.x

[30] Sjödin B, Westing YH, Apple FS. Biochemical mechanisms for oxygen free radical formation during exercise Sport Med. Springer International Publishing; 1990; 10(4): 236–54. Available from: http://link.springer.com/10.2165/00007256-199010040-00003 [http://dx.doi.org/10.2165/00007256-199010040-00003]

[31] Kelsey Fisher-Wellman RJ. Acute exercise and oxidative stress: a 30 year history. Dyn Med. BioMed Central 2009; 8: 1.

[32] Halliwell B. Cross1 CE.Oxygen-derived Species: Their Relation to Human Disease and Environmental Stress. 1994; 102(Suppl (10)): 5-12.

[33] Allen RG, Tresini M. Oxidative stress and gene regulation. Free Radic Biol Med 2000; 28(3): 463-99. [http://dx.doi.org/10.1016/S0891-5849(99)00242-7] [PMID: 10699758]

[34] Legge A. A. The truth about extreme exercise, oxidative stress, and your health. Available from: http://www.completehumanperformance.com/exercise-oxidative-stress/

[35] Viña J, Gomez-Cabrera MC, Lloret A, *et al.* Free radicals in exhaustive physical exercise: mechanism of production, and protection by antioxidants IUBMB Life; 50(4-5): 271–7. Available from: http://www.ncbi. nlm.nih.gov/ pubmed/ 11327321

[36] Kregel KC, Zhang HJ. An integrated view of oxidative stress in aging: basic mechanisms, functional effects, and pathological considerations 2006. AJP Regul Integr Comp Physiol. 2006; 292(1): R18–36. Available from: http://ajpregu.physiology.org/ cgi/ doi/ 10.1152/ajpregu.00327.2006 [http://dx.doi.org/10.1152/ajpregu.00327.2006]

[37] Pearl R. The rate of living being an account of some experimental studies on the biology of life duration. New York: Knopf 1928.

[38] Holliday R. (Robin) Understanding ageing. Cambridge University Press 1995. [http://dx.doi.org/10.1017/CBO9780511623233]

[39] Barja G. Mitochondrial Free Radical Production and Aging in Mammals and Birdsa Ann N Y Acad Sci. Blackwell Publishing Ltd; 1998; 854(1 TOWARDS PROLO): 224–38. Available from: http://doi.wiley.com/10.1111/j.1749-6632.1998.tb09905.x [http://dx.doi.org/10.1111/j.1749-6632.1998.tb09905.x]

[40] Harman D. Aging: A theory based on free radical and radiation chemistry. Sci Aging Knowl. Environ. 2002; 2002(37): 1-14.

[41] Gemma C, Vila J, Bachstetter A, Bickford PC. Oxidative Stress and the Aging Brain: From Theory to Prevention. CRC Press/Taylor & Francis 2007.

[42] Balaban RS, Nemoto S, Finkel T. Mitochondria, oxidants, and aging. Cell 2005; 120(4): 483-95. [http://dx.doi.org/10.1016/j.cell.2005.02.001] [PMID: 15734681]

[43] Staniek K, Nohl H. Are mitochondria a permanent source of reactive oxygen species?Biochim

Biophys Acta - Bioenerg 2000.
[http://dx.doi.org/10.1016/S0005-2728(00)00152-3]

[44] Finkel T. Oxygen radicals and signaling Curr Opin Cell Biol. Elsevier Current Trends; 1998; 10(2): 248-53. Available from: http://linkinghub. elsevier.com/ retrieve/ pii/ S0955067498801476
[http://dx.doi.org/10.1016/S0955-0674(98)80147-6]

[45] Cederbaum AI. Alcohol Drinking-Associated Factors Role of CYP2E1 in Ethanol-Induced Oxidant Stress, Fatty Liver and Hepatotoxicity. Dig Dis. 2010; 28: 802–11. Available from: www.karger.com

[46] Yan L-J. Positive oxidative stress in aging and aging-related disease tolerance. Redox Biol 2014; 2: 165-9.
[http://dx.doi.org/10.1016/j.redox.2014.01.002] [PMID: 25460727]

CHAPTER 11

Oxygen: The Future

Abstract: Oxygen has had a long history; it was the 18th element to be discovered of 118 known elements, of which 28 have been created artificially. Oxygen is essential for all aerobic life on earth and pure oxygen at various partial pressures and mixed with other gases is used as a medicine for conditions where the oxygen supply is inadequate, including for premature infants whose lungs are not completely developed. Oxygen is also used for pathologies where oxygen supply to tissues is compromised. Oxygen is a component in gas mixtures used for deep-water diving for sport, recreation, salvage, exploration and mining. Oxygen is a requirement for manned space flight and commercial air travel. In the future, it is a certainty that all of these uses and requirements will continue. It is almost as sure that things now predictable based on past and current research and things as yet unimagined are destined to happen in the future. For therapy, it is probable that artificial blood that adequately perfuses the body will be routine. For recreation, it is likely that private oxygen chambers will get fancier and be more widely used, as will oxygen as a novelty is "sports" bars. In sports, the use of oxygen in breathing mixtures for deep-sea diving and by free divers will continue although it is difficult to see how current records can be much extended. Artificial gills will never become practical but by mimicking sea mammals, man may become as free as dolphins to explore the deeps. For endurance sports and sprints, oxygen supplementation, particularly for training recovery and its psychological effect, will become common and contribute to new records. Ways to manufacture oxygen cheaply and reliably will make possible undersea habitats that will include underwater cities. As future therapy, antioxidants and other oxidant stress protector molecules will allow expanded therapies for many conditions which are limited by the toxicity of oxygen. The most spectacular advance will be in artificial photosynthesis that will remove carbon dioxide from the atmosphere, provide oxygen and carbon-based fuels and power cells that will far outstrip solar cells.

Keywords: Artificial blood, Artificial photosynthesis, Aquarius, Biosphere 2, *Chlorobiaceae*, CRISPR, Deep water diving, Dubai underwater hotel, Gills,

Habitats, Hyperbaric chambers, ISSHL, Lagrangian point, Myoglobin, Ocean spiral, Oxidative stress, Perfusion, Space cities, Sports bars, Underwater mining, Whales.

INTRODUCTION

Oxygen is one of either 90 or 92 (depending on who is counting) "natural" elements found on the earth. There are additional elements that are man-made and which have very short half-lives, that bring the total to currently 118. Oxygen was the eighteenth element to be officially discovered (which occurred in 1774). Several elements including gold, silver, iron and copper (credited as having been known longer than any other element) were known from antiquity with no specific date of "discovery" recorded [1, 2].

Writing about the future of oxygen necessarily involves thoughts about: what we may expect to learn about the chemistry and biology of the molecule, new approaches and technologies for oxygen therapy, oxygen as it relates to the nano world, and breathing gas requirements for space and undersea habitats. Predicting these discoveries, applications and uses is only safely done if they are the immediate future and based on small steps beyond the present perception, or if they are so far in the future and so fanciful as to not be considered seriously or detrimentally to the credit of the dreamer, should they never happen.

FRONTIER OF PHYSICS, CHEMISTRY AND BIOLOGY OF OXYGEN

The basic chemistry and biology of oxygen has been thoroughly explored. The mechanisms within the cell for the metabolism of oxygen have been described. Oxidant stress and biological defenses against harmful effects are known in considerable detail. Formation of oxygen radicals in biological systems is understood. The details of second messengers and subsequent consequences in terms of damage and repair are only beginning to emerge. There will be new discoveries of how the oxidation-reduction state of the cell is controlled including in different tissues in physiologically normal and diseased states. Since the work of Otto Warburg, it has been appreciated that oxidative metabolism may hold one of the keys to defeating cancer cells and the current state of knowledge is ripe for such research. Some xenobiotics are toxic *via* a free radical mechanism involving oxygen. There is the likelihood that new drugs that are protective and new ways to stimulate anti-oxidant cellular defenses will advance both cancer chemotherapy and antidotes for poisons that work by this means.

Areas Ripe for New Discoveries

A most exciting prospect for the near future are discoveries, at the margins of physics and chemistry, relating to the quantum behavior of the electrons set free from oxygen and their participation in various biological systems. These systems include photosynthesis, where artificial photosynthesizing components have already been designed [3]. It is reasonable that these designed systems will be partly derived from biological systems and partly artificial. Since oxygen is produced, and carbon dioxide is consumed, they have great ecological potential. They also result in generation of reduced organic molecules and these have various uses as food, fuel, and starting chemical for synthesis including plastics. The potential for artificial photosynthesis is truly enormous. Initial efforts will be toward making the systems more durable and more efficient at transferring energy from light into carbon-based chemistry. This is where greater understanding of the quantum behavior of oxygen's electrons is required and it will be the focus of research that brings together biologists, chemists and theoretical physicists. Increasing the efficiency of energy transfer is one key to successful applications and resonance effects of entangled electrons will be a focus of discovery designs. A second essential will be durability and extended temperature of operation and this is where improvements will be sought in the biological components of systems. New advances in 3-D printing will be ideally suited to applications of this system for artificial photosynthesis to remove carbon dioxide from the atmosphere and for creating useful energy that will be more efficient than current solar energy devices. In the future, many new carbon-based by-products will be produced, some as starting chemicals for a variety of chemical syntheses. Biodegradable plastics will likely be an early, successful product application.

Synthetic photosynthesis will produce oxygen as a by-product which together with the energy and carbon-based compounds produced will have applications for the undersea and outer space habitats that I shall subsequently address.

It is predictable that mechanisms to selectively and specifically make point mutations, as by CRISPR, will be applied to alter the cellular production of antioxidant, especially enzyme-mediated, defense mechanisms.

RECREATIONAL USES OF OXYGEN

Sports bars, since the mid-1970s, have offered adventuresome spirits in some larger cities in the U. S. the opportunity to breathe elevated oxygen [4, 5]. Because pure oxygen is used medicinally, it is technically a drug, and it is therefore regulated by the FDA [6, 7]; its use without a prescription is not legal. I know of no cases where the FDA or law enforcement has intervened because it is a low-priority offense. Apparently, breathing oxygen is popular among some who claim it increases energy and provides a pleasant experience. This is reminiscent of the various uses made of oxygen therapy by physicians more than a century ago without a proven mechanism. Indeed, the oxygen delivered to the tissues of a healthy person by breathing a gas mixture that contains 40% or more of oxygen (which likely is the case in sports bars) will not result in a measurable change in tissue oxygenation.

Historically, few people ventured beneath the sea for time periods that required provisions for a breathing gas. Until fairly recently, few people dived more deeply than snorkeling and performing shallow breath-holding diving. Exceptions were the Ama (see Chapter 2) Japanese pearl divers. Bold adventurers, mostly employed in underwater salvage operations, and some early navy recruits, went deeper into the sea and for extended periods in hard-hat diving suits, diving bells, and caissons that required breathing compressed air, or various mixtures of oxygen that exposed them to pressure effects as well as oxygen toxicity and nitrogen narcosis (and narcosis from other gases). Submarines are an exception in that personnel are at atmospheric pressure breathing air or an artificially generated air-equivalent in terms of composition.

Underwater breathing devices have advanced to re-breathers, to SCUBA gear, and the most sophisticated versions are used by U.S. Navy divers including frog men, and SEAL team members. Future developments will be focused on allowing divers to go deeper in the ocean, stay longer, perform more work and return to the surface with less danger of decompression sickness. I recall in the Viet Nam era, visiting with a young Naval Officer who impressed me greatly. He expressed great concern at having to order young men (frog men) to engage in underwater missions requiring extreme physical dangers. The officer said he spent every minute he could spare when in port at the local University library trying to learn the latest information about the physical limits of humans relative to diving (depth, temperature, exertion, and breathing mixtures). He feared for his divers. Today, we know more and in the future, the focus will be on development of ways to employ underwater habitats that do not require divers to decompress for long missions. There is a need for lighter gear with increased oxygen supplies or even regeneration or extraction from sea water. Extraction of oxygen from sea water by "artificial gills" will become feasible but will not solve the problems. This will require cooperation of biologists, with knowledge gained by studying fish and whales and other diving mammals, and chemists and material scientists merging biology and technology to overcome current human physiological limitations. A chief problem, for which no solution is now apparent to me, is the presence in humans of air-filled chambers in the head, torso (lungs) and digestive system. Whales can dive to depths of at least one mile and stay for an hour or more. They have increased hemoglobin concentrations and also greatly increased myoglobin concentrations (Fig. **1**). Thus, they are able to "store" oxygen [8, 9].

Future research and technology aimed at freeing humans for deep and sustained living under sea would be better directed towards the example of the whale than toward devising artificial gills, in my opinion. Whales are mammals and have lungs and breathe air. Yet they are designed to withstand pressures, to store oxygen in their tissues, and to replenish their oxygen supply by freely returning to the surface without decompression effects. The gills of fish serve them admirably for extracting oxygen efficiently but the function of gills is not readily adaptable to interface with the human lung. It is totally impractical to think that humans can safely breathe seawater with any technology. The problem lies not in mimicking the function of gills by an artificial technology; the problem is that this system must interface with the lung and circulatory system of the human to be applied to diving. An artificial gill may be designed to be light and energy efficient (although this is enormously difficult with today's technology) but the problem is not one of extracting oxygen from sea water (although this is also difficult technologically). The problem is in getting this oxygen into the human circulatory system which requires the human lung to be interfaced physically with the artificial gill which is not feasible because of the insurmountable biological limitations.

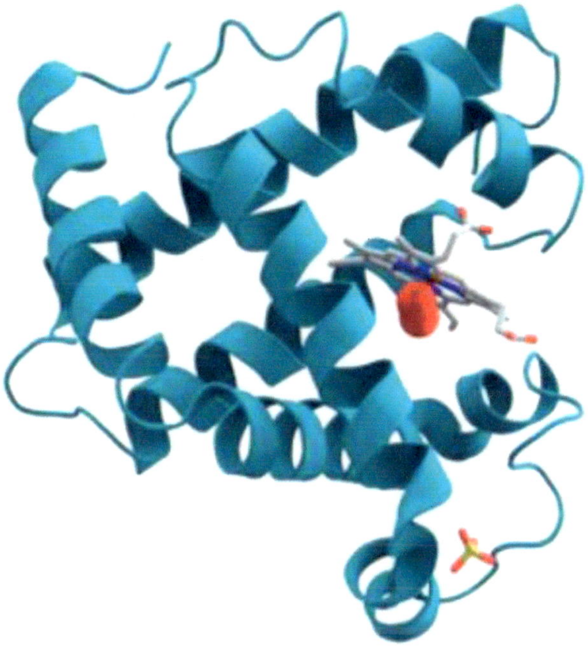

Fig. (1). Myoglobin structure showing bound oxygen molecule (red balls).

The problem of humans being able to exist in the deep seas is, however, not an insurmountable problem, either biologically or technically. We must look to the biology of the whale and other diving mammals. The importance of the designation "mammal" must not be overlooked. They share much with humans, relative to breathing air, but they are designed for the sea. Science must overcome the limitations of the collapsible human lung, relative to the whale, and the problem of myoglobin storage of oxygen. The former seems harder than the latter.

It is problematical whether any human would desire to undergo the genetic modification required to permit increased myoglobin in muscles, or whether this would have damaging or even lethal consequences. The future of humans living in the sea requires solving the problems of making the human more like the whale in terms of respiration, and not more like the fish.

Decompression

There is a need to advance the technology of decompression. Current technology is advanced but more will be possible based on artificial means for delivering oxygen to tissues. Hemoglobin is extremely efficient. It will require amazing developments in new materials to outdo it in terms of allowing oxygen to be safely carried at increased amounts dissolved in the blood. The medically scary prospect of divers with "artificial blood" seems unlikely as a routine, but will be necessary to explore as demands for underwater "living" increase in the future. The limits of "inert" gases to supply the pressure necessary for safe delivery of oxygen at high pressures, probably has been reached. The problem caused by the time and inconvenience of decompression, has been approached by simply having divers remain at depth and under pressure. The physiological consequences of this are manageable for reasonable time, but the consequences for truly long term as required for pleasurable existence in undersea habits or undersea cities of the future, is an unknown.

Aquarius (Fig. **2**) is an underwater laboratory and home, owned and operated by Florida International University [10]. It is 4.5 kilometers offshore near Key Largo, Fl and rests 60 feet deep in the Ocean. It is occupied by scientists investigating the undersea environment. Using "saturation" diving technology, scientists are able to remain underwater 24 hours per day for typically 10 days at a time. In saturation diving, rather than returning to the surface after each free dive, the divers return to Aquarius. Since the divers remain at pressure, they do not have to decompress. However, at the end of a mission, before leaving the ocean, they must undergo 15 hours and 45 minutes of decompression from the 2.5 atmospheres pressure back to surface pressure, inside the chamber. The divers then scuba dive back to the surface. Aquarius has a habitat module, an 85-ton chamber that is 43 feet long and 9 feet in diameter. Aquarius has a living space of 1,400 cubic feet. The system includes other components. Such habitats will become routine and extended for recreation and adventure.

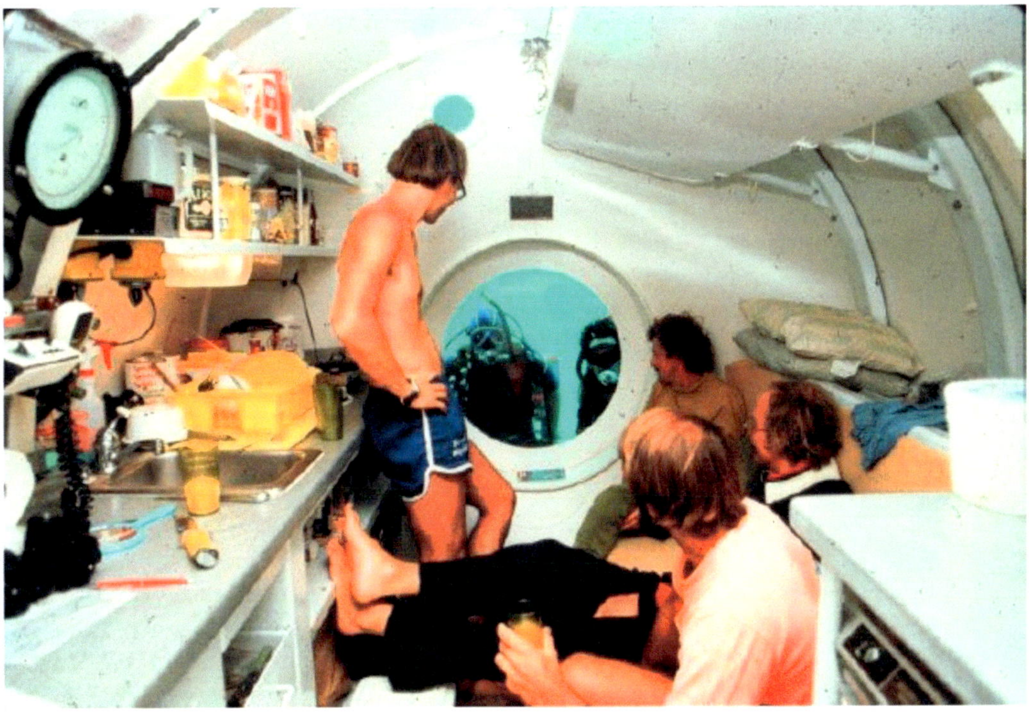

Fig. (2). A view inside Aquarius. Divers remain typically for 10 days of saturation diving before having to undergo decompression to return to the surface *via* scuba free-dives.

OXYGEN AS THERAPY

The use of personal hyperbaric chambers may be considered as recreational or perhaps it may be done as preventative, or as therapy. The best publicized example is probably Michael Jackson in the 1970s. Hyperbaric chambers are often designed for use by a single person lying down (mono-place chamber) and typically have a clear Plexiglass® structure and some are designed to meet a specific patient's needs or desires. They are usually pressurized with essentially 100% oxygen and a mask is not required. Multi-place chambers are also produced that hold from two to ten people.

Since short-term and low level oxygen exposures do not produce measurable harm, and people easily make the connection that it sounds beneficial, the use of these chambers will increase and they will become popular for various conditions that are not treated successfully by today's medical practices. One example of the use of personal hyperbaric chambers (Fig. **3**) is for treating autism for which there have been clinical trials [11]. The connection between autism and any mechanism relating to lack of oxygen delivery has not been established, and the use of oxygen therapy is not warranted. The weight of evidence is against oxygen becoming a "cure" in the future for autism.

Fig. (3). A personal hyperbaric chamber being used experimentally for treatment of autism.

In modern times, there have been proposed uses of hyperoxia for various conditions. The Board of Directors of the Undersea Hyperbaric Medicine Society on October 8, 2011 approved the use of hyperbaric oxygen for the treatment of idiopathic sudden sensorineural hearing loss (ISSHL) [12]. The incidence of this condition is estimated to be 5 to 20 cases per 100,000 annually in the U.S. (however, many cases may be unreported). The rationale for hyperbaric oxygen therapy (although, as the name implies, the cause is unknown) is based on the known high metabolic rate and paucity of vascularity of the cochlea. Perilymph is the primary source of oxygen and its oxygen tension is decreased significantly in patients with ISSHL. The use of hyperbaric oxygen therapy was determined to have an "A" level of evidence from multiple randomized clinical trials [12]. Recommended therapy is 100% oxygen at 2.0 to 2.5 atmospheres absolute for 90 minutes daily for 10 to 20 sessions [12]. The Editors of "Wound Source", citing a publication in 2008, listed 13 conditions that they say are approved by the FDA for treatment by hyperbaric oxygen therapy [13]. It is difficult to predict which additional conditions will be approved for hyperbaric therapy. It is safe to say that additional conditions will in the future be approved for treatment and that these will be almost entirely conditions requiring oxygen at elevated pressures, *i.e.*, conditions where there is a pathologically low perfusion of oxygen into a vital tissue.

Rossignol *et al.* have reviewed the use of hyperoxia for the treatment of traumatic brain injury and several other disorders where there is evidence of decreased perfusion with oxygen [11]. The rationale for oxygen seems to lie only in pathologies where there is impaired delivery of oxygen to tissues, since normal oxygenation supplies the normal needs of tissues. Rather than breathing elevated concentrations of oxygen, it will be possible to apply oxygen *via* surface exposure to certain types of injuries and chronic conditions where capillary supply of oxygen is not adequate. New modalities for the treatment of burns and skin grafts are likely to be an area of future advances. It is not predictable that hyperoxia or hyperbaric oxygen will provide successful treatment of dementia, Parkinson's, Alzheimer's, or Huntington's diseases. The already successful use of oxygen by various pathologies where the delivery of oxygen by the lung is impaired will see improved, and more portable delivery of oxygen with on-site extraction or generation.

The means for delivering therapeutic oxygen, from cannulas, to oxygen masks, and to mono-place, and multi-patient chambers for therapy range from adequate to sophisticated. Future developments will likely only be in the means of oxygen supply. Complex storage and delivery systems in hospitals are required and it seems likely that on-site generation of oxygen by extraction system will be future developments that also will find sophisticated applications for home use.

OXYGEN: UNDERSEA HABITATS AND UNDERSEA CITIES

For many years there have been plans and experiments for undersea habitats. Currently, China has plans, but lacks the required funding, to proceed with the boldest and most extensive current plan to allow undersea living (Fig. **4**). Rachel Nuwer has written an interesting piece with the provocative title: "Will we ever ... live in underwater cities?" [14]. Current at the time of writing in 2013, Nuwer stated that in her view, compared to 1972, interest has ebbed for experimenting with humans underwater for extended periods of time. Nuwer stated that in 1972 "Ian Koblick opened La Chalupa, then the largest and most advanced underwater habitat and research facility in the world, and of more than a dozen underwater habitats that once existed, just three remain, all in the Florida Keys". Koblick and colleagues own and operate the interesting "Jules Verne Lodge" which provides patrons with education and training, undersea weddings, and luxury romantic getaways at $675 per night, according to Nuwer [14]. Nuwer wrote that Koblick believes "… the technology already exists to create colonies supporting up to 100 people– the few bunker-like habitats in operation today providing a blueprint". Nuwer quotes Koblick as saying:

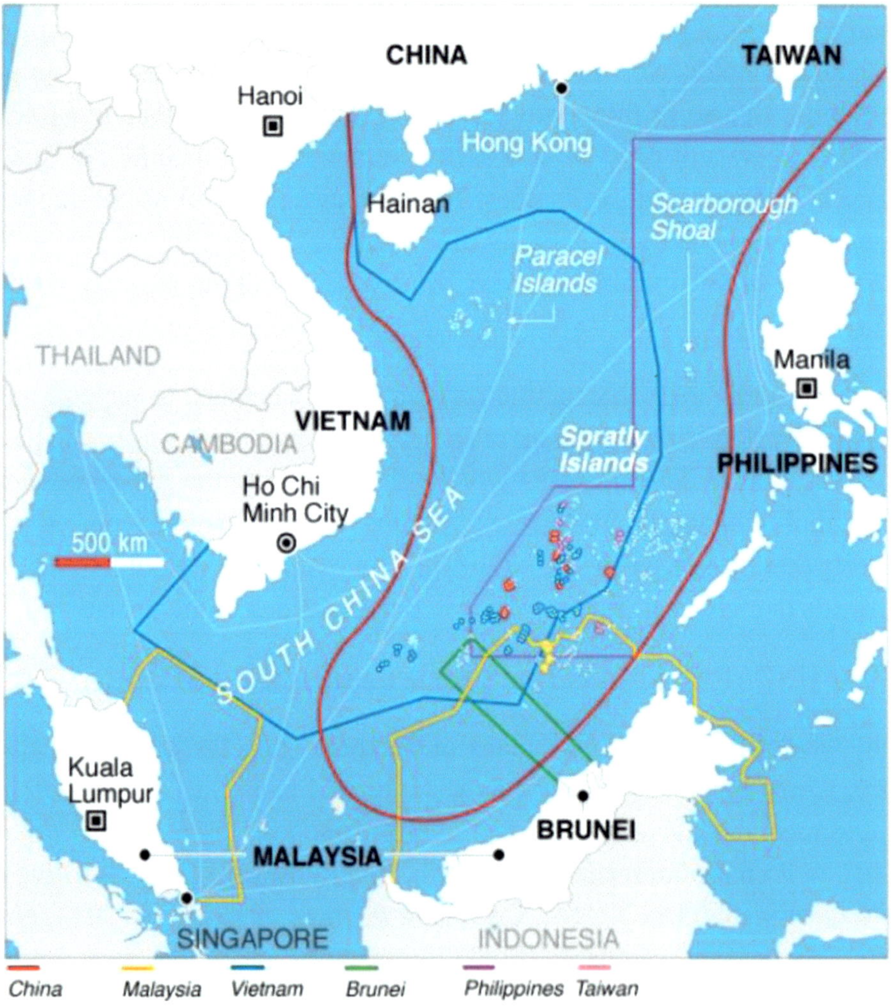

Fig. (4). China is planning extensive undersea facilities in the South China Sea.

"There are no technological hurdles… If you had the money and the need, you could do it today. Beyond that number, technological advances would be needed to deal with emergency evacuation systems, and environmental controls of air supply and humidity."

Nuwer stated:

"Making larger habitats with multiple modules made of steel, glass and special cement used underwater would be simpler than trying to create one giant bubble… we would not want to build any deeper than 1,000 ft (300 m), because the pressures at such depths would require very thick walls and excessive periods of decompression… To become self-sustaining, future habitats could potentially grow plants using natural or artificial light to generate a fresh supply of oxygen, or develop other methods to produce their own oxygen… Many marine biologists are enthusiastic about the possibility of being able to live underwater… There is growing interest in deep sea mining for minerals and metals, especially around island nations such as the Cook Islands, the Seychelles and Tonga. The Chinese… have been investing in deep sea expeditions to investigate the viability of mining manganese nodules and rocks and that contain nickel, copper, cobalt, manganese, gold and also valuable rare earth minerals".

Japan has announced plans for futuristic underwater cities to be created by 2030 [15]. For such a giant technological and financial undertaking this seems a short lead time. Hideo Imamura, a spokesperson for the Shimizu Corporation, in an interview[2] stated:

"This is a real goal, not a pipe dream. The Astro Boy cartoon character had a mobile phone long before they were actually invented– In the same way, the technology and knowhow we need for this project will become available."

Plans call for a \$25 to \$26 billion underwater "eco-city" called "Ocean Spiral" with 5,000 people. Energy will be "drawn from the seabed". The claim was made that the Ocean Spiral "… will capitalize on the infinite possibilities of the deep sea to accommodate human life, as rising sea levels endanger the survival of island communities". Plans call for a research and residential station some 10 miles in length that begins just below the surface of the sea and extends beneath the ocean floor. There are three parts to the Ocean Spiral: (a) the floating surface topped by a 500-meter sphere; (b) a central structure that is a spiral 15 km long with room for 5,000 people; and (c) at the bottom (3000 to 4000 meters under the sea) a system for developing the needed energy source for the city. It was stated that the "… spiral structure would descend as much as 9 miles… [and] produce methane from carbon dioxide by using micro-organisms". The methane will be useful fuel.

The financial costs of the endeavors proposed by China and Japan are extraordinary, but advanced planning and feasibility studies apparently have established technical feasibility for both. Challenges include making the components structurally able to withstand high winds and high seas and even earthquake activity. Currently, there are many political ramifications that are likely driving this as well as economic considerations. The technological problems with respect to providing a sustained, breathable atmosphere for its inhabitants is not a matter of public record. It is reasonable that in the future this or its equivalent will actually be constructed. It is likely to be preceded by more modest endeavors. Such efforts have been underway for some time with underwater habitats (Fig. **5**). These pressurized facilities have equipment and housing facilities that permit several scientists to remain on the bottom of deep seas at considerable pressure. They can exit and reenter the habitat and perform work and make observations while remaining at depth without decompressing. Current technologies provide for up to 10 days by several divers and the record of safety has been good. There is good reason that in the future these will become more prevalent. They will house more people and offer an experimental bridge to many of the factors that will be needed, but with more extreme parameters in larger habitats, underwater cities and underwater recreational and novel habitats. In the relatively near future, the latter will likely to grow in number, in safety, in luxury, and in the convenience necessary for commercial successes. It is this commercial success that likely will expand most in the near future.

Fig. (5). An underwater habitat; a diver can been seen above the structure.

Underwater Hotel

One of these future developments is the luxury underwater hotel which is likely to be first exploited as a novelty for the venturesome wealthy. By stressing luxury, it will lead the way toward technologies that are reliable. One factor that will likely result in breakthrough technologies and innovation, is new ways of generating oxygen economically and reliably. The biggest threat to safety, other than structural reliability against sea pressure and storms, is the reliability of the breathable atmosphere necessary for the large number of people that will be involved. It is difficult to predict whether such facilities will seek to generate oxygen by technologies at the site, or whether they will find it more feasible to deliver compressed oxygen, perhaps as liquid oxygen. It is my belief, that the development of artificial and hybrid photosynthesis (combined biological and engineered components) to generate oxygen will become available, and may even drive development of underwater facilities including luxury hotels. It seems probable that there will be competition in promoting venturing into space and venturing under the sea and the former seems to have, perhaps, a temporary advantage based on recent developments in reusable space vehicles by commercial enterprises.

An underwater hotel has been proposed in Dubai (Fig. **6**). Three areas are planned: a land station where guests will arrive, a connecting tunnel that will transport guests by underwater train to the underwater hotel.

Fig. (6). Futuristic underwater hotel, Dubai.

Underwater Mining

One driving force for technology to maintain people deep under the sea is the presence of valuable minerals. These deposits include precious metals, scarce construction metals, and rare-earth metals. These metal deposits are sometimes found on sea beds covered by extremely deep water, and they are found near volcanic activity expressed as "smokers" (Fig **7**) [16]. Oddly, a species of bacteria was found living near a black smoker off the Mexican coast at a depth of about 8,200 feet where no significant sunlight penetrates [17]. The bacterium, belonging to the family *Chlorobiaceae,* is capable of using the faint light from the black smoker to drive photosynthesis. This provides a basis for hope that human ingenuity can create marvelous synthetic and semi-synthetic photosynthesis technology to fit the needs of undersea and underwater habitats. The future of this endeavor will require the cooperative efforts of biologists, biochemist, molecular biologists, theoretical physicists, and engineers. The benefits of artificial photosynthesis are hard to overestimate. It has ramifications for space colonies, underwater habitats, and for what otherwise may become unreasonable efforts to reduce global atmospheric carbon dioxide levels. Improving the efficiency of light harvesting will make this more than competitive compared to solar energy cells, in my opinion. There is great potential for this process to create carbon-based byproducts of many kinds that will be useful as energy sources and as starting materials for biodegradable plastics and other structural materials, including those useful for 3-D printing which is already developing needs for new chemical substances to serve as the "ink" for printing complex new devices of almost unlimited potential.

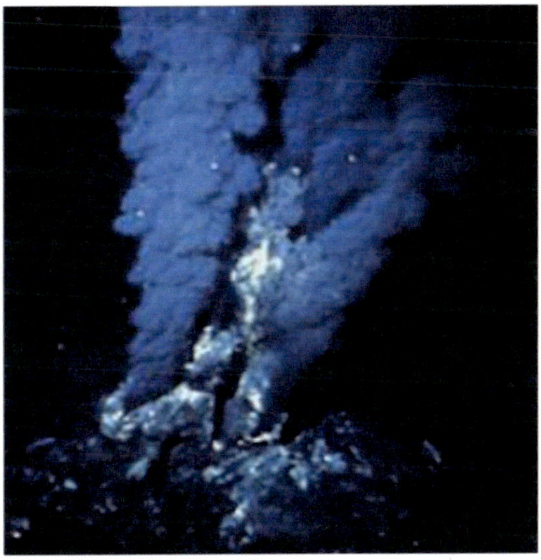

Fig. (7). Underwater "smoker".

As early as the late 1950s, there already was successful mining of diamonds underwater; however, humans underwater activity was not a significant part of the process which involved dredging diamond-containing sediment to the surface by an American company from about 1961 to 1970 [16]. The foreseeable future of underwater mining likely will not require extensive human undersea activity and the relationship to our subject, oxygen, will continue to be serendipities like the unusual and unexpected photosynthetic ability of the *Chlorobiaceae.*

SIMULATION ON EARTH: BIOSPHERE 2

Failure to produce adequate oxygen was a significant problem for Biosphere 2, an experiment on earth at living in a contained, self-sustaining habitat [18]. A lack of ability to generate adequate oxygen will be a serious limitation for all underwater and outer space facilities designed for long-term occupancy. In some circumstances resupply of oxygen is not feasible or will make the cost unacceptable.

Biosphere 2 (the Earth itself is designated Biosphere 1) was built in the Sonora desert outside Tucson, AZ and completed in September 1991 (Fig. **8**). The first occupancy was a mission from September 1991 to September 1993 with four men and four women inside the sealed environment of Biosphere 2. Oxygen levels fell so low halfway through the first year, that for safety, additional oxygen had to be added [19].

Fig. (8). Biosphere 2, in the Sonora desert.

Soil microbes were blamed for rapidly changing the atmosphere in the sphere. Oxygen was absorbed from the air and large amounts of CO_2 were released which exceeded the photosynthetic capacity of plants to assimilate and to regenerate oxygen [18].

The failure of the system to remain in balance with regard to the atmosphere, with the critical failure to maintain adequate oxygen, is a testament to the absolute necessity, and intimate relationship of oxygen in the atmosphere. It certainly was not anticipated that microbes in the soil would create the severe imbalance that was observed. Currently, knowledge about mechanisms of photosynthesis including artificial photosynthesis with biological and non-biological components provides encouragement that this problem is solvable. It seems reasonable that a balanced ecosystem is achievable that can provide recycling of oxygen and CO_2 as required for maintaining habitable environments both in space and undersea artificial environments. The problems (aside from pressure) are similar in space and undersea except that light energy is abundant in the former and poor in the latter. It is reasonable that fiber optic light guides can be used to pipe light to underwater cities. Artificial photosynthetic processes, therefore, can adequately supply and recycle the oxygen essential for under-water cities and space cities of the future.

ARTIFICIAL PHOTOSYNTHESIS

The potential use of photosynthesis to generate oxygen for undersea and outer space communities is an area of research with great promise. Recently, progress has been made in artificial photosynthesis and a separate section is warranted. Research is progressing with a bright future of realistic, but also fantastic, advancements that are reasonable and even likely. Traditionally, the focus has been on capturing solar energy for: producing carbon-based chemicals for use as fuel (or for other purposes), for removing carbon dioxide from the atmosphere, and for generating oxygen. Green plants and certain other life forms do all three with enviable efficiency.

Progress has been made in combining light-collecting components from green plants with artificial components to capture the energy of light and generate electricity. Fig. (**9**) shows an artificial leaf against a backdrop of a real leaf. These devices are inefficient compared to the leaf, and the use of plant light-capturing antennae and pigments is not yet sufficiently durable. Fig. (**10**) diagrammatically shows the idea that sunlight can be captured to provide electrons that can drive biosynthetic processes naturally in leaves to make sugar, but that can be diverted in the artificial leaf to synthesizing a hydrocarbon fuel. The type of fuel synthesized depends only on the ingenuity of man.

Fig. (9). Artificial leaf with a real leaf in the background.

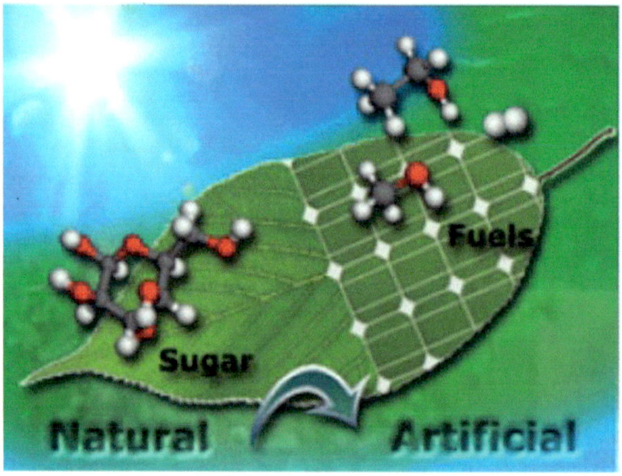

Fig. (10). Natural leaf making sugar and artificial leaf making a hydrocarbon fuel.

Experiments are focused on designing durable, efficient light collecting devices. A variety of enzyme-catalyzed processes, theoretically, could make various organic compounds with various uses and applications (Fig **11**). The simple idea is to capture the energy of sunlight, use the electrons to reduce carbon obtained from carbon dioxide and with the aid of water, to create oxygen and organic compounds. Thus, "free" energy (sunlight) and a waste product that some desire to remove from the atmosphere (carbon dioxide) and water can be turned into desired organic compounds of considerable complexity and usefulness.

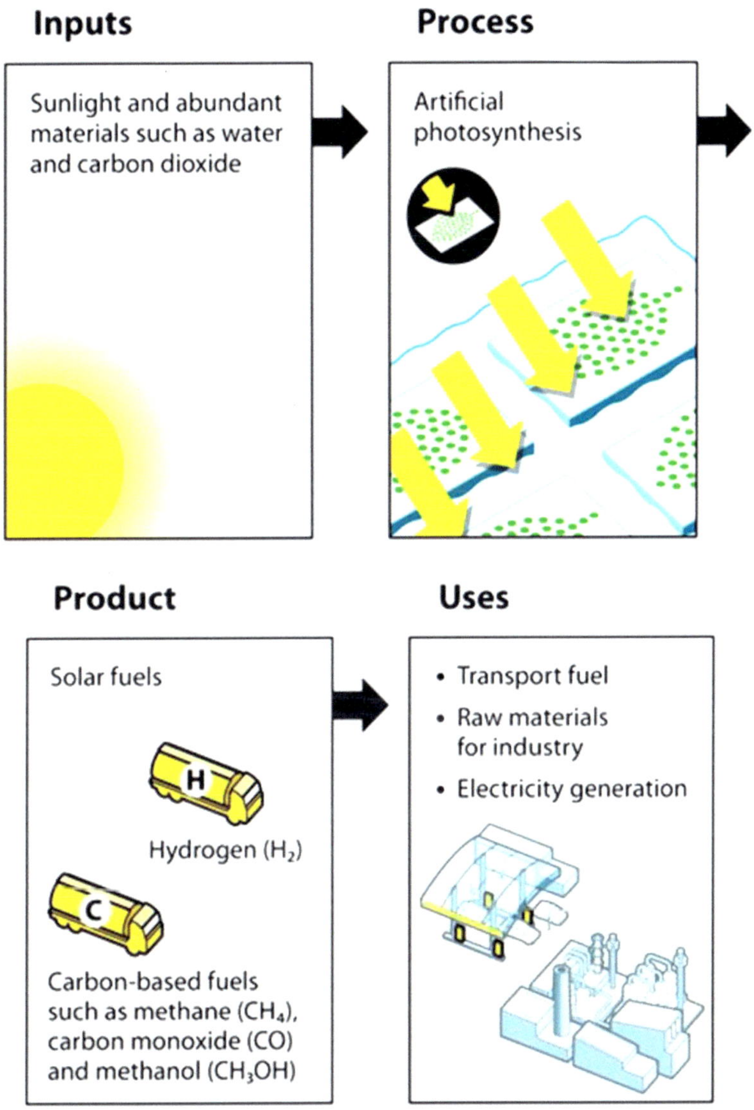

Fig. (11). Sunlight, water and carbon dioxide can be captured by processes that combine natural and artificial means and devices to make products with a variety of uses.

Delving into nature's secrets that make this process not only possible but efficient, have led to the discovery that the magic in a leaf involves physics at the quantum level. The distances involved in the extraction and transfer of electrons by chlorophyll and the related components of the "leaf antennae" are so short that quantum physics takes over. Quantum dot technology is a possible answer and biologists, chemists, physicists, and material scientists are collaborating in seeking practical, durable, miniature solutions (Fig. **12**).

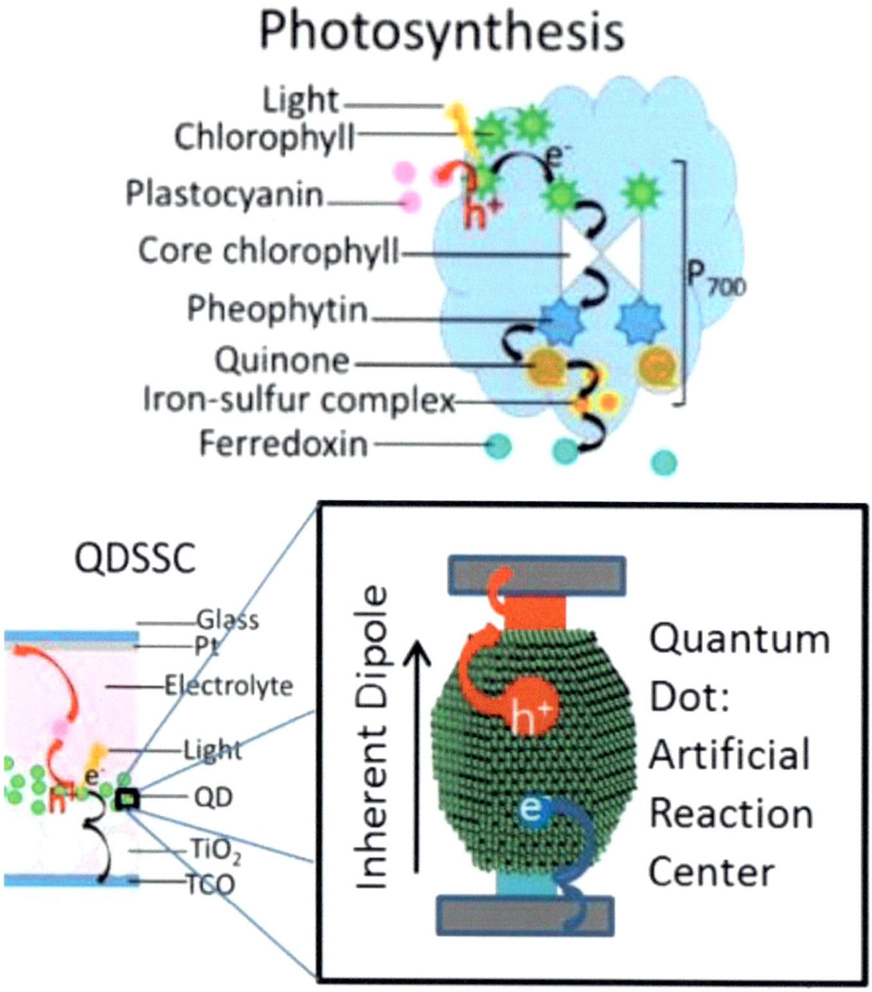

Fig. (12). Plant photosynthesis compared to quantum dot artificial technology.

As far back as 2010, researchers in Spain, the U.S., Ireland, France, and Russia had developed technology involving the concept of the quantum dot [20]. The quantum dot is the name given to an inorganic nanocrystal that is designed to "harvest" light which is accomplished in nature by an organic system that involves chlorophyll. The group of researchers just named have combined the photosynthetic center from a bacterium, with the light harvesting inorganic nanocrystal (quantum dot). The initial step in photosynthesis is the capture of light by an "antenna" which is a complex of proteins and pigments that is raised to an electronically-excited state by absorption of photons of light. This energy, as electrons, is transferred to chlorophyll and other pigments in the "reaction" center of the leaf chloroplast. In the reaction center, the energy is used to make ATP and to synthesize sugar. During its transfer between components, the energy must have electron states in which the sending and receiving element are in resonance: this is called Főrster resonance energy transfer (FRET) [20]. Of course, any artificially-made systems must have an efficient antenna system that is able to send the energy along through FRET. Pigment molecules were initially used but they generally capture a very narrow range of light wavelengths, corresponding to the pigment color, and they are comparatively unstable [20]. Thus the concept: design and replace this system with artificial quantum dots as antennae. These dots are so small that they behave like molecules and smaller particles and obey the laws of quantum mechanics. The electronic and optical properties of the dots can be designed to specifications based on composition, shape and size [20]. The researchers have been able to couple their artificially antennae (the quantum dots) with the reaction center (biosynthetic mechanism) of a bacterium, which performs the organic biosynthesis reactions.

Currently (in 2016), a group in Russia is doing exciting research and has published results using a purple bacterium, *Rhodobacter sphaeroides* as the source for the reaction center, and quantum dots that provide hybrid complexes that retain functional activity for at least a few months [21]. These stable hybrids of natural reactions centers (for biosynthesis of organic compounds) and artificial quantum dots (as components in an antennae system for harvesting sunlight energy) are a promising direction for the future.

The future promise of hybrid systems for harvesting sunlight and linking it to biosynthesis of useful organic compounds is significant. It is now fanciful, but in the future these systems will be combined with micromotors (described next) that use the energy of sunlight for motion and for even more complex, and directed processes.

MICROMOTORS PROPELLED BY OXYGEN BUBBLES

Microscopic particles have been designed that move about in solutions by the force generated asymmetrically by micro bubbles of oxygen derived from platinum-catalyzed cleavage of hydrogen peroxide. The basic system uses asymmetry created by particles (hemispheres when particles are spherical) that have chemically different sides (halves). These particles were named Janus particles [22] because they were "two-faced". Janus was an ancient Roman composite god associated with doorways, beginnings, and transitions– two-faced; he looks to the future and to the past (Fig. **13**).

Fig. (13). Ancient Roman god Janus: two faced, he was the god of doorways and new beginnings.

The microparticles rely on asymmetric chemical reactions occurring on their surfaces to provide driving forces that propel the particles. In one example (Fig. **14**), a spherical particle was coated on one hemisphere with platinum which catalyzes the conversion of hydrogen peroxide to water and oxygen molecules, and the other hemisphere is a different substance. When the particle (motor) is immersed in a solution containing hydrogen peroxide, oxygen bubbles are formed on the hemisphere with the platinum catalyst and not on the other hemisphere. This creates an asymmetric force of propulsion for the particle.

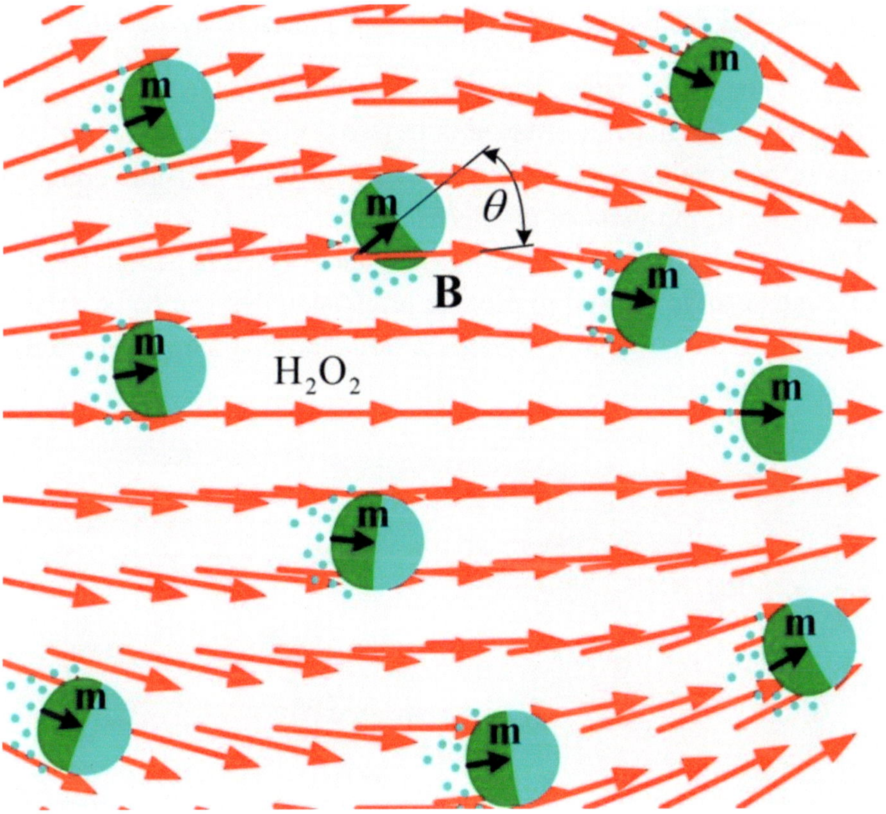

Fig. (14). Platinum spheres coated only on one hemisphere with platinum (colored green) which is a catalyst that converts hydrogen peroxide to water and microscopic oxygen bubbles (blue dots) that propel the spheres (red arrows) in a water solution containing hydrogen peroxide in the direction opposite to the platinum-coated hemisphere by the force of the oxygen bubbles.

In 2013 [23] researchers in Japan found that a structure with two "sides" with different materials was not necessary. Spheres made only of platinum spontaneously formed clumps (aggregates) of hundreds or thousands of particles with different shapes which behaved differently. Tear-dropped shaped aggregates moved in primarily one direction; aggregates shaped like windmills spun; and those shaped like a boomerang travelled in circles. Asymmetric drag controlled the direction of motion. Tiny particles have large surface-to-mass relationships and can experience large anisotropic, viscous forces in water. The authors attributed regular motion, in the solution containing a reactant (in this case hydrogen peroxide), to random forces on the tiny particles where torque, translation, spin and rotation could be controlled precisely by design.

Microscopic motors may have been inspired by the wonderfully complex "motors" that drive the flagella of bacteria to create movement (Fig. (**15**), panel A). Flagella of bacteria such as *Escherichia coli* have microscopic components that are analogous to the parts of an electric motor (Fig. (**15**), panel B). These parts are inserted in the bacterial membrane and the attached flagellum rotates and a group of flagella act in concert to provide propulsion that can be controlled by the bacterial cell. Not shown in the diagram is the driving force of protons that cause the rotor to spin.

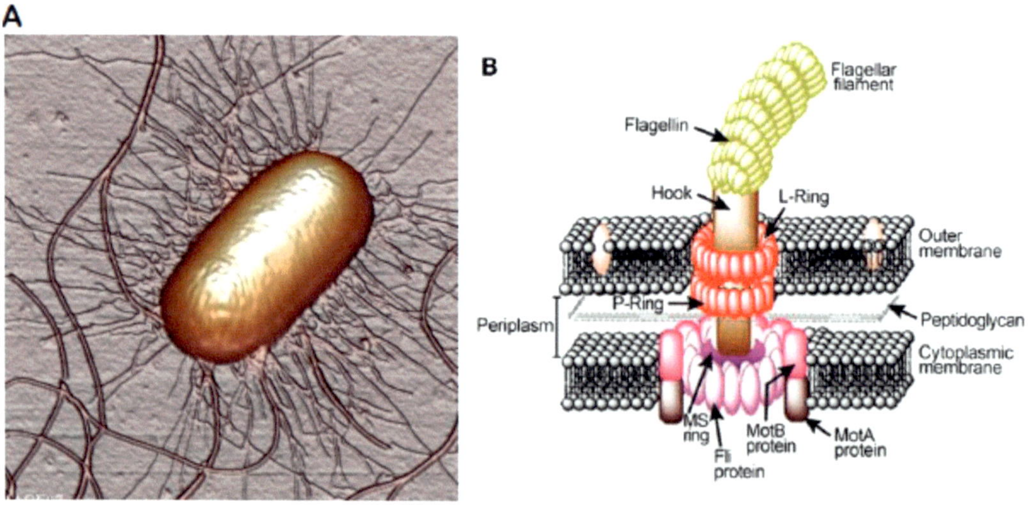

Fig. (15). Electron Microscope view (panel A) of rod-shaped bacteria (*Escherichia coli*) with many pili (smaller protein structures used in gene transfer and adhesion) and several flagella (larger, protein structure used for motion), and a diagram of the structures responsible for rotating a flagellum and creating motility.

These bacterial "motors" are capable of creating motion using isothermal conditions very different from the internal combustion engine. Therefore, investigation of their mechanisms is of great relevance to the potential development of such microscopic, directed motion with applications in medicine as conveyers of particles to transport pharmaceuticals to specific sites and possibly even to perform microsurgery (perhaps at the molecular level) in the future. Indeed, Isaac Asimov wrote a science fiction story, made into a movie, about a fanciful miniaturization (shrinking) of a human crew and submarine which was then injected into the bloodstream of a patient with a cerebral blood clot. The future will not see this come to pass (shrinkage of a crew and submarine) but it will see microscopic (nano-scale) devices that can be directed to specific sites to deliver drugs or carry out mechanical functions (including surgery and to physically attach to specific cells) very precisely.

Self-propelled Janus micromotors were described in a 2015 [24] report to remove a "wide-range of organic and inorganic pollutants from water". Activated charcoal with asymmetric deposition of a catalytic platinum "patch" on the surface as a catalyst was used to produce oxygen bubbles to propel the particles with speeds reported to be over 500 micrometers per second. This corresponds to approximately 10 particle-lengths per second by the oxygen bubbles emitted from the platinum catalytic surface. The authors claim they have combined the absorptive capacity of activated carbon microspheres with the dynamic movement of the Janus micromotors to create efficient "moving absorption platforms" that removed metals and various oxygen-containing organics from water [24]. They used commercially available carbon microparticles with a large surface area of 100 square meters per gram. Propulsion was provided by coating the particles with a 60 nanometer catalytic layer of platinum. The same authors were quoted in the article [24] to have previously used similar particles to remove oil droplets from sea water, detoxify organophosphorus nerve agents from water, and also to have designed reactive iron to degrade organic pollutants using the Fenton reaction and oxygen (previously described in Chapter 7).

It is also possible that nanomotors will be developed that can be released into the oceans to trap and fix carbon dioxide. In 2015 [25], it was reported that scientists at the University of California, San Diego, had designed: "enzyme-functionalized" micromotors that move about in water and remove carbon dioxide by converting it into solid calcium carbonate *via* the enzyme carbonic anhydrase.

It was reported that: "Within five minutes, the micromotors removed 90% of the carbon dioxide from a solution of deionized water" (and 88% from sea water) [25]. To fuel the micromotors in water, they added hydrogen peroxide which reacted with the platinum catalyst as previously described to generate a stream of oxygen bubbles that propelled the microparticles. They reported speeds up to 100 micrometers per second with hydrogen peroxide of 2 to 4%. It appears to me that the requirement for hydrogen peroxides makes the current system impractical. A possible approach to surmount this difficulty is to design a carbon-capturing micromotor that will use a pollutant reaction for propulsion.

Recently, in 2016 [26] progress has been made that has great potential for the future toward making micromotors that are biocompatible. Researchers described Janus (two-faced) spherical microspheres made of silica, coated on one hemisphere with magnetic iron particles and on the other hemisphere an enzyme, urease. Urease decomposes urea (present in human body fluids) to produce carbon dioxide micro bubbles which causes chemo-osmotic propulsion. The microspheres are only 2.3 microns in diameter (they are the size of bacteria), and they can be designed to carry drugs. The iron present on one hemisphere is magnetic and the microspheres can be guided by the placement of external magnets. The researchers [26] state that the 2 mM to 7 mM urea concentration found in human blood is sufficient to move the microspheres.

The future for this approach is immense. One can envision, not only drugs delivered precisely to targets with consequent lower toxicity to the body, but delivery to tumors specifically and perhaps even to hunt and destroy individual cancer cells in the blood or even in tissues. Micro-surgery is a possibility with control at the cellular level for delicate operations that target disease and leave the healthy cells unaffected. Today's surgery, by comparison, is a gross attack that destroys or removes whole organs in some cases.

OXYGEN AND OUTER SPACE COLONIES

It is obvious that colonies in space will require that breathable atmospheres be provided. In any reasonably imagined or foreseeable future, the fundamental qualities of humans as air-breathing mammals will remain. There will be no need to deal with pressure, as required for undersea colonies on earth; pure oxygen at lower pressure, to save on containment structures, seems to be the obvious choice.

If the space colony is large enough, an artificial gravity with an artificial atmosphere is conceivable, especially if the colony is not an artificial satellite, but a colony on an existing orbiting body. The means to generate oxygen from oxygen bound in a mineral will be the most practical way to provide oxygen. Because of the dearth of water to be anticipated in this environment, electrolysis will not likely contribute. A long-term solution to generating oxygen, when the colony is on a carbonaceous body, will be advanced forms of the artificial photosynthesis that is under development today [3, 27].

Fig. (16). Artist depiction of an outer space colony, the Stanford Taurus, from 1975, courtesy of NASA.

The Stanford Torus was the design, for a habitable space colony facility, selected for consideration by NASA in a 1975 "summer study" conducted in coordination with Stanford University[3]. The design featured a torus (do-nut shaped ring) one mile in diameter that was designed to provide normal earth gravity on the inside of the outer ring (Fig. **16**). It was planned for a space city of 10,000 people [28]. It is interesting to see the state of interest in and development of this concept 40+ years ago. Manned space travel has been practically stalled since the last Moon mission. Unmanned probes (Juno just arrived to orbit Jupiter), satellites, and the earth-orbiting space station, are the extent of man's more recent efforts. Several artist conceptions of the Stanford Taurus were made by Donald Davis. Princeton physicist Gerald O'Neill is credited with conceptualizing what was called the "O'Neill Cylinder" (Fig. **17**). The structure has two 20-mile-long cylinders that "float" side-by-side at the Lagrangian point between the earth and the moon. The cup-shaped structures in circles around one end of each tube are intended for agricultural uses [28].

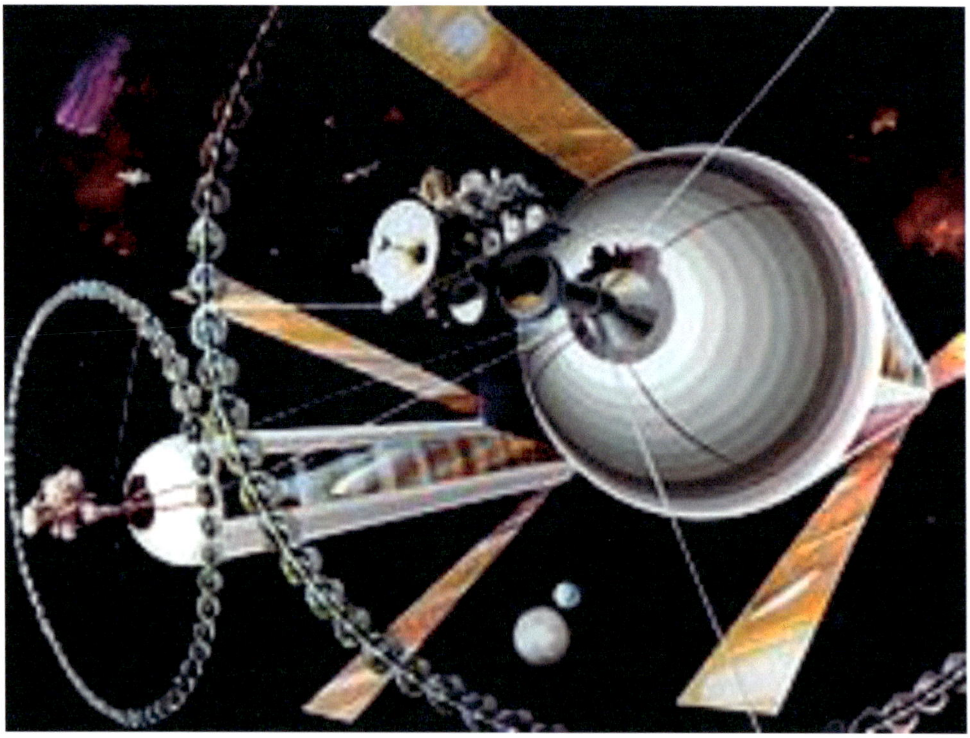

Fig. (17). One of three "imaginary" cities in a painting by Donald Davis, relating to the 1975 "Stanford Taurus", courtesy of NASA.

OXYGEN AND LONGEVITY

It is commonly considered that the cumulative effects of breathing oxygen contribute to a shortening of lifespan. Perhaps this results from "oxidative stress" that accumulates at various levels within cells: the DNA, proteins, and membranes. In a crude way, this is analogized to "rusting away". This leads to the thesis that striking just the right balance of oxidation-reduction in cells and perhaps even with a magical anti-oxidant artificially added, the lifetime of cells might be extended, or at least, the period of physiologically healthy functioning could be extended. Shortening, with time, of teleomers on chromosomes is proposed to contribute to ageing, and this does not seem to be related to oxidant stress mechanisms, and thus not related to our subject of interest, oxygen. It is perhaps a surprising conclusion to realize that germ-line cells, unlike somatic cells, do not have a built-in mortality (in terms of number of divisions they ordinarily can undergo). Certain cancer cells also in tissue culture at least, have become "immortal". At the intracellular level, the concept of mortality in terms of longevity does not seem to be a fixed outcome. There seems to be no inherent reason why genes, DNA, RNA, and proteins, with proper maintenance and repair, should intrinsically "age". However, here, we can see that "oxidative" damage (and thus – a link to oxygen) can accumulate and over time, the Second Law will exert its effects. But, why are germ line cells different?

However, even if technology can change the biology at the intracellular level, the greater challenges to mortality at the cellular level, and the even greater challenges at the organ and whole animal level, remain as towering limitations. Modestly extending lifetime however is likely to be a realized consequence of specific advances in understanding "oxidative stress". It is reasonable that such advances may extend the average lifespan to that now experienced by the few. Human longevity in the future, relating to knowledge of the biology, chemistry, and physics of oxygen, likely will not surge forward dramatically. Change to increased longevity will have small steps and understanding of oxidant stress and managing it genetically and medicinally will progress, albeit slowly.

NOTES

[1]*The Anatomy of Revolution*, 7.3, 1952.

[2] Imamura statement as told to *The Guardian*, as reported in: https://www.theguardian.com/world/2014/nov/20/ocean-spiral-japan-underwater-city.

[3] Space Settlements: A Design Study, NASA Publication SP-413.

REFERENCES

[1] Chemical elements listed by the discovery year. Llentech. Available from: http://www.lenntech.com/periodic-chart-elements/discovery-year.htm

[2] How many natural elements are there? Available from: https://www.reference.com/science/many-natural-elements-675491d9adc985f

[3] Liu C, Gallagher JJ, Sakimoto KK, *et al.* Nanowire–Bacteria hybrids for unassisted solar carbon dioxide fixation to value-added chemicals. Available from: http://pubs.acs.org/doi/abs/10.1021/acs.nanolett.5b01254 2015.
[http://dx.doi.org/10.1021/acs.nanolett.5b01254]

[4] The rise of oxygen bars. Available from: http://www.webmd.com/ balance/features/rise-of-oxyg-n-bars [Internet]. [cited 2016 Jun 11]

[5] Oxygen bars: Is a breath of fresh air worth it? Available from: http://www.medicinenet.com/script/main/art.asp?articlekey=21800 [Internet]. [cited 2011 Jun 11]

[6] Oxygen is a nutritional drug. Available from: http://drsircus.com/ medicine/oxygen-is-a-nutritio-al-drug/ [Internet]. [cited 2016 Jun 13]

[7] Clearance and approvial. Available from: http://www.hyperbariclink. com/hyperbaric-oxyge--therapy/clearance-and-approval.aspx#.V168frsrIdU [Internet]. [cited 2016 Jun 13]

[8] The evolution of diving in mammals and myoglobin. Available from: http://indianapublicmedia.org/amomentofscience/myoglobin-evolution-diving-mammals/ [Internet]. [cited 2016 Jun 13]

[9] Noren SR, Williams TM. Body size and skeletal muscle myoglobin of cetaceans: adaptations for maximizing dive duration. Comp Biochem Physiol A Mol Integr Physiol 2000; 126(2): 181-91.
[http://dx.doi.org/10.1016/S1095-6433(00)00182-3] [PMID: 10936758]

[10] NOAA Ocean Explorer: Technology: Diving: Aquarius

[11] Rossignol DA, Bradstreet JJ, Van Dyke K, *et al.* Hyperbaric oxygen treatment in autism spectrum disorders. Med Gas Res 2012; 2(1): 16.
[http://dx.doi.org/10.1186/2045-9912-2-16] [PMID: 22703610]

[12] Steven M. 2011. Tracy Leigh LeGros;Heather Murphy-Lavoie. Idiopathic Sudden Sensorineural Hearing Loss (New! approved on October 8, 2011 by the UHMS Board of Directors). Available from: https://www.uhms.org/14-idiopathic-sudden-sensorineural-hearing-loss-new-approved-on-october-8-2011-by-the-uhms-board-of-directors.html

[13] Gesell LB. Hyperbaric oxygen therapy indications : the Hyperbaric Oxygen Therapy Committee report. Undersea and Hyperbaric Medical Society 2008.

[14] Nuwer R. Will we ever... live in underwater cities? Available from: http://www.bbc.com/future/story/20130930-can-we-build-underwater-cities [Internet]. [cited 2016 Jun 12]

[15] Kathrine J. Tobal. Japan releases plans for futuristic underwater cities by 2030. Available from: https://www.collective-evolution.com/2014/11/25/plans-for-future-underwater-cities-in--apan-by-2030/[Internet]. [cited 2016 Jun 12]

[16] Undersea mining. Available from: ocean link.island.net[Internet]. [cited 2016 Jun 11]

[17] Beatty JT, Overmann J, Lince MT, *et al.* An obligately photosynthetic bacterial anaerobe from a deep-sea hydrothermal vent. Proc Natl Acad Sci USA 2005; 102(26): 9306-10. [http://dx.doi.org/10.1073/pnas.0503674102] [PMID: 15967984]

[18] Allen J, Blake AG. 1991. (Anthony GE. Biosphere 2 : the human experiment. Penguin Books; 1991. 156 p

[19] The world's largest earth science experiment: Biosphere 2. Available from: http://ecowatch.com/2015/10/16/biosphere-2-video/ [Internet]. [cited 2016 Jun 13].

[20] Nabiev I, Rakovich A, Sukhanova A, *et al.* Fluorescent Quantum Dots as Artificial Antennas for Enhanced Light Harvesting and Energy Transfer to Photosynthetic Reaction Centers 2010. Angew Chemie Int Ed. 2010; 49(40): 7217–21. Available from: http://doi.wiley.com/10.1002/anie.201003067

[21] Lukashev EP, Knox PP, Oleinikov IP, Seifullina NK, Grishanova NP. Investigation of stability of photosynthetic reaction center and quantum dot hybrid films 2016. Biochem; 2016; 81(1): 58–63. Available from: http://link.springer.com/ 10.1134/S0006297916010065 [http://dx.doi.org/10.1134/S0006297916010065]

[22] Casagrande C, Fabre P, Raphaël E, Veyssié M. 1989. "Janus Beads": Realization and Behaviour at Water/Oil Interfaces. Europhys Lett; 1989; 9(3): 251–5. Available from: http://stacks.iop.org/0295-5075/9/i=3/a=011? key=crossref. d6373125d0f2d9546e0d59e0ab8c052e

[23] Yamamoto D, Mukai A, Okita N, Yoshikawa K, Shioi A. 2013. Catalytic micromotor generating self-propelled regular motion through random fluctuation. J Chem Phys. American Institute of Physics; 2013; 139(3): 034705. Available from: http://www.ncbi.nlm.nih.gov/pubmed/23883050

[24] Jurado-Sánchez B, Sattayasamitsathit S, Gao W, *et al.* Self-Propelled Activated Carbon Janus Micromotors for Efficient Water Purification 2015. Small. 2015; 11(4): 499–506. Available from: http://doi.wiley.com/10.1002/smll.201402215 [http://dx.doi.org/10.1002/smll.201402215]

[25] Uygun M, Singh VV, Kaufmann K, Uygun DA, de Oliveira SD, Wang J. Micromotor-Based Biomimetic Carbon Dioxide Sequestration: Towards Mobile Microscrubbers. Angew Chemie. 2015; 127(44): 13092–6. Available from: http://doi.wiley.com/10.1002/ange.201505155

[26] Ma X, Wang X, Hahn K, Sánchez S. Motion control of urea-powered biocompatible hollow microcapsules 2016. Micromotor-Based Biomimetic Carbon Dioxide Sequestration: Towards Mobile Microscrubbers. Angew Chemie. 2015; 127(44): 13092–6. Available from: http://doi.wiley.com/10.1002/ange.201505155 [http://dx.doi.org/10.1021/acsnano.5b08067]

[27] Sakimoto KK, Wong AB, Yang P. Self-photosensitization of nonphotosynthetic bacteria for solar-t--chemical production 2016. Science (80-). 2016; 351(6268): 74–7. Available from: http://www.sciencemag.org/cgi/doi/10.1126/science.aad3317 [http://dx.doi.org/10.1126/science.aad3317]

[28] NASA's groovy concept art for the orbiting cities of the future. Available from: http://www.citylab.com/ design/ 2012/01/ nasas-groovy-concept-art-orbiting-cities-future/1085/ [cited 2016 Jun 12]

SUBJECT INDEX

A

B

C

Olen R. Brown

D

E

F

G

H

O

Q

R

S

T

U

V

W

Made in the USA
Las Vegas, NV
03 November 2026

57424222R00221